STATE ARCHIVES OF ASSYRIA STUDIES

VOLUME III

STATE ARCHIVES OF ASSYRIA STUDIES

Published by the Neo-Assyrian Text Corpus Project
of the Academy of Finland
in co-operation with
the Finnish Oriental Society

Project Director
Simo Parpola

Managing Editor
Robert M. Whiting

VOLUME III

Marco De Odorico

THE USE OF
NUMBERS AND QUANTIFICATIONS
IN THE ASSYRIAN
ROYAL INSCIPTIONS

THE NEO-ASSYRIAN TEXT CORPUS PROJECT

State Archives of Assyria Studies is a series of monographic studies relating to and supplementing the text editions published in the SAA series. Manuscripts are accepted in English, French and German. The responsibility for the contents of the volumes rests entirely with the authors.

Set in Times
Typography and layout by Teemu Lipasti
The Assyrian Royal Seal emblem drawn by Dominique Collon from original
Seventh Century B.C. impressions (BM 84672 and 84677) in the British Museum
Ventura Publisher format and custom fonts by Robert M. Whiting and Timo Kiippa
Electronic pasteup by Robert M. Whiting

Printed in Finland
by Vammalan Kirjapaino Oy

ISBN 951-45-7125-8 (Volume 3)
ISSN 1235-1032 (Series)

THE USE OF NUMBERS AND QUANTIFICATIONS IN THE ASSYRIAN ROYAL INSCRIPTIONS

By

Marco De Odorico

THE NEO-ASSYRIAN TEXT CORPUS PROJECT
1995

To my parents

PREFACE

This book is largely based on my Ph.D. ("Dottorato di Ricerca") dissertation written in late 1991 and titled "Studi sulle quantificazioni nelle iscrizioni reali assire," in the framework of the doctoral curriculum of Philological Sciences of Ancient Western Asia, under the directorship of Prof. Luigi Cagni (Naples) and the specific supervision of Prof. Carlo Zaccagnini (formerly Bologna, now Naples) and Prof. F. Mario Fales. The manuscript was subsequently revised and translated into English during a visit to the University of Helsinki, and reached its present form during a stay in Jerusalem in late 1994.

Funding for the preparation of this book was obtained, apart from the fellowship assigned to Italian doctoral candidates, from the Italian Ministry of Foreign Affairs for a 5 months scholarship at the University of Helsinki in connection with the Ministry of Education of Finland.

I am sincerely grateful to Prof. Zaccagnini for his constant advising from the very beginning of this research and for the encouragement and the scientific stimuli that led to this work. I also wish to thank Professors Luigi Cagni and F. Mario Fales for their kind support and their many suggestions during the preparation of the doctoral dissertation. I extend my heartfelt thanks also to Prof. Giovanni B. Lanfranchi (Padua) for according me access to his research archives.

Prof. Hayim Tadmor deserves my utmost gratitude for his warm hospitality and kind advice during my six months stay in Jerusalem in 1990, and again during the final stages of the writing of this work. The hours he spent reading and discussing with me its final version resulted in many improvements, and in the addition of some supplementry material. His knowledge, interest in the subject, and sense of humor have been greatly appreciated. I am also grateful to Prof. Israel Eph'al (Jerusalem) for his kindness and his constructive suggestions. Special thanks also are deserved by Eckart Frahm (Göttingen), who sent me in advance large portions of his doctoral thesis on Sennacherib's inscriptions, and Elnathan Weissert (Jerusalem), for discussing many aspects of the late neo-Assyrian inscriptions.

My deep gratitude goes to Prof. Simo Parpola (Helsinki) for allowing me full access to the facilities of the Department of Asian and African Languages and the State Archives of Assyria Project, for his kind interest and support and for the acceptance of the manuscript in the SAAS Series. I also wish to express my gratitude to Dr. Robert Whiting for his precious technical and scientific assistance in the production of the book. Raija Mattila and Laura Kataja contributed to the successful completion of this work with friendship and knowledge; I thank them heartily. The English version of this book, prepared during my stay in Helsinki, greatly benefited from the kind and competent aid of Prof. Margot Stout Whiting, who undertook the not easy task of revising my English text and style, and to whom I wish to express my sincere gratitude.

Since this study aims at examining thoroughly a specific aspect (the quantity) and facing it from different points of view, as a result, specific points are dealt with in several places, and the number of repetitions and, especially, of cross references is quite high.

In general, the needs of those who are not cuneiform scholars have been kept in mind, and this involved giving multiple references to the same text (copy - edition - translation). The copies are quoted only if numbers are discussed and/or the transliteration of a passage is given *in extenso*. In effect, in many cases it is not possible to establish how a number is written without having recourse to the copy, because of the inaccuracy of the transcription systems employed. The edition in copy is first mentioned, followed (usually within parenthesis) by the latest edition in transliteration, and one in translation (either ARI, TUAT, ARAB or ANET). Semicolons in such cases (within parentheses) separate different editions of the same text. In certain cases, when some passages are mentioned *en passant* reference only to an edition *or* a translation is made. The translations have been, at the expense of English style, rendered as literally as possible — to have an idea of the exact position of a numeral within a sentence is quite important. A parenthesis with three dots (...) indicates an omission (mine). Transliterations appearing in the footnotes are somewhat simplified (i.e. the logograms have not been "solved") to avoid having excessively crowded footnotes.

Note that, throughout the work, any number, real or hypothetical, appearing in an ancient source will be printed in italics (except in transliterations), and, contrary to the usage in SAA volumes, this does not mean it is uncertain.

July 1995 Marco De Odorico

CONTENTS

LIST OF TABLES

BIBLIOGRAPHY AND ABBREVIATIONS

PERIODICALS AND SERIES CITED IN ABBREVIATED FORM

AAA	(Liverpool) *Annals of Archaeology and Anthropology*
AfO	*Archiv für Orientforschung*
AnSt	*Anatolian Studies*
AOAT	*Alter Orient und Altes Testament*
ARRIM	*Annual Review of the R. I. M. Project*
BaF	*Baghdader Forschungen*
BaM	*Baghdader Mitteilungen*
BASOR	*Bulletin of the American Schools of Oriental Research*
BiOr	*Bibliotheca Orientalis*
IEJ	*Israel Exploration Journal*
JAOS	*Journal of the American Oriental Society*
JCS	*Journal of Cuneiform Studies*
JHS	*Journal of Hellenic Studies*
JSS	*Journal of Semitic Studies*
JNES	*Journal of Near Eastern Studies*
MAOG	*Mitteilungen der altorientalischen Gesellschaft*
MDOG	*Mitteilungen der Deutschen Orient Gesellschaft*
MVAG	*Mitteilungen der Vorderasiatisch(-Aegyptisch)en Gesellschaft*
OLZ	*Orientalistische Literaturzeitung*
Or	*Orientalia* (*Nova Series*)
RA	*Revue d'Assyriologie*
RIMA	*The Royal Inscriptions of Mesopotamia, Assyrian Periods*
SAA	*State Archives of Assyria*
SAAB	*State Archives of Assyria Bulletin*
WO	*Die Welt des Orients*
WVDOG	*Wissenschaftliche Veröffentlichungen der Deutschen Orient-Gesellschaft*
ZA	*Zeitschrift für Assyriologie*
ZDPV	*Zeitschrift des deutsches Palästina-Vereins*

BIBLIOGRAPHICAL ABBREVIATIONS

ABL	R.F. Harper, *Assyrian and Babylonian Letters belonging to the K collection of the British Museum*, Parts 1-14, Chicago 1892-1914 [by no.].
ABZ	R. Borger, *Assyrisch-babylonische Zeichenliste* (AOAT 33), Neukirchen-Vluyn 1978.
AD	(see Smith, AD).
Adams, *Heartland*	R.McC. Adams, *Heartland of Cities*, Chicago 1981.
ADD	C.H.W. Johns, *Assyrian Deeds and Documents recording the transfer of Property*, I-IV, Cambridge 1898-1923 [by no.].
Ahmed, *Southern Mesop.*	S.S. Ahmed, *Southern Mesopotamia in the Time of Ashurbanipal*, The Hague 1968.
AHw	W. von Soden, *Akkadisches Handwörterbuch*, I-III, Wiesbaden 1959-1981.
AKA	E.A. Wallis Budge - L.W. King, *Annals of the Kings of Assyria ... in the British Museum*, I, London 1902.
Amiaud - Scheil, *Salm.*	A. Amiaud - V. Scheil, *Inscriptions de Salmanasar II, roi d'Assirie*, Paris 1890.
Andrae, FWA	W. Andrae, *Die Festungswerke von Assur* (WVDOG 23), Leipzig 1913.
ARAB 1/2	D.D. Luckenbill, *Ancient Records of Assyria and Babylonia*, I-II, Chicago 1926-27 [by §].
ARE 2/3/4/5	J.H. Breasted, *Ancient Records of Egypt*, II-V, Chicago 1906-07 [by §].
ARI 1/2	A.K. Grayson, *Assyrian Royal Inscriptions*, I-II, Wiesbaden 1972-76 [by §].
ARINH	F.M. Fales (ed.), *Assyrian Royal Inscriptions: New Horizons* (Orientis Antiquis Collectio XVII), Roma 1981.
Aynard, *Prisme*	J.-M. Aynard, *Le Prisme du Louvre A.O. 19939*, Paris 1957.
BAL^2	R. Borger, *Babylonisch-Assyrische Lesestücke* (2. Neubearbeitete Auflage, Analecta Orientalia 54), I-II, Rom 1979.
Bauer, IWA	Th. Bauer, *Das Inschriftenwerk Assurbanipals*, 1/2, Leipzig 1933. [Tf. refers to the 1st vol.]
Becking, *Samaria*	B. Becking, *The Fall of Samaria* (Studies in the History of the Ancient Near East 2), Leiden 1992.
Böhl, MVLS 3	F.M.Th. Böhl, *Mededeelingen uit de Leidsche Verzameling van Spijkerschrift-Inscripties, III - Assyriasce en nieuw-babylonische Oorkonden (1100-91 V.Chr.)* (Mededeelingen der Kon. Akademie van Wetenschappen 82/2), Amsterdam 1933.

Boissier, RT 25	A. Boissier, "Deux fragments des annales de Salmanasar II," *Recueil de Travaux relatifs a la philologie et a l'archéologie égyptiennes et assyriennes* 25 (1903), 81-86.
Borger, *Asarh.*	R. Borger, *Die Inschriften Asarhaddons Königs von Assyrien* (AfO *Beiheft* 9), Graz 1956 [by page].
Borger, AfO 18	R. Borger, "Die Inschriften Asarhaddons (AfO Beiheft 9). Nachträge und Verbesserungen," AfO 18 (1957-58), 113-118.
Borger, ARRIM 6	R. Borger, "König Sanheribs Eheglück," ARRIM 6 (1988), 5-12.
Borger, BiOr 21	R. Borger, "Zu den Asarhaddon-Texten aus Babel," BiOr 21 (1964), 143-148.
Börker-Klähn, *Bildstelen*	J. Börker-Klähn, *Altvorderasiatische Bildstelen und vergleichbare Felsreliefs* (BaF 4), Mainz 1982.
Botta	P.E. Botta, *Monument de Ninive*, vols. III-IV, Paris 1849.
Brinkman, PKB	J.A. Brinkman, *A Political History of Post-Kassite Babylonia 1158-722 B.C.* (Analecta Orientalia 43), Roma 1968.
Brinkman, *Prelude to Empire*	J.A. Brinkman, *Prelude to Empire, Babylonian Society and Politics 747-626 BC*, Philadelphia 1984.
Budge, PEFQS 1890	T. Hayter Lewis, "An Assyrian Tablet from Jerusalem," *Quarterly Statement of the Palestine Exploration Fund* 1890, 265-266.
Burn, *Persia and the Greeks*	A.R. Burn, *Persia and the Greeks*, New York 1962 - London 1984[2].
CAD	*The Assyrian Dictionary of the University of Chicago*, Chicago-Glückstadt 1956-.
CAH	*The Cambridge Ancient History* [older edition as CAH, second and third edition marked as such].
Cameron, *Sumer* 6	G.C. Cameron, "The Annals of Shalmaneser III, King of Assyria," *Sumer* 6 (1950), 6-26, pl. I-II.
Cat.	C. Bezold, *Catalogue of the Cuneiform Tablets in the Kouyunjik Collection*, I-IV, London 1889-1896.
Cat. Suppl.	L.W. King, *Catalogue of the Cuneiform Tablets in the Kouyunjik Collection. Supplement*, London 1914.
Cazelles, CRAIBL 1969	H. Cazelles, "Une nouvelle stèle d'Adad-nirari d'Assyrie et Joas d'Israel," *Comptes rendus des séances Académie des inscriptions et belles-lettres* 1969, 106-117.
Cogan, HHI	M. Cogan, "Omens and Ideology in the Babylonian Inscription of Esarhaddon," in HHI (q.v.), 76-87.
Cogan, JCS 32	Mordechai Cogan, "Ashurbanipal Prism Inscriptions once Again," JCS 32 (1980), 147-150.
Cogan - Tadmor, JCS 40	M. Cogan - H. Tadmor, "Ashurbanipal Texts in the Collection of the Oriental Institute, University of Chicago," JCS 40 (1988), 84-96.

Cogan - Tadmor, Or 46 — M. Cogan - H. Tadmor, "Gyges and Ashurbanipal: a Study in Literary Transmission," Or 46 (1977), 65-84.

Cogan - Tadmor, Or 50 — M. Cogan - H. Tadmor, "Ashurbanipal's Conquest of Babylon: The First Official Report - Prism K," Or 50 (1981), 229-240.

Cogan - Tadmor, *2 Kings* — M. Cogan - H. Tadmor, *II Kings, A New Translation with Introduction and Commentary* (The Anchor Bible 11), New York 1988.

Craig, *Monolith Inscr.* — J.A. Craig, *The Monolith Inscription of Salmaneser III*, New Haven 1887.

Cross, HHI — F.M. Cross, "The Ammonite Oppression of the Tribes of Gad and Reuben: Missing Verses from 1 Samuel 11 Found in 4QSamuel[a]," in HHI (q.v.), 148-158.

CT 13 — [L.W. King] *Cuneiform Texts from Babylonian Tablets, &c., in the British Museum*, Part XIII, London 1910.

CT 26 — [L.W. King] *Cuneiform Texts from Babylonian Tablets in the British Museum*, Part XXVI, London 1909.

CT 35 — [A.W.A. Leeper] *Cuneiform Texts from Babylonian Tablets in the British Museum*, Part XXXV, London 1918.

CT 53 — S. Parpola, *Cuneiform Texts from Babylonian Tablets in the British Museum*, Part 53, London 1979.

Dalley, *Iraq* 47 — S. Dalley, "Foreign Chariotry and Cavalry in the Armies of Tiglath-Pileser III and Sargon II," *Iraq* 47 (1985), 31-48.

Dandamaev, *Achaemenid Empire* — M.A. Dandamaev, *A Political History of the Achaemenid Empire* (transl. by W.J. Vogelsang), Leiden 1989.

Delbrück, *Geschichte* — H. Delbrück, *Geschichte der Kriegskunst im Rahmen der politischen Geschichte* - 1. *Das Alterthum*, Berlin 1900.

Delbrück, *Numbers* — H. Delbrück, *Numbers in History*, London 1913.

Delitzsch, BA 6 — F. Delitzsch, "Die Palasttore Salmanassars II von Balawat (Zweiter Teil: Die Inschrift)," *Beiträge zur Assyriologie* (Hrs. F. Delitzsch - P. Haupt), Band 6/1, 129-155.

Deller - Fadhil - Ahmad, BaM 25 — K. Deller - A. Fadhil - K.M. Ahmad, "Two New Royal Inscriptions dealing with construction work in Kar-Tukulti-Ninurta," BaM 25 (1994), 459-472, Tf. 13-17.

Dombrowski, RA 67 — W.W. Dombrowski, "The Original of British Museum Tablets 90,984, 90,979 and 90,985 and the Oldest Part of the Main Inscriptions of Ashurnasirpal II," RA 67 (1973), 131-145.

Driver, *Semitic Writing* — G.R. Driver, *Semitic Writing*, London 1948.

Durand, *Documents cunéiformes* — J.-M. Durand, *Documents cunéiformes de la IVe Section de l'Ecole pratique des Hautes Etudes* (*Tome I: Catalogue et copies cunéiformes*), Genève-Paris 1982.

EAK 1	R. Borger, *Einleitung in die assyrischen Königsinschriften*, I (Handbuch der Orientalistik, Erg. Bd. V, 1/1), Leipzig-Köln 1961[1], 1964[2].
EAK 2	W. Schramm, *Einleitung in die assyrischen Königsinschriften*, II (Handbuch der Orientalistik, Erg. Bd. V, 1/2), Leipzig-Köln 1973.
Ebeling, ArOr 17	E. Ebeling, "Beschwörungen gegen den Feind und den Bösen blick aus dem Zweistromlande," *Archiv Orientální* 17 (1949), 172-211.
Edzard, *Sumer* 20	D.O. Edzard, "A New Inscription of Adad-narari I," *Sumer* 20 (1964), 49-51.
Elat, AfO *Bh.* 19	M. Elat, "The Impact of Tribute and Booty on Countries and People within the Assyrian Empire," AfO *Beiheft* 19 (Vorträge der 28 R.A.I., Wien 1981), Horn 1982, 244-251.
Elat, IEJ 25	M. Elat, "The Campaigns of Shalmaneser III against Aram and Israel," IEJ 25 (1975), 25-35.
Elat, OLA 6	M. Elat, "The Monarchy and the Development of Trade in ancient Israel," in E. Lipiński (ed.), *State and Temple Economy in The Ancient Near East* (Orientalia Lovanensia Analecta 6, Leuven 1979), 527-546.
Ellis, FD	R.S. Ellis, *Foundation Deposits in Ancient Mesopotamia*, New Haven-London 1968.
Engels, *Alexander*	D.W. Engels, *Alexander the Great and the Logistics of the Macedonian Army*, Los Angeles 1978.
Eph'al, *Ancient Arabs*	I. Eph'al, *The Ancient Arabs*, Jerusalem and Leiden 1982.
Eph'al, JAOS 94	I. Eph'al, "«Arabs» in Babylonia in the 8th Century B.C.," JAOS 94 (1974), 108-15.
Evetts, ZA 3	B.T.A. Evetts, "On Five unpublished Cylinders of Sennacherib," ZA 3 (1888), 311-31.
Fales, RAI 25	F.M. Fales, "The Enemy in the Assyrian Royal Inscriptions: «The Moral Judgement»," in H.-J. Nissen - J. Renger (eds.), *Mesopotamien und seine Nachbarn,* [Actes du] XXV Rencontre Assyriologique Internationale (Berliner Beiträge zum vorderen Orient 1), Berlin 1982, 425-35.
Fales, SAAB 4	F.M. Fales, "Grain Reserves, Daily Rations, and the Size of the Assyrian Army: a Quantitative Study," SAAB 4 (1990), 23-34.
Farbridge, *Studies*	M.H. Farbridge, *Studies in Biblical and Semitic Symbolism*, London 1923.
Faulkner, JEA 39	R.O. Faulkner, "Egyptian Military Organization," *Journal of Egyptian Archaeology* 39 (1953), 32-47.
Ford, JCS 22	M. Ford, "The Contradictory Records of Sargon II of Assyria and the Meaning of *palû*," JCS 22 (1968-69), 83-84.
Forrer, MVAG 20/3	E. Forrer, *Zur Chronologie der neuassyrischen Zeit* (Mitteilungen der Vorderasiatischen Gesellschaft 20/3), Leipzig 1916.

Freedman, *St. Louis*	R.D. Freedman, *The Cuneiform Tablets in St. Louis*, Diss., Columbia Univ. 1975.
Fs Landsberger	H.G. Güterbock - Th. Jacobsen (eds.), *Studies in Honor of Benno Landsberger on his 75th Birthday* (Assyriological Studies 16), Chicago 1965.
Fs Tadmor	M. Cogan - I. Eph'al (eds.), *Ah Assyria... Studies (...) Presented to Hayim Tadmor*, Jerusalem 1991.
Gadd, *Iraq* 10	C.J. Gadd, "Two Assyrian Observations," *Iraq* 10 (1948), 19-25, pl. V-VI.
Gadd, *Iraq* 16	C.J. Gadd, "Inscribed Prisms of Sargon II from Nimrud," *Iraq* 16 (1954), 173-201, pl. XLIII-LI.
GAG	W. von Soden, *Grundriss der akkadischen Grammatik* (Analecta Orientalia 33/47), Roma 1952 - 1969².
Galil, SAAB 6	G. Galil, "Conflicts between Assyrian Vassals," SAAB 6 (1992), 55-63.
Galter, JCS 40	H.D. Galter, "28.800 Hethiter," JCS 40 (1988), 217-35.
Galter, *Paradies*	H.D. Galter, "Paradies und Palmetod. Ökologische Aspekte im Weltbild der assyrischen Könige," in B. Scholz (ed.), *Beiträge zum 2. Grazer Morgenländischen Symposion*, Graz 1989, 235-53.
Galter - Levine - Reade, ARRIM 4	H.D. Galter - L.D. Levine - J.E. Reade, "The Colossi of Sennacherib's Palace and their Inscriptions", ARRIM 4 (1986), 27-32.
Garelli, RA 68	P. Garelli, "Remarques sur l'administration de l'empire assyrien," *Revue d'Assyriologie* 68 (1974), 129-140.
Gelb, JNES 32	I.J. Gelb, "Prisoners of War in early Mesopotamia," JNES 32 (1973), 70-98.
Gelio, RBI 32	R. Gelio, "Fonti mesopotamiche relative al territorio palestinese (1000-500 a.C.)", *Rivista Biblica Italiana* 32 (1984), 121-151.
Gelio, VO 1	R. Gelio, "Caratterizzazione ideologica e politica del cilindro Rassam," *Vicino Oriente* 1 (1978), 47-63.
Gerardi, JCS 40	P. Gerardi, "Epigraphs and Assyrian Palace Reliefs: the Development of the epigraphic Text," JCS 40 (1988), 1-35.
Gonçalves, *L'expédition de Sennachérib*	F.J. Gonçalves, *L'expédition de Sennachérib en Palestine dans la littérature hébraïque ancienne*, Louvain 1986.
Götze, *Iraq* 25	A. Goetze, "Warfare in Asia Minor," *Iraq* 25 (1963), 124-130.
Grayson, ABC	A.K. Grayson, *Assyrian and Babylonian Chronicles*, Locust Valley 1975.
Grayson, AfO 20	A.K. Grayson, "The Walters Art Gallery Sennacherib Inscription," AfO 20 (1963), 83-96, Tf. I-IV.
Grayson, ARINH	A.K. Grayson, "Assyrian Royal Inscriptions: Literary Characteristics," in ARINH (q.v.), 35-48.

Grayson, *Fs Tadmor*	A.K. Grayson, "Old and Middle Assyrian Royal Inscriptions - Marginalia," in *Fs Tadmor* (q.v.), 264-66.
Grayson, Or 49	A.K. Grayson, "Assyria and Babylonia [History and Historians of the ancient Near East]," Or 49 (1980) 140-94.
Grayson, ZA 70	A.K. Grayson, "The Chronology of the Reign of Ashurbanipal," ZA 70 (1980), 227-45.
Hallo, ErIs 14	W.W. Hallo, "Assyrian Historiography Revisited," *Eretz Israel* 14 (1978), 1*-7*.
Hallo, JNES 15	W.W. Hallo, "Zariqum," JNES 15 (1956), 220-222.
Hammond, *Alexander*	N.G.L. Hammond, *Alexander the Great, King, Commander and Statesman*, Bristol 1989^{2}.
Hammond, JHS 100	N.G.L. Hammond, "The Battle of the Granicus River," JHS 100 (1980), 73-88, pl. Ia.
Harper, *Hebr.* 4	R.F. Harper, "Some Unpublished Esarhaddon Inscriptions (Cylinder C; 80,7-19; PS. and K. 1679)," *Hebraica* 4 (1887-88), 18-25.
Harrak, *Hanigalbat*	A. Harrak, *Assyria and Hanigalbat*, Hildesheim 1987.
Hawkins, *Fs Mellink*	J.D. Hawkins, "Royal Statements of ideal Prices: Assyrian, Babylonian, and Hittite" in *Ancient Anatolia, Aspects of Change and Cultural Development* (Essays in Honour of M.J. Mellink), Madison 1986, 93-102.
Hehn, LSS 2/5	J. Hehn, *Siebenzahl und Sabbat bei den Babyloniern und im Alten Testament* (Leipziger Semitistische Studien II, 5), Leipzig 1907.
Heidel, *Sumer* 9	A. Heidel, "The octogonal Sennacherib Prism in the Iraq Museum," *Sumer* 9 (1953), 117-87.
Heidel, *Sumer* 12	A. Heidel, "A new hexagonal Prism of Esarhaddon," *Sumer* 12 (1956), 9-37, pl. I-XII.
Henshaw, *Palaeologia* 16	R.A. Henshaw, "The Assyrian Army and Its Soldier, 9th - 7th C., B.C.," *Palaeologia* 16 (1969), 1-24.
HHI	H. Tadmor - M. Weinfeld (eds.), *History, Historiography and Interpretation: Studies in Biblical and Cuneiform Literature*, Jerusalem 1982.
HKL 1 / 2	R. Borger, *Handbuch der Keilschriftliteratur*, I-II, Berlin 1967-75.
Hignett, *Xerxes*	C. Hignett, *Xerxes' Invasion of Greece*, Oxford 1963.
Hulin, *Iraq* 25	P. Hulin, "The Inscription on the Carved Throne-base of Shalmaneser III," *Iraq* 25 (1963), 48-69, pl. X.
IAK	E. Ebeling - B. Meissner - E.F. Weidner, *Die Inschriften der Altassyrischen Könige* (Altorientalische Bibliothek 1), Leipzig 1926.
ICC	A.H. Layard, *Inscriptions in the cuneiform Character from assyrian Monuments*, London 1851.

Jankowska, *Ancient Mesopotamia*	N.B. Jankowska, "Some Problems of the Economy of the Assyrian Empire," [transl. G.M. Sergheyev] in I.M. Diakonoff (ed.), *Ancient Mesopotamia, Socio-Economic History*, Moscow 1969, 253-76.
Jaritz, *Adeva Mitt.* 8	K. Jaritz, "Geheimschriftsysteme im alten Orient," *Adeva Mitteilungen* 8 (1966), 11-15.
KAJ	E. Ebeling, *Keilschrifttexte aus Assur juristischen Inhalts* (WVDOG 50), Leipzig 1927.
KAH 1 / 2	*Keilschrifttexte aus Assur historischen Inhalts*, I (L. Messerschmitt) and II (O. Schroeder) (WVDOG 16 and 37), Leipzig 1911-1922 [by no.].
KB 1 / 2	E. Schrader (ed.), *Keilinschriftliche Bibliothek*, I-II, Berlin 1889-90.
Kessler, WO 8	K. Kessler, "Die Anzahl der assyrischen Provinzen des Jahres 738 v. Chr. in Nordsyrien," WO 8 (1975-76), 49-63.
Kinnier-Wilson, *Iraq* 24	J.V. Kinnier-Wilson, "The Kurba'il Statue of Shalmaneser III," *Iraq* 24 (1962), 90-115, pl. XXX-XXXV.
Kitchen, *Third Int.*	K.A. Kitchen, *The Third Intermediate Period In Egypt*, Warminster 1973.
Knudsen, *Iraq* 29	E.E. Knudsen, "Fragments of historical Texts from Nimrud - II," *Iraq* 29 (1967), 49-69, pl. XIV-XXIX.
König, HCI	F.W. König, *Handbuch der chaldischen Inschriften*, (AfO *Beiheft* 8), I-II, Graz 1955-57.
König, *Ktesias*	F.W. König, *Die Persika des Ktesias von Knidos* (AfO *Beiheft* 18), Graz 1972.
Kühne, BaM 11	H. Kühne, "Zur Rekonstruktion der Feldzüge Adad-Nīrārī II., Tukulti-Ninurta II. und Aššurnaṣirpal II. im Hābūr-Gebiet," BaM 11 (1980), 44-70.
Lackenbacher, PSR	S. Lackenbacher, *Le palais sans rival. Le récit de construction en Assyrie*, Paris 1990.
Lackenbacher, RB	S. Lackenbacher, *Le Roi Bâtisseur. Les récits de construction assyriens des origines à Teglatphalasar III* (Études Assyriologiques, cah. n° 11), Paris 1982.
Lambert, AfO 18	W.G. Lambert, "Two Texts from the early Part of the Reign of Ashurbanipal," AfO 18 (1957-58), 382-87, Tf. XXIII-XXV.
Lambert, AnSt 11	W.G. Lambert, "The Sultantepe Tablets: VIII. Shalmaneser in Ararat," AnSt 11 (1961), 143-58, pl. XIX.
Lambert, *Ladders*	W.G. Lambert, "Portion of inscribed Stela of Sargon II, King of Assyria," in O.W. Muscarella (ed.), *Ladders to Heaven* (Toronto 1981), 125.
Landsberger, JCS 8	B. Landsberger, "Assyrische Königsliste und 'Dunkles Zeitalter'," JCS 8 (1954), 31-73, 106-33.
Læssøe, *Iraq* 21	J. Læssøe, "A Statue of Shalmaneser III from Nimrud," *Iraq* 21 (1962), 147-57, pl. XLf.

LCA	F.M. Fales - G.B. Lanfranchi, *Lettere dalla corte assira*, Venezia 1992.
van Leeuwen, OTS 14	C. van Leeuwen, "Sanchérib devant Jérusalem," *Oudtestamentische Studiën* 14 (1965), 245-72.
Le Gac, *Asn.*	Y. Le Gac, *Les Inscriptions d'Aššur-naṣir-aplu III*, Paris 1907.
Lehmann-Haupt, *Materialen*	C.F. Lehmann-Haupt, *Materialen zur älteren Geschichte Armeniens und Mesopotamiens*, Berlin 1907.
Levine, ARINH	L.D. Levine, "Manuscripts, Texts, and the Study of the Neo-Assyrian Royal Inscriptions," in ARINH (q.v.), 49-70.
Levine, HHI	L.D. Levine, "Preliminary Remarks on the historical Inscriptions of Sennacherib," in HHI (q.v.), 58-75.
Levine, *Stelae*	L.D. Levine, *Two Neo-Assyrian Stelae from Iran*, Toronto 1972.
Lie, *Sargon*	A.G. Lie, *Inscriptions of Sargon II, King of Assyria: Part I, the Annals*, Paris 1929.
Ling-Israel, *Fs Artzi*	P. Ling-Israel, "The Sennacherib Prism in the Israel Museum - Jerusalem," in J. Klein - A. Skaist (eds.), *Bar-Ilan Studies in Assyriology dedicated to Pinhas Artzi*, Ramat Gan 1990, 213-48, pl. IV-XVI.
Liverani, *Ideology*	M. Liverani, "The Ideology of the assyrian Empire," in M.T. Larsen (ed.), *Power and Propaganda, a Symposium on ancient Empires* (Mesopotamia 7, Copenhagen 1979), 297-317.
Luckenbill, *Senn.*	D.D. Luckenbill, *The Annals of Sennacherib* (Oriental Institute Publications 2), Chicago 1924.
Luckenbill, AJSL 41	D.D. Luckenbill, "The Black Stone of Esarhaddon," *American Journal of Semitic Languages and Literatures* 41 (1925), 165-73.
Lyon, *Sargon*	D.G. Lyon, *Keischrifttexte Sargons II, Königs von Assyrien*, Leipzig 1883.
MacGinnis, SAAB 6	J. MacGinnis, "Tablets from Nebi Yunus," SAAB 6 (1992), 3-19.
Machinist, JAOS 103	P. Machinist, "Assyria and Its Image in the First Isaiah," JAOS 103 (1983), 719-37.
Mahmud - Black, *Sumer* 44	M. Mahmud - J. Black, "Recent Work in the Nabu Temple, Nimrud," *Sumer* 44 (1985-86), 135-55.
Malbran-Labat, *L'armée*	F. Malbran-Labat, *L'armée et l'organisation militaire de l'Assyrie*, Paris 1982.
Mallowan, *Nimrud*	M.E.L. Mallowan, *Nimrud and its Remains*, London 1966.
Manitius, ZA 24	W. Manitius, "Das stehende Heer der Assyrerkönige und seine Organisation," ZA 24 (1910), 97-149, 185-224.
Marsden, *Gaugamela*	E.W. Marsden, *The Campaign of Gaugamela*, Liverpool 1964.
Martin, *Tribut*	W.J. Martin, *Tribut und Tributliestungen bei den Assyrern* (Studia Orientalia 8/1), Helsingforsiae 1936.

Mayer, MDOG 112	W. Mayer, "Sargons Feldzug gegen Urartu - 714 v. Chr.: Eine militärhistorische Würdigung," MDOG 112 (1980), 13-33.
Mayer, MDOG 115	W. Mayer, "Sargons Feldzug gegen Urartu - 714 v. Chr.: Text und Übersetzung," MDOG 115 (1983), 65-132.
Melikišvili, UKN	G.A. Melikišvili, *Urartskie klinoobraznye nadpisi*, Moskva 1960.
Michel, WO	E. Michel, "Die Assur-Texte Salmanassars III. (858-824)," WO 1 (1947-52) 5-20, 57-71, 205-22, 255-71, 385-96, 454-75, Tf. 1-9, 12f, 18-24; 2 (1954-59), 27-45, 137-57, 221-33, 408-15, Tf. 2-4, 6f, 11; 3 (1964), 146-55, Tf. 1f; 4 (1967-68), 29-37.
Millard, *Eponyms*	A. Millard, *The Eponyms of the Assyrian Empire 910-612 BC* (SAA Studies 2), Helsinki 1994.
Millard, *Fs Tadmor*	A.R. Millard, "Large Numbers in the Assyrian Royal Inscriptions," in *Fs Tadmor* (q.v.), 213-22.
Millard, *Iraq* 30	A.R. Millard, "Fragments of Historical Texts from Nineveh: Ashurbanipal," *Iraq* 30 (1968), 98-11, pl. XIX-XXVII.
Millard, *Iraq* 32	A.R. Millard, "Fragments of Historical Texts from Nineveh: Middle Assyrian and later Kings," *Iraq* 32 (1970), 167-76, pl. XXXII-XXXVII.
Millard - Tadmor, *Iraq* 35	A.R. Millard - H. Tadmor, "Adad-Nirari III In Syria, another Stele Fragment and the Dates of his Campaigns," *Iraq* 35 (1973), 57-64, pl. XXIX.
Na'aman, BASOR 214	N. Na'aman, "Sennacherib's "Letter to God" on his Campaign to Judah," BASOR 214 (1974), 25-39.
Na'aman, *Iraq* 46	N. Na'aman, "Statements of Time-spans by Babylonian and Assyrian Kings and Mesopotamian Chronology," *Iraq* 46 (1984), 115-23.
Na'aman, *Tel Aviv* 3	N. Na'aman, "Two Notes on the Monolith Inscription of Shalmaneser III from Kurkh," *Tel Aviv* 3 (1976), 89-106.
Na'aman, ZA 81	N. Na'aman, "Chronology and History in the Late Assyrian Empire (631-619 B.C.)," ZA 81 (1991), 243-67.
Nassouhi, AfO 2	E. Nassouhi, "Prisme d'Assurbânipal daté de sa trentième année, provenant du temple de Gula à Babylone," *Archiv für Keilschriftforschung* 2 (1925), 97-106.
Nassouhi, MAOG 3	E. Nassouhi, "Textes divers relatifs à l'historie de l'Assyrie," MAOG 3/1-2, Leipzig 1927.
ND	(number of the tablets excavated at Nimrud).
Neugebauer, *Sciences*	O. Neugebauer, *The exact Sciences in Antiquity*, Providence 1957[2].
Noble, SAAB 4	D. Noble, "Assyrian Chariotry and Cavalry," SAAB 4 (1990), 61-68.
Oded, *Deportations*	B. Oded, *Mass Deportations and Deportees in the Neo-Assyrian Empire*, Wiesbaden 1979.

Olmstead, *Historiography*	A.T.E. Olmstead, *Assyrian Historiography, A Source Study,* Columbia 1916.
Olmstead, *History*	A.T.E. Olmstead, *History of Assyria*, New York 1923 (repr. 1975).
Olmstead, JAOS 38	A.T.E. Olmstead, "The Calculated Frightfulness of Ashur Nasir Apal," JAOS 38 (1918), 209-63.
Olmstead, JAOS 41	A.T.E. Olmstead, "Shalmaneser III and the Establishment of the Assyrian Power," JAOS 41 (1921), 345-82.
Oppenheim, JNES 19	A.L. Oppenheim, "The City of Assur in 714 B.C.," JNES 19 (1960), 133-47.
Oppenheim, *Letters*	A.L. Oppenheim, *Letters from Mesopotamia*, Chicago 1967 [by no.].
Oppenheim, *Propaganda and Communication*	A.L. Oppenheim, "Neo Assyrian and Neo-Babylonian Empires", in H.D. Lasswell - D. Lerner - H. Speier (eds.), *Propaganda and Communication in World History, Vol. I: The Symbolic Instrument in Early Times*, Honolulu 1979, 111-44.
Paley, *Aššurnaṣirpal*	S.M. Paley, *King of the World: Aššur-nāṣir-pal II of Assyria, 883-859*, New York 1976.
Page, *Iraq* 30	S. Page, "A Stela of Adad-Nirari III and Nergal-ereš from Tell al Rimah," *Iraq* 30 (1968), 139-53, pl. XXXIX.
Page, VT 19	S. Page, "Joash and Samaria in a New Stela excavated at Tell al Rimah, Iraq," *Vetus Testamentum* 19 (1969), 483f.
Parpola, ARINH	S. Parpola, "Assyrian Royal Inscriptions and Neo-Assyrian Letters," in ARINH (q.v.), 117-42.
Parpola, JSS 21	S. Parpola [Review of J.V. Kinnier-Wilson, *The Nimrud Wine Lists*], JSS 21 (1976), 165-74.
Parpola, LASEA 1 / 2	S. Parpola, *Letters from Assyrian Scholars to the Kings Esarhaddon and Assurbanipal*, I-II (AOAT 5/1-2), Neukirchen-Vluyn 1970-1983.
Parpola, *Toponyms*	S. Parpola, *Neo-Assyrian Toponyms* (AOAT 6), Neukirchen-Vluyn 1970.
Pearce, *Cryptography*	L.E. Pearce, *Cuneiform Cryptography, Numerical Substitutions for Syllabic and Logographic Signs*, Diss., Yale Univ. 1982.
Pecorella - Salvini, *Tra lo Zagros e l'Urmia*	P.E. Pecorella - M. Salvini, *Tra lo Zagros e l'Urmia*, Roma 1984.
Pedersén, *Archives*	O. Pedersén, *Archives and Libraries in the City of Assur*, I, Uppsala 1985.
Peñuela, *Sefarad* 9	J.M. Peñuela, "El registro de tributos de los principes sirios en la estela de Kurḫ (857 a.C)," *Sefarad* 9 (1949), 3-25.
Pettinato, *Semiramide*	G. Pettinato, *Semiramide*, Milano 1985.
Pfeiffer, SLA	R.H. Pfeiffer, *State Letters of Assyria*, New Haven 1935 [by no.].

Piepkorn, *Asb.*	A.C. Piepkorn, *Historical Prism Inscriptions of Ashurbanipal*, I (Assyriological Studies 5), Chicago 1933.
Pinches, *Balawat*	Th. Pinches, *The Bronze Ornaments of the Palace Gates of Balawat*, London 1902.
Pitard, *Ancient Damascus*	W.T. Pitard, *Ancient Damascus*, Winona Lake 1987.
Poebel, JNES 1	A. Poebel, "The Assyrian King List from Khorsabad," JNES 1 (1942), 247-306 and 460-92.
Porter, IPP	B. Nevling Porter, *Images, Power, and Politics. Figurative Aspects of Esarhaddon's Babylonian Policy*, Philadelphia 1993.
Postgate, FNALD	J.N. Postgate, *Fifty Neo-Assyrian Legal Documents*, Warminster 1976.
Postgate, GPA	J.N. Postgate, *The Governor's Palace Archive* (*Cuneiform Texts from Nimrud* II), London 1973.
Postgate, *Sumer* 29	J.N. Postgate, "The Inscription of Tiglath-Pileser III at Mila Mergi," *Sumer* 29 (1973), 47-59, fig. 1-7.
Postgate, TCAE	J.N. Postgate, *Taxation and Conscription in the Assyrian Empire* (Studia Pohl S.M. 3), Rome 1974.
R	H.C. Rawlinson (ed.), *The Cuneiform Inscriptions of Western Asia*, I-V, London 1861-1891 [as 1 R, 2 R etc.].
Rasmussen, *Salm.*	N. Rasmussen, *Salmanasser den II's Indskrifter.* I - *Kileskrift, Transliteration og Translation, ...*, Kjøbenhavn 1897.
Reade, *Iraq* 34	J.E. Reade, "The Neo-Assyrian Court and Army: Evidence from the Sculptures," *Iraq* 34 (1972), 87-112, XXXIII-XXXIX.
Reade, *Iraq* 37	J.E. Reade, "Aššurnaṣirpal I and the White Obelisk," *Iraq* 37 (1975), 129-50, pl. XXVIII-XXXI.
Reade, JCS 27	J. Reade, "Sources for Sennacherib: the Prisms," JCS 27 (1975), 189-96.
Reade, RA 72	J. Reade, "Studies in Assyrian Geography. Part I: Sennacherib and the Waters of Nineveh," RA 72 (1978), 47-72.
Reade, ZA 68	J.E. Reade, "Assyrian Campaigns, 840-811 B.C., and the Babylonian Frontier," ZA 68 (1978), 251-60.
Reade - Walker, Af0 28	J. Reade - C.B.F. Walker, "Some Neo-Assyrian Royal Inscriptions," AfO 28 (1981-82), 113-22.
Reiner, JNES 26	E. Reiner - M. Civil, "Another Volume of Sultantepe Tablets," JNES 26 (1967), 177-211.
Renger, *Sargon*	J. Renger, *The Inscriptions of Sargon II, King of Assyria*, unpub. manuscript [1978 - the introduction (cited as *Introd.*) has a separate pagination].
Rigg, JAOS 62	H.A. Rigg, Jr., "Sargon's 'Eighth Military Campaign'," JAOS 62 (1942), 130-38.
RIMA 1	A.K. Grayson, *Assyrian Rulers of the Third and Second Millennia BC* (RIMA 1), Toronto 1987.

RIMA 2	A.K. Grayson, *Assyrian Rulers of the Early First Millennium BC*, I (RIMA 2), Toronto 1991.
RlA	*Reallexikon der Assyriologie*, 1-6, Berlin 1932-.
Rost, *Tigl. III*	P. Rost, *Die Keilschrifttexte Tiglat-Pilesers III*, I-II, Leipzig 1893 [Pl. refers to the plates of the 2nd vol.].
Russell, AnSt 34	H.F. Russell, "Shalmaneser's Campaign to Urartu in 856 B.C. and the Historical Geography of Eastern Anatolia according to the Assyrian Sources," AnSt 34 (1984), 171-201.
Russell, *Iraq* 47	H.F. Russell, "The Historical Geography of the Euphrates and Habur according to the Middle- and Neo-Assyrian Sources," *Iraq* 47 (1985), 57-74.
Russell, *Sennacherib's Palace*	J.M. Russell, *Sennacherib's Palace without Rival at Nineveh*, Chicago 1991.
SAA 1	S. Parpola, *The Correspondence of Sargon II*, Part 1 (SAA I), Helsinki 1987 [by no.].
SAA 3	A. Livingstone, *Court Poetry and Literary Miscellanea*, (SAA III), Helsinki 1989 [by no.].
SAA 5	G.B. Lanfranchi - S. Parpola, *The Correspondence of Sargon II*, Part 2 (SAA V), Helsinki 1990 [by no.].
SAA 11	F.M. Fales - J.N. Postgate, *Imperial Administrative Records*, Part 2 (SAA XI), Helsinki 1995 [by no.].
Sader, *États araméens*	H.S. Sader, *Les états araméens de Syrie, depuis leur fondation jusqu'à leur transformation en provinces assyriennes*, Beirut 1987.
Safar, *Sumer* 7	F. Safar, "A Further Text of Shalmaneser III from Assur," *Sumer* 7 (1951), 3-21, pl. I-III.
Saggs, *Assyria*	H.W.F. Saggs, *The Might that was Assyria*, London 1984.
Saggs, *Iraq* 21	H.W.F. Saggs, "The Nimrud Letters, 1952 - Part V," *Iraq* 21 (1959), 158-179, pl. XLIII-IL.
Saggs, *Iraq* 25	H.W.F. Saggs, "Assyrian Warfare in the Sargonid Period," *Iraq* 25 (1963), 145-54.
Saggs, *Iraq* 36	H.W.F. Saggs, "The Nimrud Letters, 1952 - Part IX," *Iraq* 36 (1974), 199-221, pl. XXXIV-XXXVII.
Saggs, *Iraq* 37	H.W.F. Saggs, "Historical Texts and Fragments of Sargon II of Assyria, I. The «Aššur Charter»," *Iraq* 37 (1975), 11-20, pl. IX.
Salonen, *Jagd*	A. Salonen, *Jagd und Jagdtiere im alten Mesopotamien* (Ann. Acad. Scientiarum Fennicæ, Ser. B, 196), Helsinki 1976.
Saporetti, ANLR 8/25	C. Saporetti, "Una deportazione al tempo di Salmanassar I," *Accad. Naz. dei Lincei: Rendiconti*, Serie VIII vol. 25 (1970), 437-53.
Sasson, *Mari*	J.M. Sasson, *The Military establishments at Mari* (Studia Pohl 3), Rome 1969.

Sauren, WO 16	H. Sauren, "Sennachérib, les Arabes, les déportés Juifs," WO 16 (1985), 80-99.
Scheil, *Prisme S*	V. Scheil, *Le prisme S d'Assaraddon*, Paris 1914.
Schneider, *Shalm.*	T.J. Schneider, *A new analysis of the royal annals of Shalmaneser III*, Diss., Univ. of Pennsylvania 1991.
Schott, MVAG 30/2	A. Schott, *Die Vergleiche in den akkadischen Königsinschriften*, (MVAG 30.2), Leipzig 1926.
Schramm, BiOr 27	W. Schramm, "Die Annalen des assyrischen Königs Tukulti-Ninurta II (890-884 v. Chr.)," BiOr 27 (1970), 147-50, pl. I-V.
Segal, JSS 10	J.B. Segal, "Numerals in the Old Testament," JSS 10 (1965), 2-20.
Seidmann, MAOG 9/3	J. Seidmann, *Die Inschriften Adadnirâris II* (MAOG 9/3), Leipzig 1935.
Seux, *Épithètes*	M.-J. Seux, *Épithètes royales akkadiennes et sumériennes*, Paris 1967.
Shea, JCS 30	W.H. Shea, "Adad-Nirari III and Jehoash of Israel," JCS 30 (1978), 101-113.
Smith, AD	G. Smith, *Assyrian Discoveries*, London 1875.
Smith, *First Camp.*	S. Smith, *The First Campaign of Sennacherib, King of Assyria, B.C. 705-681*, London 1921.
Smith, *Senn.*	G. Smith, *History of Sennacherib translated from the Cuneiform Inscriptions*, London 1878.
von Soden, *Fs Stier*	W. von Soden, "Sanherib vor Jerusalem 701 v. Chr.," in R. Stiehl - G.A. Lehmann (Hrsg.), *Antike und Universalgeschichte* (*Festschrift H.E. Stier*), Münster 1972, 43-51.
von Soden, *Iraq* 25	W. von Soden, "Die Assyrer und der Krieg," *Iraq* 25 (1963), 131-144.
Sollberger, *Iraq* 36	E. Sollberger, "The White Obelisk," *Iraq* 36 (1974), 231-38, pl. XLI-XLVIII.
Speiser, *Idea of History*	E.A. Speiser, "Ancient Mesopotamia," in R.C. Dentan (ed.), *The Idea of History in the Ancient Near East*, New Haven 1955.
Stohlmann, *Scripture in Context* 2	S. Stohlmann, "The Judaean Exile after 701 B.C.E.," in W.W. Hallo - J.C. Moyer - L.G. Perdue (eds.), *Scripture in Context II, More Essays on the Comparative Method*, Winona Lake 1983, 147-75.
Streck, *Asb.*	M. Streck, *Assurbanipal und die letzten assyrischen Könige ...*, I-III (Vorderasiatische Bibliothek 7), Leipzig 1916.
STT 1	O.R. Gurney - J.J. Finkelstein, *The Sultantepe Tablets* I, London 1957 [by no.].
STT 2	O.R. Gurney - P. Hulin, *The Sultantepe Tablets* II, London 1964 [by no.].

Tadmor, AAC	H. Tadmor, "Monarchy and the Elite in Assyria and Babylonia: The Question of Royal Accountability," in S.N. Eisenstadt (ed.), *The Origins and Diversity of Axial Age Civilizations*, New York 1986, 203-24.
Tadmor, ARINH	H. Tadmor, "History and Ideology in the Assyrian Royal Inscriptions," in ARINH (q.v.), 13-34.
Tadmor, *Fs Finkelstein*	H. Tadmor, "Observations on Assyrian Historiography," in M. de Jong Ellis (Ed.), *Ancient Near Eastern Studies in Memory of J.J. Finkelstein*, Hamden 1977, 209-213.
Tadmor, HHI	H. Tadmor, "Autobiographical Apology in the Royal Assyrian Literature," in HHI (q.v.), 36-57.
Tadmor, IEJ 11	H. Tadmor, "Que and Muṣri," IEJ 11 (1961), 143-50.
Tadmor, *Intr. Remarks*	H. Tadmor, "Introductory Remarks to a new Edition of the Annals of Tiglath-Pileser III," *Proceedings of the Israel Academy of Sciences and Humanities* 2/9 (1968), 168-87.
Tadmor, *Iraq* 35	H. Tadmor, "The Historical Inscriptions of Adad-nirari III," *Iraq* 35 (1973), 141-50.
Tadmor, JCS 12	H. Tadmor, "The Campaigns of Sargon II of Assur: a chronological Study," JCS 12 (1958), 22-40 and 77-100.
Tadmor, *Scr.Hier.* 8	H. Tadmor, "Azriyau of Yaudi," *Scripta Hierosolymitana* 8 (1961), 232-71.
Tadmor, *Tigl. III*	H. Tadmor, *The Inscriptions of Tiglath-Pileser III King of Assyria*, Jerusalem 1994.
Tadmor, *Unity and Diversity*	H. Tadmor, "Assyria and the West: The Ninth Century and Its Aftermath," in H. Goedicke - J.J.M. Roberts (eds.), *Unity and Diversity* (Baltimore-London 1975), 36-47.
Tallqvist, APN	K.L. Tallqvist, *Assyrian Personal Names*, Helsingfors 1914
TCL 3	(cf. Thureau-Dangin, TCL 3).
Thompson, AAA 19	R.C. Thompson - R.W. Hamilton, "The British Museum Excavations on the Temple of Ishtar at Nineveh, 1930-31," AAA 19 (1932), 55-116; pl. XLVI-XCII.
Thompson, AAA 20	R.C. Thompson - M.E.L. Mallowan, "The British Museum Excavations at Nineveh, 1931-32," AAA 20 (1933), 71-186; pl. XXXV-CVI.
Thompson, *Arch.* 79	R.C. Thompson - R.W. Hutchinson, "The Excavations on the Temple of Nabû at Nineveh," *Archaeologia* 79 (1929), 103-48, pl. XLI-LXXIX.
Thompson, *Iraq* 7	R.C. Thompson, "A Selection from the Cuneiform Historical Texts from Nineveh (1927-32)," *Iraq* 7 (1940), 85-111, fig. 1-20.
Thompson, PEA	R.C. Thompson, *The Prisms of Esarhaddon and Ashurbanipal*, London 1931.
Thureau-Dangin, JA 1909	F. Thureau-Dangin, "L'U, le qa et la mine, leur mesure et leur rapport," *Journal Asiatique* X. 13 (1909), 79-111.

Thureau-Dangin, RA 22 — F. Thureau-Dangin, "La grande coudée assyrienne," RA 22 (1925), 30.

Thureau-Dangin, RA 24 — F. Thureau-Dangin - Ch.-F. Jean, "Les annales de la salle II du palais de Khorsabad," RA 24 (1927), 75-80.

Thureau-Dangin, TCL 3 — F. Thureau-Dangin, *Une Relation de la Huitième Campagne de Sargon* (Textes Cunéiformes, Musée du Louvre, 3), Paris 1912.

TUAT — R. Borger - W. Hinz - W.H.Ph. Römer, *Historisch-chronologische Texte* I (Texte aus der Umwelt des Alten Testaments, Bd. I Lf. 4), Gütersloh 1984 [only this volume is taken into consideration].

Turner, *Iraq* 32 — G. Turner, "Tell Nebi Yūnus: the *ekal māšarti* of Nineveh," *Iraq* 32 (1970), 68-85, pl. XV.

Unger, *Balawat* — E. Unger, *Zum Bronzetor von Balawat*, Leipzig 1912.

Unger, FF 7 — E. Unger, "Altorientalische Zahlensymbolik," *Forschungen und Fortschritte* 7 (1931), 263.

Unger, MDAIA 45 — E. Unger, "Die wiederherstellung des Bronzetors von Balawat," *Mitteilungen der Deutschen Archäologischen Instituts - Athenische Abteilung* 45 (1920), 1-105, Tf. I-III.

Unger, PKOM 2 — E. Unger, *Reliefstele Adadniraris III. aus Saba'a und Semiramis* (Publicationen der Kaiserlich Osmanischen Museen 2), Konstantinopel 1916.

Ungnad, OLZ 21 — A. Ungnad, "Der Gottesbrief als Form assyrischer Kriegsberichterstattung," OLZ 21 (1918), 72-75.

Ungnad, ZAW 59 — A. Ungnad, "Die Zahl der von Sanherib deportierten Judäer," *Zeitschrift für die alttestamentliche Wissenschaft* 59 (1942-43), 199-202.

Van der Spek, JEOL 25 — R.J. van der Spek, "The Struggle of King Sargon II of Assyria against the Chaldaean Merodach-Baladan," *Jaarbericht Ex Oriente Lux* 25 (1977-78), 56-66.

VS 1 — L. Messerschmidt - A. Ungnad, *Vorderasiatischen Schriftdenkmäler der Königlichen Museen zu Berlin*, I, Leipzig 1907.

Weidner, AfO 3 — E.F. Weidner, "Die Annalen des Königs Aššurdân II. von Assyrien," AfO 3 (1926), 151-61.

Weidner, AfO 6 — E.F. Weidner, "Die Annalen des Königs Aššurbêlkala von Assyrien," AfO 6 (1930-31), 75-94.

Weidner, AfO 9 — E.F. Weidner, "Die Feldzüge Šamši-Adads V. gegen Babylonien," AfO 9 (1933-34), 89-104.

Weidner, AfO 10 — E.F. Weidner, "Aus den Tagen eines assyrischen Schattenkönigs," AfO 10 (1935-36), 1-48.

Weidner, AfO 12 — E.F. Weidner, "Neue Bruchstücke des Berichtes über Sargons achten Feldzug," AfO 12 (1937-39), 144-48.

Weidner, AfO 14 — E.F. Weidner, "Šilkan(ḫe)ni, König von Muṣri, ein Zeitgenosse Sargons II.," AfO 14 (1943-44), 40-53.

Weidner, AfO 15	E.F. Weidner, "Bemerkungen zur Königliste aus Chorsābād," AfO 15 (1945-51), 85-102.
Weidner, AfO 18	E.F. Weidner, "Die Feldzüge und Bauten Tiglatpileser I.," AfO 18 (1957-58), 342-360, Tf. XXVI-XXX.
Weidner, AfO 21	E.F. Weidner, "Assyrische Itinerare," AfO 21 (1966), 42-46, Tf. VIII.
Weidner, *Tn.*	E.F. Weidner, *Die Inschriften Tukulti-Ninurtas I. und seiner Nachfolger* (AfO *Beiheft* 12), Graz 1959.
Weippert, WO 7	M. Weippert, "Die Kämpfe assyrischen Königs Assurbanipals gegen die Araber," WO 7 (1976), 39-85.
Weippert, ZDPV 89	M. Weippert, "Menahem von Israel und seine Zeitgenosse in einer Steleinschrift des assyrischen Königs Tiglatpileser III. aus dem Iran," ZDPV 89/1 (1973), 26-53.
Winckler, AOF 2	H. Winckler, *Altorientalische Forschungen*, II, Leipzig 1899.
Winckler, *Sargon*	H. Winckler, *Die Keischrifttexte Sargons*, I-II, Leipzig 1889 [number (no.) or plate (Tf.) refers to the 2nd vol.].
Winckler, ZA 2	H. Winckler, "Studien und beiträge zur babylonisch-assyrischen geschichte," ZA 2 (1887), 148-178, 299-315, Tf. I-III.
Wiseman, *Iraq* 13	D.J. Wiseman, "Two Historical Inscriptions from Nimrud," *Iraq* 13 (1951), 21-26, pl. XI-XII.
Wiseman, *Iraq* 14	D.J. Wiseman, "A New Stela of Aššur-naṣir-pal II," *Iraq* 14 (1952), 24-44, pl. II-VI.
Wiseman, *Iraq* 18	D.J. Wiseman, "A fragmentary Inscription of Tiglath-pileser III from Nimrud," *Iraq* 18 (1956), 117-129, pl. XXIf.
Yadin, *Warfare*	Y. Yadin, *The Art of Warfare in Biblical Lands* (transl. by M. Pearlman), London 1963.
Yamada, ZA 84	S. Yamada, "The Editorial History of the Assyrian King List," ZA 84 (1994), 11-37.
YOS 9	F.J. Stephens, *Votive and Historical Texts from Babylonia and Assyria* (*Yale Oriental Series* 9), New Haven 1937.
Zaccagnini, *Le tribut*	C. Zaccagnini, "Prehistory of the Achaemenid Tributary System," in P. Briant - C. Herrenschmidt (ed.), *Le tribut dans l'empire perse*, Paris 1989, 193-215.
Zaccagnini, *Opus* 3	C. Zaccagnini, "La circolazione dei beni di lusso nelle fonti neo-assire (IX-VII sec. a.C.)," *Opus* 3 (1984), 235-52.
Zawadzki, *Fall of Assyria*	S. Zawadzki, *The Fall of Assyria and Median-Babylonian relations in light of the Nabopolassar Chronicle*, Poznań 1988.
Zimansky, *Ecology and Empire*	P.E. Zimansky, *Ecology and Empire: the Structure of the Urartian State*, (Studies in Ancient Oriental Civilizations 41), Chicago 1985.

1. INTRODUCTION

The use of quantifications[1] in the Assyrian royal inscriptions[2] has never been the subject of a systematic study, even if often it is affirmed that these numbers have been, or could have been, inflated for propaganda purposes.[3] In some cases single numbers have been considered, or suspected of being, exaggerated, already during the edition of the texts.[4] Even, the neo-Assyrian royal inscriptions have been cited as examples of gross exaggerations or inventions, especially for their numbers, and mostly by those outside the field of specialists in Assyrian history.[5] On the other hand, sometimes these numbers are taken into account for statistics and calculations, being considered "facts" that, until they can be demonstrated as wrong, cannot be ignored on the basis of simple suspicions or suppositions. In the case of internal inconsistency between the texts, usually the lower numbers are taken

---

1 Note that for "quantification" I will mean any representation of quantity, both through a number or through other more generic indications, such as "many," "all," etc. However, only the numerical quantifications are the subject of this study, the other ones being taken into consideration as far as they represent an alternative to the number, and as far as they are subject to the same compositive rules. As a matter of principle when speaking of a "quantification" (singular) I will mean a single representation of quantity (a single "accounting") even if it is given by several inscriptions (or by several copies of a certain inscription) with diverging digits/quantity. For instance, the camels carried off to Assyria by Sennacherib at the end of his first campaign are quantified *5,230, 5,233* or "without measure (*ša lā nībi*)" (cf. p. 58 no. 67). Now, this is a single quantification: the number of camels carried off to Assyria on that occasion (the number is, in a sense, part of the quantification). Thus, it is then possible to state that the royal inscriptions of Sennacherib quantify the camels only once: even if it will be possible to find several quantities of camels, all these are referring to a single instance.

Certain numbers are not quantifications, not being the result of an "accounting". Thus, the expression "four parts of the world," having the meaning of oecumene, is not subject to the same editing rules and processes applied to the other quantifications, being moreover employed in the royal "titulary" and not within the narrative sections of the inscriptions; it will therefore be excluded from the analysis. Other numbers having an evident symbolic significance, instead, will be considered, being after all the result of an "accounting": think for instance of the "*7* kings of Cyprus living at *7* days of travel in the midst of the sea" (cf. p. 133).

2 These documents belong to a well-defined literary genre (for some remarks on their classification, cf. beginning of chapter 2). In this work, also some texts of "literary" nature (e.g. epics or the like) will be taken into consideration — mostly for purposes of comparison — even if quite rarely since these texts show a very low number of quantifications.

3 Cf. e.g. Luckenbill, *Senn.*, 1; Gelb, JNES 32, 72.

4 Cf. e.g. Weidner, *Tn.*, 26 nn. 27-30; *id.*, AfO 9, 104; Michel, WO 2, 41 n. 10.

5 For example, Chester G. Starr's *A History of the Ancient World*, a one-volume general introduction to the ancient western and eastern cultures (incl. China, India, etc.), in two places outlines, in a very brief fashion, the most notable characteristics of the neo-Assyrian annalistic inscriptions, without forgetting to mention the "incredible inflation of the numbers of the captives and killed" or to point out that "the numbers of cattle captured and other details are increased amazingly from the earlier to the later version" (respectively p. 136 and 142 of the New York 1974 edition).

into consideration.[6] It is unnecessary to stress how either to accept or to refuse *en bloc* all the quantifications could lead to error. Each case should be individually analyzed, and the truthfulness of each number will be based on the context and on a very small number of rules always valid (for all the quantifications of the *corpus*), obviously using, as far as possible, also other sources besides the royal inscriptions (administrative documents, etc.). It is not surprising that no systematic studies on the truthfulness of the data reported by these inscriptions have ever been undertaken: they would not make much sense. The (very few) works on this topic are, as a matter of fact, centered on some specific quantifications, even if conclusions of a more general value are sometimes drawn: see the very short article by Ungnad in ZAW 59 (*ca.* 3 pages) or the brief remarks made *en passant* by Olmstead in his *Historiography*. In both cases certain exaggerations are pointed out, thanks to the presence of older inscriptions with much lower numbers, and general indications on the ways the editors of the royal inscriptions made these exaggerations are deduced.[7] More recently, in a short article H.D. Galter discussed an episode (and a quantification) relative to the reign of Tukulti-Ninurta I,[8] while A.R. Millard (in *Fs Tadmor*) confronted the argument "numbers" in a broader manner — even if not undertaking a systematic analysis, in view of the brevity of the article — yet reaching conclusions of general significance.[9]

A systematic study of the quantifications should be conceived as the study of each single quantification, to be evaluated for itself and within its documentary and historical context. Also a comprehensive, "statistical" examination could be useful, being possible in this manner to stress the general tendencies, to be used simply as a general background to the evaluation of each single quantification.

The studying of the Assyrian royal inscriptions has revived in recent years, with a special interest in their aims and their "messages"; this, however, is not a matter for this work.[10] Yet the possibility exists of analyzing the quantifications for their ideological significance and for their narrative functions. In such ideologically characterized ("propaganda") documents, the use of quantifications to stress one's achievements corresponds to definite rules, made of devices aimed at presenting the data (more or less manipulated) in such a way as to confirm the message carried by the text and/or the context. However, since the Assyrian royal inscriptions are not war bulletins nor

6 Cf. Oded, *Deportations,* 21 n. 6. He remarks (p. 18f) that calculations based on these figures allow one to obtain what he calls "a factual picture," *i.e.* a picture as drawn by the records, or the scribes. Thus, Liverani, *Asn.*, 133ff, takes into account all the data given by Ashurnasirpal II's "annals," to draw a number of conclusions on the demography and the economy of the Assyrian empire.

7 Cf. below chapter 3 §4.

8 JCS 40; cf. pp. 150ff.

9 The article by Millard appeared quite recently (1991), and arrived when the present work was already in a rather advanced phase. Some conclusions of Millard are to be shared: for instance the function of the "round numbers," the copying errors as the cause of most of the variants, cf. p. 214 and 218, or the arguments on the truthfulness of the numerals on p. 217f. Also the general conclusions (p. 221) can be accepted, but the fact that the "exact" numbers are the result of careful accounting is decidedly disputable (cf. below, p. 86). There are also other points of disagreement: cf. below p. 88 and p. 166 n. 36.

Millard's article has been recently supplemented by D.M. Fouts (in JNES 53), who stressed the occurrence of "literary hyperbole" with the numerals reported in the Assyrian royal inscriptions.

10 Some contributions are collected in ARINH and in HHI, but several others have appeared, by Cogan, Grayson, Levine, Liverani, Reade, Tadmor, etc. (cf. the bibliography). The present work owes much, from the point of view of the methodology, to Liverani's contribution in ARINH.

simple booty lists, but, generally speaking, fluctuate between these two forms and "free narration"; one has to conceive of the quantifications as a non-indispensable complement of the text, an amplification whose use is not imposed by any compositive or stylistic rule. In this respect, thus, the *absence* of the number is the rule, and its presence is a simple option. The use of quantifications cannot give, therefore, a 360° view of the narrative, ideological, etc. characteristics of the inscriptions.[11] At most, some editing processes determining which quantifications to use, where and when, can be pointed out. As for the number itself, one may note that for a "correct" propagandistic usage it — obviously — must be as much quantitatively characterized (i.e. "high") as possible, and if it is not, it can be altered or passed over in silence.

The ideologically distorted, propagandistic use of the number is certainly not an exclusivity of the Assyrian royal inscriptions, and has been the subject of research in relation to other historical contexts.[12] However, works concerning other ambits, as well as quantifications appearing in other ancient Near Eastern historical texts,[13] will be referred to only when discussing the feasibility of the figures, for purposes of comparison,[14] but they will not be taken into consideration when analyzing the editorial rules employed in the Assyrian chancellery, in view of the particular characteristics shown by the texts under examination.

NUMERICAL NOTATION

The technical aspects of the numeral notation are not the subject of this work. To the non-specialist reader, however, a short description of the system(s) employed will be useful.[15] Weights and measures, the latter quite numerous, will be discussed *ad hoc*.[16]

---

11 The fact that the number represents, within these texts, a sort of "additional" information has dissuaded me from making a "discourse analysis" of the narrative context in which the quantifications appear.

12 I have no knowledge of systematic works on a *corpus* of documents comparable to the Assyrian royal inscriptions in vastness, ideological relevance etc. Thus, the figures given by the classical authors for the Persian armies, often characteristically exaggerated (cf. p. 109), have been repeatedly discussed, but usually limiting themselves to one or a few quantifications, as for instance the size of Xerxes' or Alexander's armies (cf. below, p. 110 n. 268f and Engels, *Alexander*, 144ff and the bibliography there quoted). Note, however, that the Greek sources are later and indirect, while here we are dealing with primary and direct texts.

13 E.g. the Urartian royal inscriptions, for which see below p. 187.

14 The size of certain armies, the number of captives or dead in a given battle or war, etc. have often been discussed by military historians, mostly basing themselves on the method of *Sachkritik*, a criticism based on "reality" or "facts" (e.g. considerations of logistics, supply, space and time). This method was developed and extensively used by Hans Delbrück, Professor of Military History at the University of Berlin around the beginning of this century. He faced repeatedly the problem of the truthfulness of the figures reported by historical documents, his interests ranging (temporally) from the Persian wars to the beginning of the 20th century (thus excluding the ancient Orient). In any case, he showed himself very skeptical about the quantifications given by the ancient sources: see especially his book on *Numbers*, p. 22f, 47ff and *passim*. His monumental 7-volume *Geschichte* opens with a chapter on "Heereszahles" (pp. 7-24), and the topic is taken up again in several instances (pp. 33ff, 149, 166f, etc. — cf. also below, p. 109 n. 265).

15 In general, on the Mesopotamian numerical notation, cf. A. Deimel, *Šumerische Grammatik der Archaischen Texte mit Übungstücken* (Orientalia, Series Prior, 9-13, Roma 1923ff) §§43 and 45; cf. also Neugebauer, *Sciences*, and J. Friberg in RlA 7, *s.v.* "Mathematik."

16 For the weights cf. here below, n. 70, and for the most frequently used measures, the cubit and the double-hour of travel (*bēru*), nn. 98 and 103.

The Mesopotamian numerical system, as employed within the *corpus*, represents a compromise between the decimal and the sexagesimal systems.[17] Two basic signs were employed: the vertical wedge, 𒁹 (DIŠ), = *1*, and the corner wedge (*Winkelhaken*), 𒌋 (U), = *10*. Obviously, to obtain digits such as *2, 3, 4* etc. the sign was repeated, up to 9 times.[18] These signs could thus serve to write the numbers up to 99. Beyond this number, the ideograms 𒈨 ME (*me'at,* "hundred") and 𒇷 LIM (*līmu,* "thousand," actually 10 × ME) were used, preceded by the necessary number of vertical or corner wedges. However, very frequently the values between 70 and 99 were represented through a sexagesimal system: in this case, the vertical wedge took up the value *60*, its value being clear since it precedes, reading from left to right, one or more *Winkelhaken*. Thus, the writing 𒌋𒁹 meant *11* while 𒁹𒌋 meant *70*.[19] Sexagesimal notations based on the term *šuššu* ("soss"), with the meaning "sexantine," are also found. In this case we may have 1 *šu-ši* = 60, 2 *šu-ši* = 120, etc. — the latter number could thus be written either as 1 *me* 20 or 2 *šu-ši*.[20] Sexagesimal and decimal digits could be employed simultaneously, e.g. writing the number "160" with 1 *me* 1 *šu-ši*.[21]

For the values *600* or *3,600* some other sexagesimal signs could be used, as 𒁹𒌋 (DIŠ+U, *nēru*; combination of 60 and 10) = *600*; 𒊹 (ŠÁR, *šāru*) = *3,600*, but they appear rather rarely in these inscriptions.[22]

Note that, throughout the work, numbers will be consistently "translated" into our notation system. Special notation, such as "1,10" for *70*,[23] will be used only when necessary, to clarify certain numerical relations.

SOME "CATEGORIES" OF NUMBERS

Certain quantifications can be conveniently arranged in "categories" — "round," "exact" numbers, etc. — bearing in mind especially the practical purposes. These categories are not applied to the formal aspect only, but also to the function the number takes within the narrative context in which it appears. Any discussion on the relations between the various "categories," on their uses and significance, etc. is postponed to the later chapters.

---

17 "Mixed-decimal system" in Friberg's words, RlA 7, 537.

18 The digit "9" may also be written with three overlaid vertical strokes (as in the case of the 9 *me* horses mentioned below, p. 93, Shalmaneser III's "totals"), while the numerals *1, 2*, or *4*, may be written with a proper number of wedges placed horizontally (thus readable AŠ, ḪAL or LÍMMU) — cf. e.g. 3 R, 8:29 (and cf. below, p. 35 n. 72 for the number *3*).

19 Cf. p. 140 and p. 23 n. 15. Writings such as "1,05" (one vertical wedge = 60 + five vertical wedges = 5, giving 65) are not found in the *corpus*. Place-value notation, in which the position of a digit in a number determines its value (as in our system), is thus not consistently used.

20 The *šuššu*, or *šūši* (cf. CAD Š/3 *s.v.*), may be written also UŠ (cf. e.g. nn. 112 on p. 140 and 157 on p. 148).
Diacronically, the trend is that the sexagesimal system appears to be declining in usage throughout the *corpus*: cf. below Table 1 on pp. 42-43 and chapter 2 in general.
It is possible, however, to find 3 *me* and 5 *šu-ši* within the same inscription (cf. below p. 30 n. 53).

21 Weidner, AfO 21, Tf. VIII (VAT 9968) Vs. 7.

22 Certain inscriptions of late epoch, in the "building" sections prefer the use of sexagesimal numbers: cf. e.g. the 3 *šuššu tibkē* of Borger, *Asarh.*, 87 Rs. 2 (or cf. the perimeter of Dur-Sharruken, below, p. 140, or other examples at p. 56 no. 62, or at p. 152 n. 173).

23 See Neugebauer, *Sciences*, and cf. also RlA 7, 534 end ("compact format").

"Round" numbers

Frequently, in the Assyrian royal inscriptions, we meet quantities — sometimes even very high — represented by numbers formed by one or two digits only, such as 1 *lim*, 1 *me* (= *1,000, 100*) etc. Since the graphic system did not allow specification of the low-order digits (our "zeros"), these numbers may represent either exact quantities, in a form that only by coincidence is "round," or they may be an approximation simply intended to represent a magnitude (or even an intentionally indefinite reality). Conventions and context make it possible to distinguish between the two alternatives: in a business record we will always regard it as a determined quantity, while in a literary text we will rather favor the other alternative. Which solution has to be applied within a royal inscription is debatable. The former — i.e. that we have to understand in such cases an exact quantity — is rather unlikely. Actually, it is difficult to believe that, e.g., the enemies killed during a battle or the sheep looted in a given zone really added up to a round number.[24] Therefore, in the case "round" numbers are given for such occurrences, we may safely assume we are dealing either with a "rounding up," with an approximation or with an indefinite quantity. Then, we should consider as "round" — and by no means representing an exact quantity — *all* the numbers formed by high-order digits only (from the hundreds on) and deprived of digits on the order of tens or units, especially in the cases where they quantify people (prisoners, losses, military contingents), booty or cities.

Actually, most of the text editions, in translation or in transcription, simply report (or "translate") the numbers in "Arabic" digits, possibly being misleading, since they exclude the possibility of giving the second of the above mentioned interpretations, leading indeed to the first one.[25] This applies also and especially to such sexagesimal writings as 4 ŠÁR (= 4 × *3,600*) that, transcribed "14,400," may seem a relatively exact number.[26] It has therefore been suggested to "translate" such a number as "15,000,"[27] which would give the impression of its "roundness," while approximating its "real" value — not so negative a thing since it is rather difficult to believe that this number really expressed a well-defined quantity.

"Exact" numbers

Since all the numbers are formally exact, the creation of a special category for these numbers is necessitated by the presence in the *corpus* of such numerals as, e.g., *800,509, 200,150*, etc., formed by the sum of a round number of hundreds of thousands and a much smaller number. Most of these

---

24 Some doubt remains when speaking of tribute, inasmuch as this was possibly demanded, in advance, in a round figure.

25 Thus the recent RIMA 1 regularly transliterates 1,000, 2,000 etc. (instead of 1 LIM, 2 LIM), which does not allow us to understand how the numbers are written (if, for instance, *120* is written 1 *me* 20 or 2 *šu-ši*). The subsequent RIMA 2, however, remedied this inconvenience.

26 Cf. the remarks of Poebel, JNES 1, 298 n. 138.

27 R. Borger in EAK 1, 57.

numbers thus clearly show their "round" origin.[28] The addition of the low order digits clearly has the aim of excluding the interpretation of the numeral as an approximation or an indefinite quantity (cf. here above), in favor of that of "exact quantity." One should also take into consideration the fact that in many cases the "exact" numbers quantify objects that most often, in different epochs but in similar contexts, are indicated with "round" numbers.

The larger a number is the more its exactness will be relevant: thus, *25, 34* or *16* do not appear "exact," at least not as much as *16,020* or *100,225* do — and, to the writers of these inscriptions, appearance had its importance. Then, all the "exact" numbers, as well as the "round" ones, have to be "high."[29]

"High" and "low" numbers

These categories, differently from the previous ones, concern not only the number but the quantified object as well. A number is never absolutely "high" or "low," being placed within a value scale with two ends (the "low" and the "high") whose "center" depends on the quantified object and on the general context. Therefore, "high number" will indicate particularly relevant quantifications in relation to what is quantified.[30]

The concept of "high" number is not completely a "modern" category, as it might seem, since the texts may sometimes underscore this characteristic: "with *208,000* people, a heavy booty (*šallat nišē kabittu*), *7,200* horses ... I returned to Assyria,"[31] or:

> [*it*]-*ti* 20 *lim um-ma-na-te-šu-nu* DAGAL^{MEŠ}*(rapšāte) i-na* ^{KUR}*Ta-la lu am-da-ḫi-iṣ*
>
> "with their *20,000* extensive troops near mount Tala I fought" (Tiglath-pileser I).[32]

Most "low" numbers concern lengths of time[33] and are used to concretely demonstrate how certain achievements have been carried out in a short period of time. A couple of interesting examples of "low" numbers come from Sennacherib's Bavian stela inscriptions. On line 15 of the text, the king states that he dug the "Sennacherib channel," carrying water to the city of Nineveh,

---

[28] That these "exact" numbers come from a "round" number is possibly demonstrated by the variation no. 27 (cf. p. 50) in which an original quantification of 16 *lim* (*16,000*) is given by a later document as 16 *lim* 20 (*16,020*).

Actually, one may remark, the writing system did not permit the perception of the roundness of the former ("round") numeral so clearly as it happens in our system, since the zeros were not represented — thus, *100,225* was written simply "1 hundred thousand 2 hundred 25." However, in the spoken language the zeros are not pronounced in any case, and yet, in the above example, nothing (= other digits) stands between "hundred" and "thousand," directly juxtaposed to express the multiplication (and thus the sense of "roundness").

[29] Consequently, throughout this work, all the numbers lower than *100* will never be classified "round" nor "exact."

[30] In practice, this definition will include any really very high number (e.g. *100,000* or more), since the objects quantified by the Assyrian royal inscriptions have in any case an intrinsically quite relevant value (cf. the tendency to quantify preferably the objects having a higher value, pp. 120ff).

[31] Luckenbill, *Senn.*, 55:60 (ARAB 2, 267), cf. p. 58. Cf. also below pp. 177f for other examples.

[32] AKA, 77:87f (RIMA 2, 24; ARI 2, 37). According to RIMA 2, 33:18 he once faced "*12,000* troops of the extensive Mušku (KUR *Muš-ki*.MEŠ DAGAL.MEŠ)" (cf. p. 46 no. 5).

[33] Cf. here below §6.

with the help of (only) *70* workers.[34] Then, a few lines later (23-26), the king swears to any future prince who might wonder how he could have accomplished such a task "with these few workers,"[35] that he has indeed dug it with these people and completed it in *one* year and *3* months.[36]

"Totals"

The quantification may either refer to a single episode or to a repeating series of homologous episodes. There might be computed the prisoners of a city, or of a group of neighboring (or in some way connected) cities, or the prisoners captured during a series of military actions, or a whole campaign, or the whole reign. An account, besides, is always the "total" of a certain category of objects. It is for the accountant to establish the size of this category. Thus, the quantifications of cities are always referring to several episodes, since it is difficult to conquer or to destroy many cities simultaneously. The single episodes are then grouped, presenting in the narrative the conquest of a given number of cities as a single deed ("*X* cities I conquered"). But, rather than "totals," these are "sub-totals." For "total" I will mean, on the contrary, quantifications relating to events that are not directly connected, or anyway that are temporally distant. Thus, according to this separation or distance there could be totals encompassing more or less extensive periods and/or areas.

In the Assyrian royal inscriptions we find quantifications expressing totals, whatever their time-span might be, without the nature of "total" being declared, or in any manner explained, especially when, as it often happens, they refer to a limited series of contiguous events. Thus, for instance, most of the quantifications of prisoners or booty even from a first examination appear as "sub-totals," referring to specific geographical areas or sections of campaigns.[37] "Totals" are the number of prisoners captured in neighboring cities (actually conquered over the course of time): "the cities of Daiakanša, Sakka, Ippa ... (names some others) ... [I captured] and I defeated. *900* people, *150* oxen, *1,000* sheep, horses, mules, asses I carried off" (Tiglath-pileser III).[38] Or: "*8* fortresses of Tuaiadu, a province of Telušina of Andia, I captured. *4,200* people, together with their possessions, I carried off" (Sargon).[39]

What distinguishes these "totals," in substance, is that they refer to temporally distant events or to events that could be narrated, or are actually narrated by other inscriptions, as separate episodes. In these cases the "total" encompasses "quantities" pertaining to a more or less wide span of time, but presented as referring to a narrower context.

---

34 According to E. Weissert's forthcoming new edition of the text(s), based on L.W. King's handcopies made on the spot, line 15 (cf. Luckenbill, *Senn.*, p. 80) should be read (after *ušraddi*): *i-na* ŠÀ 1,10 LÚ.ERÍN ÍD [(*šu-a-tu*)] *aḫ*-[*re*⌉-*ma*, "with the help of 70 workers I dug [(that)] canal."

35 ⌈*i*⌉-*na* [Ø!] LÚ.ERÍN^MEŠ *an-nu-ti e-ṣu-ti* (line 24).

36 *i-na* MU.AN.NA 3 ITI (line 26). I thank E. Weissert for according me access to his notes on the subject.

37 Cf. Oded, *Deportations,* 19 n. 2.

38 The passage is quoted here below at p. 9.

39 Lie, *Sargon*, 18:106f and Winckler, *Sargon,* 104:44f (resp. Fuchs, *Sargon*, 107 and 204; ARAB 2, 13 and 56).

Note that some totals — especially those referring to the whole reign — can undergo "updating," thus augmenting as time goes on.[40]

QUANTIFIED OBJECTS

We will here review the quantifications according to the numbered objects.[41] However, only some general indications will be pointed out, since giving a more or less complete "inventory" of the various combinations would result in a voluminous and rather useless list.[42]

1. People

The quantifications of enemy soldiers, warriors — that in most cases are killed or captured — are frequent from the reign of Shalmaneser I on, with the exception of the last rulers.[43]

Starting with Tiglath-pileser I also the "civilians" are enumerated, even if in most of the cases the terms employed do not allow a clear distinction between them and the "military" men.[44] This is not surprising, if one considers that in wartime — the ancient Near East being no exception — almost every male capable of holding a weapon is seen as a "soldier."[45] Actually, even distinguishing between "things" and "people" could be difficult, considering that they are mingled in the expression *šallatu*, "booty," a term that may indicate either people or things (especially livestock), or both.[46] Thus the deportees, from Tiglath-pileser III on, mostly are simply labeled as *amēlû* (LÚ.MEŠ), "men," or *nišē* (UN.MEŠ), "people":

---

40 The argument will be taken up again, chapter 3 §12.

41 For the quantifications of beasts killed or captured during the hunting expeditions see below, chapter 4 §9.

42 One should consider that such a list should include almost any physical object mentioned by the inscriptions (perhaps it would be easier to give a list of what is never quantified). Even to establish which terms are more frequently quantified could be quite difficult and complicated (for some indications see chapter 2).

43 For Shalmaneser I cf. pp. 150ff. At p. 194 will be found an almost complete list of quantifications of people.

44 For this reason, I decided, for this section discussing the people, to distinguish the quantifications according to the type of action, grouping the verbs having similar meaning.

A subdivision of the quantifications according to the quantified terms would be complicated and confused. The texts sometimes use different terms to indicate the subject of the same quantification, or more generally the same events: for example, Tiglath-pileser III's annals sometimes do not specify any term for the deported people, not even LÚ — (cf. e.g. Tadmor, *Tigl. III*, Ann. 9:10 (3 quantifications) and Summ. 8:24. Thus, the soldiers killed by Shalmaneser III during his 18th campaign are sometimes called ERÍN, at others *mundaḫṣu* (comp. e.g. 3 R, 5 no. 6:10 with Kinnier-Wilson, *Iraq* 24, 94:23).

45 The neo-Assyrian empire had a standing professional army (cf. below p. 111 n. 280), but this does not appear to be the case of many unorganized or semi-tribal bordering political entities (according to the Assyrian texts). One should also consider that, especially in later times, open field battles were quite rare, and the wars mostly consisted of more or less prolonged sieges by the Assyrians: in this case, obviously, the distinction between military and civilian men did not make much sense, at least for the besiegers.

46 Cf. Oded, *Deportations*, 7. In this work the quantifications of *šallatu* will always be regarded as referring to people, since it is rather difficult to imagine a quantification of "things." However, generally speaking, the Assyrian inscriptions are careless in distinguishing between people or (inanimate) things, since the former often appear at the top or at the bottom of the booty or tribute lists, quantified and treated like any other physical object.

> 9 *me* UN^{MEŠ}*(nišē)* 1 *me* 50 GU$_4$.NÍTA^{MEŠ}*(alpē)* 1 *lim* UDU.NÍTA^{MEŠ}*(immerē)* ANŠE.KUR.RA^{MEŠ}*(sīsē)* ANŠE.GÌR.NUN.NA^{MEŠ}*(kudanī)* ANŠE.NÍTA^{MEŠ}*(imērē)* *áš-lu-la*
>
> "*900* people, *150* oxen, *1,000* sheep, horses, mules, asses I carried off."[47]

This example is typical of the quantifications of people in the inscriptions of the last rulers.

Subdividing the quantifications of people according to the kind of action, three situations are the most frequent:

Killing (*maqātu* Š, *dâku, napāṣu, zaqāpu*, etc.). At the time of Ashurnasirpal II and Shalmaneser III this lot fell normally to the "military" men, usually indicated with ERÍN(.ḪÁ)^{MEŠ} (GAZ): 1 *me* 72 ERÍN^{MEŠ}*(ṣābē)* GAZ*(tidūki)-šú-nu a-dúk* ("I killed *172* of their combat troops"),[48] and this is a very frequent quantification — actually, in consideration of the verb employed, the most frequent quantification appearing in the inscriptions of Ashurnasirpal II is that of soldiers killed (verb *maqātu* Š).

Capture (*ṣabātu*). This fate concerns, in Ashurnasirpal II's time, the "military" men who escaped from the killing. In fact the "captured" are quantified only after the "killed" were quantified (and always in a lower measure):

> 6 *me* ERÍN^{MEŠ}*(ṣābē) mun-daḫ-ṣi-šú-nu ina* GIŠ.TUKUL^{MEŠ}*(kakkē) ú-šam-qit* SAG.DU^{MEŠ}*(qaqqadē)-šú-nu ú-na-kis* 4 *me* ERÍN^{MEŠ}*(ṣābē)* TI.LA^{MEŠ}*(balṭūti) ina* ŠU^{MEŠ}*(qātē) ú-ṣab-bi-ta* 3 *lim šal-la-su-nu ú-še-ṣi-a*
>
> "I felled *600* of their fighting men with the sword (and) cut off their heads. I captured *400* soldiers alive, I brought out *3,000* of their captives."[49]

The quantification of captured enemies is very rare in the later reigns.[50]

In some cases the verbs *ekēmu* or *leqû* are used to indicate the capture or the deportation of military men (usually quite small contingents).[51]

---

47 3 R, 10 no. 1a:6′ (Tadmor, *Tigl. III*, Ann. 5:6; Rost, *Tigl. III, Pl.* VIIIa and p. 32:182; ARAB I, 775 — Rost's copies have 6 *me* instead of 9 *me*; but cf. photograph e.g. in Cogan - Tadmor, *2 Kings*, fig. 8).

48 AKA, 308:41 (RIMA 2, 204 and ARI 2, 557). Cf. also RIMA 2, 245:9.

49 AKA, 378 (RIMA 2, 220:106f and ARI 2, 587). For other instances of quantifications of soldiers captured at the same time as the killing of some others, cf. RIMA 2, 207:72 (*50* killed and *20* captured), 210:111 and 260:78 (*700* k. and *40* var. *50* c.), 211:115 (*1,000* k., *200* c. and *2,000* carried off), 214:32 (*470* k. and *20* var. *30* c.), 220:111 (*1,400* k., *580* var. *780* c. and *3,000* carried off). In RIMA 2, 213:19 *50 pētḫallu* and *3,000* soldiers (ERÍN.MEŠ *ti-du-ki*) are captured alive without anybody being killed — but *70* enemy soldiers fell in a river escaping the Assyrians. According to RIMA 2, 266:13′, a fragmentary passage, *174* were captured, while *12* were flayed, *153* (?) were beheaded and another *20* (?) (apparently) impaled. When it is written, within a single passage, of enemies killed, of others captured (alive) and of some others deported, and all the three groups are quantified, the first number surpasses the second while the third surpasses them both. This order (killed - captured - deported) is respected in any case, with the sole exception of the just mentioned fragmentary passage of RIMA 2, 266.

50 Shamshi-Adad V: KB 1, 180 III 13 (*1,200*, another *6,000* are killed), 186:31 (*3,000*, another *13,000* are killed), 186:43 (*2,000*, the killed are *5,000*, while *100* chariots and *200 pētḫallu* are also captured — verb *ekēmu*). Sargon: *250* var. *260* nobles captured, cf. p. 53 no. 46.

Earlier, in Tiglath-pileser I's reign: *6,000* were captured, without anyone being killed (RIMA 2, 14:84 = AKA, 37).

51 See, with *ekēmu*, Shalmaneser III: *470 pētḫallu* (Michel, WO 1, 265:11ff, plus various later texts: cf. p. 50 n. 24); Shamshi-Adad V (besides the *pētḫallu* mentioned in the preceding note): *30 pētḫallu* (Weidner, AfO 9, 93 and 92:12; another *650* soldiers killed); Sargon: *1,000* soldiers (Lie, *Sargon*, 66: 447). With *leqû*: Tiglath-pileser I, RIMA 2, 33:21f = 42:20f (*4,000* troops ERÍN.MEŠ).

Deportation (*šalālu, nasāḫu*). The deported are in most cases "civilians," *šallatu*, e.g. in 2 *lim šal-la-su-nu aš-lul* "I carried off *2,000* of their captives";[52] up to (but not including) Tiglath-pileser III this kind of quantification is quite rare, then becomes more frequent. The verb *nasāḫu* is used mostly to refer to troops: 1 *lim* 2 *me* ERÍN.HÁ^MEŠ^*(ṣābē)-šú-nu a-su-ḫa* "I uprooted *1,200* of their troops."[53] In Tiglath-pileser III's inscriptions also the resettling of prisoners is often mentioned.[54]

As concerns the relation among these three situations, it is evident that the "preference" is granted to the "killed-militarymen" over the "deported prisoners/civilians," since the former are sometimes quantified whereas the latter are not: the basilar model is "I killed *X* soldiers and I deported many / the rest / etc.," and is very common in the inscriptions of Ashurnasirpal II.[55]

Note that the survivors, who often are indicated as the "rest" who escaped the massacre or the capture, are never quantified. The enemies are killed and/or captured/deported: the "rest" is quantified only in the event it is, at a later time, in turn killed or captured.[56]

Also foreign rulers or nobles may be quantified, obviously in a much more moderate measure, starting from Tiglath-pileser I who boasts of having vied with *60* enemy kings.[57]

In addition to being defeated, captured or deported, people may be quantified also when they approach bringing gifts:

> *ina* u_4*-me-šú-ma šá* 24 MAN^MEŠ^*(šarrā)-ni šá mat Ta-ba-li i-gi-si-šú-nu am-daḫ-ḫar*

---

52 AKA, 339:115 (RIMA 2, 211 and ARI 2, 573).

53 AKA, 305:31 (RIMA 2, 204 and ARI 2, 554). Cf., for Shalmaneser III, a similar passage in Safar, *Sumer* 7, 7 II 4 = Michel, WO 2, 30 (annals "edition E," it also appears in some other editions, in which it changes from an original *17,500* to *22,000*). In the broken obelisk (credited to Ashur-bel-kala) some "captives" are uprooted (*4,000 šallatu* + *nasāḫu*: AKA, 129 II 2). In Tiglath-pileser I *nasāḫu* refers to *300* families (5 *šu-ši qi-in-na-a-te*, AKA, 81:31ff = RIMA 2, 25).

Another verb employed is *târu* D: Shamshi-Adad V brought back *120* var. *140 pētḫallu* (KB 1, 180:34; another *2,300 tidūki* were killed).

54 Cf. simply the list at pp. 194ff. Worth mentioning also are the *6,300* guilty Assyrians (*bēl ḫīṭi*, "malefactors") settled in Hamath according to Sargon's stelae (Winckler, *Sargon*, 178:61; Lambert in *Ladders*, 125:5; Renger, *Sargon*, 768:5).

55 8 *me* ERÍN.MEŠ(*ṣābē*) *mun-daḫ-ḫi-ṣi-šú-nu ina* GIŠ.TUKUL.MEŠ(*kakkē*) *ú-šam-qit* SAG.DU.MEŠ(*qaqqadē*)-*šú-nu* KUD-*is* (var. *ú-[na-kis]*) ERÍN.MEŠ(*ṣābē*) TI.LA.MEŠ(*balṭūti*) ḪÁ.MEŠ (var. *ma-ʾ-du-te*) *ina* ŠU(*qāti*) DIB-*ta* (var. *ú-ṣab-bi-ta*) "I felled 800 of their combat troops with the sword (and) cut off their heads. I captured many (other) soldiers alive" (Ashurnasirpal II, RIMA 2, 210:107ff = AKA, 336, variants from the "Kurḫ monolith," AKA, 233:25ff; cf. respectively ARI 2, 571 [giving incorrectly *300* instead of *800*] and 638). For other examples from this reign cf., in translation: ARI 2, 544, 549 (end), 560, 571, 572, 634, 635, 639; for the reign of Shalmaneser III: ARAB 1, 563. In some other cases "the others" or "the rest," not quantified, escape (cf. ARI 2, 546, 556, 557, 579) or submit (cf. ARI 2, 567). Obviously in many cases both the killed and the deported are quantified. It is not possible to establish a relation of "preference" in respect to the "captured": in a passage of ARI 2, 572 both the killed and the prisoners are not quantified, but only *50* (var. *40*) captured soldiers (cf. p. 47 no. 11).

56 The passages of the Assyrian royal inscriptions mentioning the "rest" (*sittu* or *riḫtu*) are collected in O. Carena, *Il Resto di Israele. Studio storico-comparativo delle iscrizioni reali assire e dei testi profetici sul tema del resto*, Bologna 1985, pp. 22-43.

The explanation that the killed and the prisoners were quantified as they were counted after the battle, while the "escaped" were not quantified, being impossible to count, would be an over-simplification. The inscriptions report also digits relating to enemy armies that were *not* defeated (at least this does not appear from the inscriptions themselves, cf. e.g. the battle of Qarqar, below, p. 104), while, on the other hand, certain so grossly rounded numbers of military contingents would suggest to us that not accountings but rather very inaccurate estimates had been carried out.

57 Within the "titulary," cf. p. 92 n. 190. Cf. also the *5* kings of Muški mentioned in n. 59 below.

"In those days I received the presents of *24* kings of Tabal";[58]

or when they ally with each other and advance against the king of Assyria: "In those days Adad-idri of Damascus, Irḫulina of Hamath together with *12* kings of the seacoast trusted in their own strength and advanced to give battle."[59]

In Sargon's inscriptions some quantifications of "city prefects" or enemy "chiefs" appear.[60]

2. *Objects in booty and tribute lists*

Quite often the tribute is not explicitly defined as such (*biltu, madattu*, etc.): a distinction between categories ("tribute," "gift," "booty," etc.) based on the terminology would therefore be rather problematic.[61] For practical purposes, however, bearing in mind that the context does not always allow a clear distinction, the following tripartition can be made:[62]

Tribute received (any object received, *maḫāru*, by the Assyrian king). These quantifications are the most frequent.[63] However, they do not refer to the collection (and the amount) of the "regular" tribute (*biltu, madattu*), which was an administrative matter, and not a military one, but rather to some "extraordinary" contribution which is paid upon the arrival of the Assyrian army, as a form of submission. In these cases, often the tribute is labeled as *nāmurtu*, a term indicating an "audience gift."[64]

Tribute imposed (any object established, *šakānu*, as tribute).[65] These are the most frequently quantified in later times. Generally speaking, the quantities of received tribute far surpass those of the imposed tribute, which represents a "regular" amount for the time being.[66]

---

58 Shalmaneser III, ICC, 92:105f (= Michel, WO 2, 154; ARAB 1, 579 — Michel doubts the number *24*), similar passage also in Læssøe, *Iraq* 21, 154:27f and 155:12 (with the number *20*; cf. p. 50 no. 31). Cf. another instance in ICC, 93:119f (*27* Persian kings). In the inscriptions of Ashurnasirpal II no quantification of kings appears.

59 Shalmaneser III: episode of the "*12*" Syrian kings (several versions, cf. pp. 134ff). For other instances cf. Tiglath-pileser I in AKA, 35:62ff (= RIMA 2, 14; *5,000* men and *5* kings of Muški), 67:83 (= RIMA 2, 21; *23* kings of Nairi, elsewhere *30*, cf. p. 46 no. 6).

60 Cf., in the annals (not considering the pertinent passages appearing in the other inscriptions) the "city prefects," BE URU.MEŠ(*bēl ālā*)-*ni*, of Lie, *Sargon*, 16:100 (*28*), 20:115 (*22*), 30:192 (*45*) and the "chiefs," LÚ *nasikku*, of 44:281 (*8*), 48:5 (*5*, named), 48:237 (*4*, named), 52:2 (*6*, named). In 74:2 are mentioned *7* Assyrian functionaries LÚ *šu-ut-reš-ia* (*lit.* "eunuchs").

61 On the terminology cf. Postgate, TCAE, 119ff and 146ff; Martin, *Tribut*, 13ff. More generally, on the tribute in neo-Assyrian times, see also Jankowska, *Ancient Mesopotamia*; Elat, AfO *Bh.* 19; *id., Economic Relations in the Lands of the Bible, c. 1000-539 B.C.*, Jerusalem 1977 [in Hebrew] and Zaccagnini, *Opus* 3 (especially 241f — the conclusions are summed in *id., Le tribut*, 195f).

62 See Liverani, *Asn.*, 155f for a similar classification.

63 Cf. pp. 199ff for an overview of the quantifications. The most frequently employed term is *madattu*.

64 In Tukulti-Ninurta II's annals, within the "itinerary" passage, the amount of some "gifts" from local rulers who submit to the Assyrian king are recorded: cf. below, pp. 97ff (12 tributes, all indicated as *nāmurtu* except the last one that appears in a fragmentary passage).

65 Usually, speaking of "imposed tribute," the appropriate term is cited: "*1* talent of silver, *2* talents of purple wool, *200* cedar logs as tribute (*ma-da-tu*) I imposed" (Shalmaneser III), 3 R, 7 col. II 23f (KB 1, 162 and ARAB 1, 601).

66 Cf. below, p. 99, for some examples from Shalmaneser III's monolith inscription in which both the tribute received and that imposed on the same occasion are quantified.

Booty (any object obtained as spoil, being the result of a military action, verbs: *šalālu, nešû, waṣû* Š, *leqû*, etc.).[67] This situation is not quantified very frequently.[68] The quantities are usually higher than those of the collected or imposed tributes. All these numbers, however, tend to be "high," and these categories are the most subject to "inflation" during the Sargonic period.

Various "movable" goods are subject to enumeration: the most common are horses, chariots, livestock (especially oxen and sheep), bronze or copper objects (casseroles, kettles, etc.), precious metals. In the case of metals, the quantification is mostly expressed in weight,[69] given in talents (GUN, GÚ.UN = *biltu*), minas (MA.NA = *manû*), and, rarely, in shekels (GÍN(.MEŠ) = *šiqlu* — only to specify quantities expressed in talents and/or minas).[70]

3. Lands, regions, etc.

The quantification of territories is rather rare in the Assyrian royal inscriptions. Most of the time the districts conquered during a part of a campaign are quantified:

> ^{KUR}*Ḫi-im-me* ^{KUR}*Ú-at-qu-un* ^{KUR}MAŠ-*gu-un* ^{KUR}*Sa-lu-a* ^{KUR}*Ḫa-li-la* ^{KUR}*Lu-ḫa* ^{KUR}NI-*li-pa-aḫ-ri ù* ^{KUR}*Zi-in-gu-un* 8 KUR^{DIDLI}*(matāti) ù* ILLAT^{MEŠ}*(illatē)-ši-na ak-šud*
>
> "The lands of Ḫimme, Uatqun, Mašgun (or Bargun), Salua, Ḫalila, Luḫu, Nilipaḫri (or Ṣ/Zallipaḫri) and Zingun (var. Zinigun), *8* lands and their fighting forces I conquered" (Shalmaneser I).[71]

In some cases the lands conquered by the king during a whole campaign or the whole reign are quantified (being thus typical "totals").[72]

In one instance "enemy" mountains are quantified: "(across) *16* mighty mountains (I rode) in my chariot over smooth terrain, and I hacked out the rough one with copper picks" (Tiglath-pileser I).[73]

---

67 The (material) booty is not indicated by a specific term, except in rare cases: cf. e.g. the *kišittu* from Karkemish at p. 52 n. 36. Thus, in KB 1 182:41ff, the Assyrian troops brought out (*mašā'u*) as "tribute" (*ma-da-ta*) "the sons, the daughters, the goods, the properties, the oxen and the sheep," but this is apparently booty. Live booty (people and/or animals), as we saw above (n. 46), is indicated as *šallatu*.

68 For instance, Ashurnasirpal II's inscriptions, though quite long and numerous, never quantify the booty (the verbs *waṣû, nešû* and *leqû* refer to prisoners or animals captured during the hunting expeditions). In Shalmaneser III's inscriptions only a booty of chariots and horses is quantified: "*1,121* of his chariots and *470* of his horses, together with his camp, I took from him (*ekēmu*)"; cf. p. 50 n. 24. The booty, in many cases, is composed only of military implements: cf. the table at pp. 194ff.

69 Only rarely are they received in the form of "ingots" or bars: cf. ARI 2, 471 for *18* tin bars. In ARI 2, 25, however, *30* talents of copper bars are recorded (cf. RIMA 2, 19).

70 The mina corresponds to ca. 500 g, the talent (= 60 minas) to ca. 30 kg, while the shekel is a sixtieth of a mina (thus ca. 8 g). Also "heavy" talents, minas and shekels existed, of double value (cf. Postgate, FNALD, 64f; Powell in RlA 7, 515ff).

71 RIMA 1, 183 I 32ff (A.0.77.1 = IAK, XXI, 1; ARI 1, 527; Text in KAH 1, 13). Other similar examples come from the reign of Sargon (cf. Lie, *Sargon*, 16:99 and 48:1; Gadd, *Iraq* 16, 177:44).

72 E.g. the *42* lands conquered by Tiglath-pileser I during his first *5* years of reign, cf. p. 88 and n. 170 there.

73 16 KUR.MEŠ *dan-nu-ti* A.ŠÀ DÙG.GA *i-na* GIŠ.GIGIR.MEŠ-*ia ù mar-ṣa i-na aq-qúl-lat* URUDU.MEŠ *lu aḫ-si* (Tiglath-pileser I), AKA, 65:65ff. (+ variants in n.) = RIMA 2, 21 and ARI 2, 30. The *16* mountains are named *per extenso* in the preceding lines. Mountains "without number" (*la mi-na*) were crossed by Sargon during his 8th campaign (Thureau-Dangin, TCL 3, 23:128).

For "enemy" land surfaces cf. Thureau-Dangin, TCL 3, 34:208 (*300 imēru*).

4. *Cities*

Cities (URU), subject to conquest, destruction, etc. are frequently quantified. The first example comes from Shalmaneser I's inscriptions, an example that may be considered typical of these quantifications:

> 41 (var. 51) URU^{DIDLI}*(ālāni)-šu-nu aq-qur aš-rup šal-la-su-nu* NÍG.GA*(namkūr)-šu-nu aš-lu-ul*
>
> "*41 / 51* of their cities I destroyed (and) burnt, their prisoners (and) their property I carried off."[74]

This model is subject to expansion and to variations: the names of the cities may be specified, or the name of the region or ruler to whom they belong, etc. Concerning the verbs employed, the most common variants are:

• with *kašādu*: the pattern is *X ālāni akšud*, almost never employed as such[75] but usually expanded specifying at least URU^{MEŠ}*-ni-šú-nu* "their cities" (refers to previously mentioned peoples or political entities),[76] or their geographic position: 50 URU^{MEŠ}*(ālā)-ni šá mat Di-ra-a* KUR*(akšu)-ud*, "*50* cities of the land of Dirru I conquered,"[77] or: 1 *me* 50 URU^{DIDLI}(^{MEŠ}, *ālāni*) *šá* URU*(ālu) La-ar-bu-sa-ai* URU BÀD*-Lu-lu-ma-ai* URU *Bu-ni-sa-ai* URU *Ba-ra-ai* KUR*(akšu)-ud*, "*150* cities out of those of the Larbuseans, Dūr-Lullumeans, Buniseans (and) Bareans I conquered."[78]

• with *napālu, naqāru* and *šarāpu*: *X ālāni abbul aqqur ina išāti ašrup*, "X cities I devastated, destroyed, burned with fire." This treatment is very common from Tiglath-pileser I on,[79] but the cities are only rarely quantified up to the time of Shalmaneser III.[80]

Obviously, other verbs may be employed, or these same verbs may be differently combined or used together: 1 *me* URU^{MEŠ}*(ālā) ni ša* ^{m}*A ra mu ak-šud ab-búl aq-qur ina* IZI*(išāti) áš-ru-up*, "*100* cities of the man Aramu I captured, devastated, destroyed, (and) burned with fire" (Shalmaneser III).[81]

Frequently, from Ashurnasirpal II on, the main (or "central") cities are distinguished from the neighboring ones, actually only small villages. The former are simply indicated with the term for city (URU), while for the latter it is specified that they are part "of its/their neighborhood," *ša limētišu(nu)* (lit. "limits"), or, more rarely, "of its/their surroundings," *ša siḫirtišu(nu)*.

---

74 RIMA 1, A.0.77.1 I 37ff (IAK, XXI, 1; ARI I, 527; text in KAH 1, 13 — cf. another quantification at line 77, shortly after). This passage follows immediately that of the *8* lands mentioned above.

75 Some schematic examples are given by the annals of Shalmaneser III, e.g.: 11 URU.MEŠ*-ni dan-nu-ti ak-šud* "(I went up to mount Kašiyari and) *11* fortified cities I captured" (ed. "C": Cameron, *Sumer* 6, 13:17 = Michel, WO 1, 462; the passage appears in the later editions too).

76 E.g. in Michel, WO 2, 230:180.

77 AKA, 337:111 (RIMA 2, 210; ARI 2, 572). Cf. p. 47 no. 10.

78 AKA, 309:44 (RIMA 2, 205; ARI 2, 558).

79 In his inscriptions the "burning" is mentioned before the devastation and the destruction — cf. CAD N/1, 273b and 330a for an overview of the occurrences.
Sometimes in *napālu - naqāru - šarāpu* the last verb is replaced by *târu* (+ *tillu u karmu*), "to turn (into heaps and ruins)."

80 Several examples appear in ICC, 97:180ff (Michel, WO 2, 230).

81 Cameron, *Sumer* 6, 14:70f (Michel, WO 1, 466 + later editions of the annals).

Very seldom they are explicitly indicated as "smaller,"[82] as seldom the main ones are indicated as such.[83]

In the inscriptions of Ashurnasirpal II and Shalmaneser III only the last ones are quantified, while the main cities are individually named:

> URU *Ar-ni-e* <URU*(āl)*> *šárru-ti-šu* KUR*(akšu)-ud a-di* 1 *me* URU^{MEŠ}*(ālā)-ni ša li-me-tu-šú ab-búl aq-qur ina* IZI*(išāti) áš-ru-up*
>
> "The city of Arnê, his royal <city>, I conquered, (and) together with *100* cities of its neighborhood I devastated, destroyed, burned (it) with fire."[84]

From Shamshi-Adad V on, both the main and the neighboring cities might be quantified:

> 11 URU^{MEŠ}*(ālā)-ni dan-nu-ti a-di* 2 *me* URU^{MEŠ}*(ālā)-ni-šú ša* ^{m}*Us-pi-na ik-šud*
>
> "*11* strong cities together with *200* of their cities of the man Uspina he captured."[85]

Yet the smaller cities continue to be quantified in most cases, while the main ones are named.[86]

In the extant inscriptions of Tiglath-pileser III cities are quantified only on a few occasions, and therefore do not allow us to draw many conclusions.[87] With Sargon the instances in which both the main cities and the small ones are quantified become more frequent, the latter, obviously, always in a higher measure.[88]

---

82 Cf. the example from Sennacherib's reign, cited here below, n. 90.

83 E.g. as "capital cities" or "strong cities" or "fortified cities." For a study on such "categories" cf. Y. Ikeda, "Royal Cities and Fortified Cities," *Iraq* 41 (1979), 75-87.

Since the term employed is the same in all cases, URU (= *ālu*) — a term that might indicate a very wide range of settlements from a big city to a small estate — I have translated throughout "*X* cities and the *cities* of their neighborhood," etc., and not "and the villages," or "the towns," or similarly.

84 Shalmaneser III; Cameron, *Sumer* 6, 14:58f (Michel, WO 1, 466). In Cameron, *Sumer* 6, 16:47f and 51f (Michel, WO 1, 470; two cases) the smaller cities are not quantified, as are the main ones: 5 URU.MEŠ-*ni dan-nu-ti a-di* URU.MEŠ-*ni šá li-me-tu-šu* KUR-*ud* "*5* fortified cities with the cities of the neighborhood I conquered" (then, again: "2 fortified cities with the cities of the neighborhood I conquered").

85 1 R, 30 and 33 col. II 24ff (Abel in KB 1, 178; cf. also Weidner, AfO 9, 92:22ff). Already a passage of Shalmaneser I distinguished between main cult centers and (other) cities: 9 *ma-ḫa-zi-šu dan-nu-ti* URU *be-lu-ti-šu lu ak-šu-ud ù* 3 *šu-ši* URU.DIDLI-*šu a-na* DU$_6$ *ù kar-me aš-pu-uk*, "*9* of his fortified cult centers (with) his capital I did conquer, and *180* of his cities I turned into heaps and ruins" (RIMA 1, 184:75-78 = IAK, XXI, 1 and ARI 1, 530; text: KAH 1, 13). Cf. also Tukulti-Ninurta I: 4 URU *be-lu-ti-šu dan-nu-ti* [*ša* ^{m}*E*]*ḫ-li-te-šub šar*$_4$ *mat Al-z*[*i*] 6 URU.DIDLI *šap-ṣu-ti šá mat A-ma-da-ni lu ak-šud*, "*4* strong capitals of Elḫi-Tešub, king of Alzu, (and) *6* rebel cities of the land Amadanu I did conquer" (RIMA 1, 236 IV 1-4 = Weidner, *Tn.*, no. 1 and ARI 1, 693; text in KAH 2, 158).

86 Cf. e.g. ARAB 1, 720ff.

87 Cf. the *591* cities mentioned below, p. 181. A fragmentary passage of the annals mentions *15* citi[es], cf. Tadmor, *Tigl. III*, Ann. 18:12′. The stela from Iran mentions *100* cities of Tarḫularu, conquered "together with the cities of the surroundings," and the Mila Mergi rock relief, *29* cities of the Ullubeans (*ibid.*, 102:38′ and 115:33 resp.).

88 The smaller cities always surpass the main ones when both are quantified, cf., in translation: ARAB I, 601 (Shalmaneser III), 717 (Shamshi-Adad V), ARAB 2, 19, 20 (3 cases), 56, 151, 163f, 165, 166, 167 (Sargon), 134, 261 (5 cases), 172 and AD, 298:37f (Sennacherib). Considering only these cases, and therefore excluding such instances as "Mandarazu and *2* cities of its environs," ARI 2, 568 and 635 or as "Ḫubuškia together with *100* cities," ARAB 1, 628, the lowest ratio is to be found in ARAB 2, 20 and 165 (Sargon) with *7* main cities *versus 30* smaller ones (1:5), the highest in Weidner, AfO 9, 92:22, with *3 vs. 250* (*circa* 1:83), a *record* obviously held by Shamshi-Adad V whose inscriptions give very high numbers for the "cities of the neighborhood" (cf. p. 181 — besides him the highest ratio is *6 vs. 200*, Shalmaneser III in ARAB 1, 601). It is never possible to establish a relation between the digits of the two numbers (e.g. a ratio 1:10), except perhaps in ARAB 1, 261 and 272, *88* (variant *89*) *versus 820* (cf. p. 57 nos. 65f).

Sometimes the "fortresses" are distinguished from the main cities:

> 55 URU^{MEŠ}*(ālāni)* KALA^{MEŠ}*(dannūti) bit* BÀD^{MEŠ}*(dūrāni) ša* 8 *na-gi-ì-šú a-di* 11 ^{URU}*ḫal-ṣu*^{MEŠ}*-šú mar-ṣa-a-ti ak-šud-ma i-na* ^{d}*Gíra aq-mu*
>
> "*55* strong fortified cities of his *8* provinces, together with their *11* inaccessible fortresses, I captured and burned."[89]

Frequently in Sennacherib's inscriptions, only the main cities are quantified, while the smaller ones are "without number":

> 46 URU^{MEŠ}*(ālāni)-šú dan-nu-ti bit* BÀD^{MEŠ}*(dūrāni) ù* URU^{MEŠ}*(ālāni)* TUR^{MEŠ}*(ṣeḫrūte) ša li-me-ti-šú-nu ša ni-ba la i-šu-ú* (...) KUR*(akšu)-ud*
>
> "*46* of his strong cities together with the small cities of their neighborhood, that were without number ... I conquered."[90]

From the reign of Sargon, sometimes the capital of a ruler is distinguished from the other main cities. The capitals are usually named, but not quantified:

> URU *Pa-ar-da* URU*(āl) šarru-ti-šú ina* ^{d}*Gíra aq-mu* 23 URU^{MEŠ}*(ālāni)* KALA^{MEŠ}*(dannūti) ša li-me-ti-šú ak-šud-ma áš-lu-la šal-la-su-un*
>
> "Parda, his royal city, I burned (and) *23* strong cities of its neighborhood I captured and carried off their booty."[91]

In these cases, instead of a distinction

(MAIN) CITIES	CITIES OF THE NEIGHBORHOOD
URU^{MEŠ} (KALA/*dannūti*)	URU^{MEŠ} *ša limētišu(nu)*

"skipping" to a higher degree, the capitals and the main cities "of the neighborhood" are distinguished:

CAPITAL	MAIN CITIES OF THE NEIGHBORHOOD
URU *ša* (NAM.)LUGAL	URU^{MEŠ} KALA/*dannūti ša limētišu(nu)*

Thus, the "smaller" cities are not mentioned. In other words, a three level distinction of the kind

CAPITAL / MAIN CITIES / (SMALLER) CITIES

is never in operation, except in one instance from the reign of Sargon and one from that of Sennacherib (cf. hereafter), where, however, the smaller cities are not quantified.

---

89 Botta, 146a:7′ff (Winckler, *Sargon*, no. 66; Fuchs, *Sargon*, 204:43ff; ARAB 2, 56); Botta, 94:3′ gives the variant "*[5]6* strong cities" (cf. below p. 53 n. 40). Various examples of quantifications of fortresses (*ḫalṣu*) come from this reign (cf. Lie, *Sargon*, lines 101, 103, 104, 108, 146, 448 and p. 48:3).

90 Luckenbill, *Senn.*, 172 and 32:19f (ARAB 2, 240; dupl. Ling-Israel, *Fs Artzi*, 228:17f). Cf. other similar examples, in translation, in ARAB 2, 248, 279 and 312 for Sennacherib and 808, 851, 920 and 942 for Ashurbanipal.

91 Botta, 146a:11 (Winckler, *Sargon*, no. 66 / p. 106:47; Fuchs, *Sargon*, 205; ARAB 2, 56).

For the Assyrian editors of later times the distinction between the various "categories" of cities was not very important, not as much as that between "central" cities and cities "of the neighborhood," regardless of whether these latter were "small" cities or not. Thus, the "*34* strong cities (URU^{MEŠ} *dan-nu-ti*) with the small cities of their neighborhood (URU^{MEŠ} TUR^{MEŠ} *ša li-me-ti-šú-nu*) which were without number" destroyed by Sennacherib together with their capitals Marubišti and Akkuddu according to the "Bellino cylinder,"[92] in the subsequent editions of the annals become "*34* small cities (URU^{MEŠ} TUR^{MEŠ}) of their neighborhood (*ša li-me-ti-šú-nu*, referred to the two named capitals),"[93] while there are no cities "without number." This alteration (which can be seen as a "correction") is indicative of the tendency to conform to a bipartite vision (CENTRAL) CITIES / CITIES OF THE NEIGHBORHOOD (= PERIPHERAL) typical of the inscriptions of the Sargonids.[94]

5. Distances and length measures

The height of the walls, measured in courses of bricks,[95] is the most frequent quantification in the inscriptions of the middle-Assyrian period, and this is not surprising, since the contents of these inscriptions are mostly of a "civil" type. Other kinds of data are taken into consideration, such as the number columns of a certain type, the number of statues set in a certain place, etc.[96] In later epochs these quantifications continue to appear with a certain frequency within the passages dedicated to building activities, even in the inscriptions of reigns rather sparing of quantifications such as those of Esarhaddon and Ashurbanipal.[97]

Sometimes the quantifications concern the external dimensions of temples or palaces, usually expressed in cubits,[98] as well as the perimeters of cities

---

92 ICC, 63:29f (Luckenbill, *Senn.*, 59:28; ARAB 2, 279). Cf. p. 57 n. 52 for the chronology of Sennacherib's inscriptions.

93 The change takes place already in cylinder "Rassam," which immediately follows the "Bellino": cf. Luckenbill, *Senn.*, 167 and 28:16ff (ARAB 2, 237; dupl. Ling-Israel, *Fs Artzi*, 224:15f). Cf. also Heidel, *Sumer* 9, col. II 29; Smith, AD, 301:43f. The passage involves also a "loss of value": *34* MAIN C. + *ša nība lā išû* SMALLER C → *34* SMALLER C.

94 Thus, Sargon conquered "Ḫubaḫna, their stronghold, together with *25* (other) cities and countless (*ša nība lā išû*) [towns of] their [neighborhood]" according to the Nimrud prisms (Gadd, *Iraq* 16, 177:47ff), a passage that does not appear in any later inscription (cf. *ibid.*, 179).

Liverani, *Asn.*, 125 notes that in the "annals" of Ashurnasirpal II the settlements are distinguished according to a three-level hierarchical pattern (royal cities - fortified cities - villages = "city in the neighborhood"). Actually, as can also be deduced from Liverani's remarks (p. 126), the inscription never mentions more than one or two "classes" simultaneously.

95 *tibku*, "course of bricks" (these latter written SIG$_4$).

96 Cf. Lackenbacher, RB, for the *récits de construction* of the inscriptions up to Tiglath-pileser III. The most frequent measures concern the digging works (p. 213), foundations (pp. 214f), wall dimensions (particularly their enlargement, pp. 216f and cf. 104ff), canals (pp. 222f), etc. To the volume add the passages RIMA 2, 296:9 and 299:20′ (basements, p. 214) and 180:2′ (enlargements, p. 217); Deller - Fadhil - Ahmad, BaM 25, 466:55 (a new text of Tukulti-Ninurta I mentioning *2* walls); Tadmor, *Tigl. III*, Ann. 28 (width of a palace).

97 For the successors of Tiglath-pileser III cf. Lackenbacher, PSR (this book is less detailed than the one mentioned in the previous note).

98 On the neo-Assyrian length measures cf. Postgate, FNALD, 70f and M.A. Powell in RlA 7, *s.v. Masse und Gewichte*, pp. 457ff. The cubit (*ammatu*, written KÙŠ) corresponds to ca. 50 cm, while the "great cubit" (*ammatu rabītu*) was ca. 75 cm (*ibid.*, 462a; cf. AHw, 44a; CAD A/2, 75a). In the neo-Assyrian period, however, the "great cubit" corresponds to the "normal" cubit (53-54 cm according to RlA 7, 476a), while a "lesser" cubit is also known (*ammatu ṣeḫertu*, ca. 40 cm — cf. Thureau-Dangin, RA 22, 30 and RlA 7, 474). The relation between the *aslu* cubit and the regular cubit is not clear (cf. CAD A/2, 337a) — RlA 7, 475b suggests that all the neo-Assyrian terms for cubit could be synonyms.

The text K 2411 (Ashurbanipal, cf. Streck, *Asb.*, 292-303) includes several measurements in cubits.

(such as those of Dur-Sharruken or Nineveh, cf. p. 140). Occasionally the measures are concerned with the length of canals or other architectural or urban structures.[99]

All these quantifications concern the building activities of the Assyrian rulers, but they sometimes may refer, within the annalistic or anyway "military" sections of the inscriptions, to the number of walls of an "enemy" city[100] or their height or thickness.[101]

Also the distances between regions, cities, etc. are sometimes reported, until the latest periods (Esarhaddon and Ashurbanipal).[102] They were measured "in time," most often in *bēru,*[103] which indicates a "double-hour" of march, a distance that could be covered in half a day at most.[104]

These measures "in time," rather approximated, may contain errors or inaccuracies committed in *bona fide.*[105]

6. *Measures of time*[106]

The numbers employed with time measures have the characteristic of being, most of the time, "low." For instance, they are used to show how certain military achievements had been fulfilled in short periods:

> *mat Ú-ru-aṭ-ri i-na* 3*(šalaš)-ti* u_4*-me a-na* GÌR^2*(šēpē) Aš-šur* EN*(bêl)-ia lu-šék-níš*

---

99 Length of canals (measured in *bēru*): cf. ARAB 2, 369, 377 and 414 (cf. one example at p. 62, reign of Sennacherib). The "broken obelisk" contains one quantification in *kumānu* (a surface measure: "an area of *63 kumānu* ...," cf. RIMA 2, 105:30).

100 3 BÀD.MEŠ-*šu-nu* GAL.MEŠ *ša i-na a-gúr-ri ra-áṣ-bu ù si-ḫír-ti* URU-*šu ap-púl aq-qur a-na* DU$_6$ *ù kar-mé ú-tir*, "their three great walls, constructed with baked bricks, and the entire city I devastated, destroyed and turned into a heap of ruins" (Tiglath-pileser I; AKA, 79:11ff = RIMA 2, 24 and ARI 2, 38).

URU *dan*^{an} *dan-niš* 3 BÀD.MEŠ-*ni la-a-bi* ERÍN.MEŠ *a-na* BÀD.MEŠ-*ni-šú-nu* KALA.MEŠ *ù* ERÍN.ḪÁ.MEŠ-*šú-nu* ḪÁ.MEŠ *it-tàk-lu-ma la-a ur-du-ni* GÌR.2.MEŠ-*a la-a iṣ-bu-tú* "The city was well fortified, surrounded by three walls. The people put their trust in their strong walls and their large numbers of troops and did not come down to me, they did not seize my feet" (Ashurnasirpal II — AKA, 293:114f = RIMA 2, 201 and ARI 2, 549). Another similar passage follows: URU *dan*^{an} *dan-niš* 4 BÀD.MEŠ *la-a-be* (col. II 98f = ARI 2, 569; cf. also RIMA 2, 259:64 = AKA, 230:15 and ARI 2, 636) and at line 104f: URU GIG (var. *mar-ṣi*) *dan-niš* 2 BÀD.MEŠ-*ni la-a-bi* (ARI 2, 571; var. from RIMA 2, 260:71f = AKA, 233:22f and ARI 2, 638).

101 Height of a wall (of enemy city): Sargon in Thureau-Dangin, TCL 3, 38:240 (*120* courses of bricks).
Thickness of a wall (of enemy city): Sargon *ibid.*, 30:179 (*8* cubits).
Width of a moat (*ḫirīṣu*): Adad-nerari II in KAH 2, 84:65 (RIMA 2, 151; ARI 2, 429; *9* cubits).

102 From the reign of Tiglath-pileser I on (RIMA 2, 37:22f = ARI 2, 81).

103 The *bēru*, written danna = KASKAL.GÍD, indicates a "double-hour" of march, and is translatable with "stage" (RlA 7, 467a) or "mile" (CAD B, 208b). It corresponds to 10,800 m (RlA 7, 467 and 477b; AHw 130a).

In the inscription of Adad-nerari II published in Weidner, AfO 21, Tf. VIII and p. 44f (ARI 2, 450f) the measures are given either in *bēru* or in MAL, for which the reading is not known.

104 In the fragmentary text published in Borger, *Asarh.*, §76 (TUAT, 399; ARAB 2, 558) the distances are quantified both in *bēru* and in "days of travel," for instance: "on *30 bēru* of land, a travel of *15* days (*ma-lak* 15 u_4*-me*), ... I advanced" (Rs. 3). The resulting ratio is 2 *bēru*s per day in each case, except in Rs. 8 where *15 bēru*s are equated to a travel of *8* days. This passage possibly appears also in a small unpublished stela fragment found at Qaqun (Israel). According to E. Weissert's preliminary transliteration, line 20′ of the new text gives *16 bēru* (instead of the expected *15). As Weissert suggested to me, this could represent a correction of the original 15 bēru = 8* days equation, which gives an odd ratio.

105 Generally speaking, both the distance and the length measures represent credible quantifications: cf. p. 81.

106 The dating formulae used in these inscriptions, even if representing quantifications, are not relevant to this work, except when they are subject to a "distorted" usage of ideological relevance (cf. esp. ch. 5 §7).

"(all of) the land of Uruaṭri in *3* days at the feet of Ashur, my lord, I subdued" (Shalmaneser I).[107]

It is clear that these numbers in many cases are not less obviously exaggerated than certain very "high" quantifications relating to prisoners, livestock, etc.[108]

But there are also examples of "high" time measures: "their *5* kings, who for *50* years had held the lands of Alzu and Purulumzu ...,"[109] or:

na-gu-u šu-a-tú ak-šu-ud ma-lak 10 *u_4-me* 5 *u_4-me ú-šàḫ-rib-ma*
šá-qu-um-ma-tú at-bu-uk

"That district I conquered, I devastated it for (a stretch of) *15* days of travel and I poured out dead silence (on it)" (Ashurbanipal).[110]

In these contexts the use of the number is almost compulsory, since a sentence like "the land *X* within a few days I conquered" would not be very effective.

The inscriptions (or the passages) relating to building enterprises, common especially during the old and middle-Assyrian periods, contain some chronological indications relative to earlier reigns, in most cases indicating the length of time having passed since the construction or the latest work carried out on a temple that is going to be restored, or has just been restored: "*641* years had passed and it had become dilapidated, when Ashur-dan, king of Assyria, ..."[111] These statements of time-spans (commonly known by the German word "Distanzangaben") are often accurate (using "exact" numbers)

---

107 RIMA 1, 183:40f (IAK XXI, 1; ARI 1, 527). Text in KAH 1, 13 (col. I).

108 Borger mantains that these typological numbers are not to be taken literally, cf. EAK 1, 56, where will be found some examples of "low" time measures referring to military events. Some other instances:
Tiglath-pileser I: RIMA 2, 23:48ff and 34:28 (AKA, 73 and 118; ARI 2, 34 and 70); in Schramm, BiOr 27, pl. I and 148:15 (Tukulti-Ninurta II), very fragmentary context, appears perhaps a similar passage (but cf. RIMA 2, 171).
Shalmaneser III: KB 1, 168:71.
Adad-nerari III: Millard - Tadmor, *Iraq* 35, 58:8; Page, *Iraq* 30, 142:4 (cf. *Iraq* 35, 143).
Of the Elamite king it is said that "he did not survive three months" ("did not fill," 3 ITU.MEŠ *ul ú-mal-li*, Luckenbill, *Senn.*, 41:12f). Other examples come from the reign of Esarhaddon, who captured (and destroyed) Memphis in half a day, Borger, *Asarh.*, 99:41f — cf. other examples at p. 43:63 ("I did not hesitate one day nor two," 1-*en u_4-me* 2 *u_4-me ul uq-qí*) and 50:34.
Ashurbanipal: Piepkorn, *Asb.*, 60:67f.

109 Tiglath-pileser I, AKA, 35:63ff (RIMA 1, 14; ARI 2, 12).
Ashurnasirpal II's inscriptions report some "high" time measures: "for *3* days the hero explored the mountain" (RIMA 2, 197 50f = ARI 2, 544); "for *6* days within the mounts Kašiyari ... I cut through the mountain with iron axes ..." (RIMA 2, 259:60 and 209:95 = ARI 2, 635 and 568). Cf. also Tukulti-Ninurta II in Schramm, BiOr 27, pl. II and 150:45 (RIMA 2, 173; ARI 2, 469). Other examples of quantifications of "days" appear in these same inscriptions (cf. e.g. ARI 2, 468, 469, 476, 638 and 571, 579).
Some other "high" measures of time come from the reigns of Sargon (Lie, *Sargon*, 62:5), Esarhaddon (cf. above, n. 104) and Ashurbanipal: cf. Streck, *Asb.*, 24:2, 56:77, 56:99 (= 160:30 and 164:73); Piepkorn, *Asb.*, 52:51 and 60:67.

110 5 R, 3:2f (Streck, *Asb.*, 24; ARAB 2, 786) - prism "A," older prisms had also "I completely destroyed it and burnt it up" after "I conquered," cf. e.g. Piepkorn, *Asb.*, 52:50f; Bauer, IWA, 15 IV 60 and Aynard, *Prisme*, 38:36f. On the extent of Ashurbanipal's conquests in Elam cf. below, p. 65 no. 89.
Obviously the texts give also time indications that are not "boosted" (to the low nor to the high), of the type "in a second day (I did this and that ...)."

111 Tiglath-pileser I, AKA, 95:64ff (RIMA 2, 28; ARI 2, 54).

and may furnish useful and reliable[112] elements for the reconstruction of Assyrian chronology.[113]

7. Repeating actions

The usage of multiplicative numbers is quite frequent. In the writing, they are represented by the numerical sign followed by the pronominal suffix *-šu*. Thus we have "three times (3-*šu* = *šalāšī-šu*) I marched to the land of Nairi" (Tiglath-pileser I).[114] The numeral adverbs, also quite frequent in the Assyrian royal inscriptions, are written according to the model X-*te-šu*, "for the Xth time" — e.g. "a second time (2-*te-šú* = *šanûtē-šu*) I marched to the land of Ḫanigalbat" (Adad-nerari II).[115] These quantifications obviously represent "totals" referring to the part of the reign taken into consideration by the inscription up to that moment. The numbers are subject to updating within a single inscription: thus in the previously cited one also the *3rd, 4th, 5th, 6th* and *7th* marches to that region are "counted."[116] These expressions often represent the only references to other parts of the inscription, structured as they are in "episodes,"[117] nearly independent units containing the narrative of a campaign or a part of it.

---

112 Not always: see e.g. Poebel, JNES 1, 297ff or below, chapter 3, nn. 91 and 99.

113 For an analysis of the *Distanzangaben* contained in the Assyrian royal inscriptions cf. Poebel, JNES 1, 290-306; Weidner, AfO 15, 87-95; Landsberger, JCS 8 and, more recently, Na'aman, *Iraq* 46 or R. Hachmann, "Assyrische Abstandsdaten und absolute Chronologie," ZDPV 93 (1977), 97-130. Those reported by the royal inscriptions up to the reign of Tiglath-pileser III are collected in Lackenbacher, RB, 180f (cf. also 195 and especially 15ff).

114 RIMA 2, 42:15 (ARI 2, 91 and n. 120). Cf. GAG, §71a on the multiplicatives.

115 RIMA 2, 149 (ARI 2, 425). Cf. GAG, §71b on these adverbs. For the crossings of the Euphrates of Shalmaneser III cf. pp. 136-138.

116 "For the *7th* time" is a correction since the text repeats "for the *5th* time": cf. p. 25 n. 31.

117 Using the definition of Borger, *Asarh*. The inscriptions of Ashurbanipal, however, have particular narrative characteristics (cf. p. 176), and the references to earlier events are much more frequent.

2. THE QUANTIFICATIONS IN THE ANNALISTIC INSCRIPTIONS

Hereafter I will analyze some inscriptions in relation to the quantifications they contain. The aim is to give a general view of the investigated material, the inscriptions and the numbers: the presence of the latter in the main annalistic inscriptions will be evaluated without taking into account the context (narrative or not) in which they appear. A rather flat picture will result but refer to the following chapters for a more careful evaluation. The limitation to only one type of inscription is due mainly to reasons of space.[1] Yet, the annals are preferable for a "statistical" analysis, being usually the most complete inscriptions available for each reign (also usually the longest), of which they give a general framework since, as the name itself indicates, as a rule they include a more or less detailed report for every regnal year. Furthermore, it is characteristic of the annalistic inscriptions that they are subjected to "editing," including the quantifications, which can be seen through a comparison of subsequent annalistic editions, and which can be viewed as an ideologically motivated "correction." Only in a few cases will other types of inscriptions be taken into consideration. Generally, they do not have many quantifications and, in most cases, the indications given by the contemporary annals apply to them as well. Some data are collected in Table 1 at the end of this chapter.

(1) **Type of inscription and quantifications** — the frequency of quantifications we find in the Assyrian royal inscriptions is obviously related to the kind of inscription we are examining.[2] Short inscriptions such as "labels" and dedicatory texts are completely devoid of numbers. The introductory sections

---

1 Certain annalistic texts are not considered because they are fragmentary or incomplete, or because the contents duplicate those analyzed (cf. the notes for further details).

Note that the data reported throughout the chapter (number of quantifications, of digits, etc.) are relative to the text of the inscription each time indicated, without taking into consideration reconstructions or emendations drawn from more or less similar versions, former or subsequent editions, etc. In many cases to decide whether two copies of an inscription represent the same text or not is quite an undertaking. One has to be very careful with editions in transcription that "reconstruct" the text from several copies and fragments of different provenance and format (tablets, slabs, etc. — this is the case with the annals of Tiglath-pileser III as published by Rost: cf. Tadmor, *Intr. Remarks,* 175 and *id., Tigl. III,* 18ff).

2 For the classification of the Assyrian royal inscriptions I have followed Grayson, Or 49, 150ff and ARINH, 37ff. Cf. also RlA *s.v. Königsinschriften.* B. *Akkadisch* [J. Renger], vol. 6, 71-77, and, for a classification according to the external characteristics (tablet, prism, cylinder, etc.), Ellis, FD, 94-124.

of the various inscriptions, including genealogy, titulary, epithets, etc.,[3] have only some rare numbers, if we except the very common title of "king of the four parts (of the world)."[4] The shortest *summary* inscriptions,[5] especially if they do not deal with military matters (and this happens very often during the old and middle Assyrian periods) also have very few quantifications. Since their main purpose is represented by building works, the digging of canals, etc., the numbers appearing there most of the time concern the height of the walls, or the number of years having passed since previous work was done on a given building.[6]

(2) Inscriptions of the **predecessors of Tiglath-pileser I** — with Shamshi-Adad I some military actions are for the first time narrated.[7] Another inscription of this same ruler is the first one to include a certain number of quantifications: they appear in a passage listing some commodities with their prices at the market of Ashur. These prices, compared with the actual ones, are very low; in fact they are "ideal prices" included in the inscription for propagandistic purposes.[8] During the subsequent period several attempts have been undertaken, not always successfully,[9] to find a suitable collocation for the military events within inscriptions the structure of which continues to be focused on "civil" matters.

A fragmentary tablet dating back to the reign of Arik-den-ili includes detailed information of a military character and several quantifications.[10] Its synthetic and rough style and its making reference to the king in the third

---

3 Sometimes the titulary might be included within an inscription (and not only at its beginning), cf. e.g. the "annals" of Ashurnasirpal II (and particularly the passage of ARI 2, 533).

4 For this title cf. Seux, *Épithètes*, 305ff (and 313f for "king of the totality of the *4* parts"), title used from Tukulti-Ninurta I onwards (including perhaps also Tukulti-Ninurta II: cf. RIMA 2, 166:34, very fragmentary passage; cf. also 165:22). According to Liverani, ARINH, 236, Sennacherib used this title only after having achieved successes in all the four cardinal directions. The rule, however, was not general: for instance Adad-nerari II employed this title already in a text of the 3rd year (incl. the year of accession; cf. RIMA 2, 143:2 and 142), as did some obscure kings, as Eriba-Adad II who ruled two years (RIMA 2, 114:2) or Ashur-etil-ilani (Seux, *Épithètes*, 308). It is difficult, in two years, to lead (victorious) campaigns in four directions. Thus the title appears in a recently discovered text of Shalmaneser III, containing the first two campaigns (year of accession and first regnal year, cf. Mahmud - Black, *Sumer* 44, 138:6, and cf. 10f). Anyway, he used it long before undertaking campaigns to the south in his 8th and 9th year (cf. e.g. 3 R, 7 I 5).
The title of "king of totality" (*šar kiššati*), on the other hand, if we except the short texts where only the most essential titles were employed, is exhibited by every Assyrian king from Shamshi-Adad I on who has left us a reasonably complete inscription, and among these are the obscure Ninurta-apil-Ekur, Eriba-Adad II, Shamshi-Adad IV, Ashurnasirpal I and Shalmaneser II (Seux, *Épithètes*, 309).

5 Also called "Prunkinschriften," or "display inscriptions." These terms are employed to indicate inscriptions usually written on stone slabs mainly intended for architectural purposes (often they accompany reliefs), in which the events are arranged according to a mentally associated geographical order (more of less from East to West), and only in suborder chronologically. Often, in earlier periods, these inscriptions simply give a "résumé" of the main annalistic texts, preserving a roughly chronological arrangement, and in this case it is perhaps better to call them "summary inscriptions" (thus Tadmor, *Iraq* 35, 141). Cf. also below, p. 92 n. 188.

6 Cf. the preceding chapter, pp. 16-17 and p. 18 for the "Distanzangaben."

7 Cf. ARI 1, 158 and Grayson, ARINH, 38. The attribution of this inscription is not certain, having reached us in bad condition (RIMA 1, 63, A.0.39.1001; cf. Tadmor, *Fs Finkelstein*, 212). However, it does not contain quantifications.

8 Cf. ARI 1, 127 and n. 64 there; EAK 1, 15; inscription published in IAK, VIII, 1, 22ff and RIMA 1, 47ff (A.0.39.1). On the "ideal prices" cf. Hawkins, *Fs Mellink*, and, for the instances included in the Assyrian royal inscriptions, cf. below, p. 64 nos. 86-87, to which add Weidner, AfO 13, 210:9ff and Lambert, AfO 18, 384:12 (cf. also RIMA 1, 127:27′).

9 This is the case with some inscriptions of the reigns of Shalmaneser I and Tukulti-Ninurta I: cf. Grayson, Or 49, 155; *id.*, ARINH, 38; ARI 2, 524 and 687; Harrak, *Hanigalbat*, 133f and 207.

10 Among the readable numbers there are many round ones: *2, 33, 90, 100* (3 times?), *600, 7,000* and *254,000*, written 2 *me* 54 *lim* and referring, it seems, to enemies killed (LÚ?, *dâku*). Cf. A.T. Clay, *Babylonian Records in the Library of J. Pierpont Morgan*, 4, New Haven 1923, pl. 46 (no. 49):29′ (RIMA 1, 125ff, A.0.75.8; Grayson, ABC, 185f; IAK, 50ff; ARI 1, 359ff; cf. also EAK 1, 31).

person, however, make us wonder if it actually represents a chronicle rather than a royal inscription.[11]

An inscription of Shalmaneser I, besides being one of the longest among those reaching us up to this reign, is the first one to include information of a military character sufficiently detailed and complete. In it there appear some of the quantifications typical of the later annalistic inscriptions (cities, soldiers and lands).[12] The inscriptions of Tukulti-Ninurta I do not contain many quantifications; the existing ones, however, are singularly similar to those of his predecessor.[13]

(3) The **annals of Tiglath-pileser I**, representing the first extant example of this literary form, at least in Assyria, contain 49 numbers.[14] Comprehensively, the most striking fact is the presence of several round numbers, a fact to which is connected the high frequency of numbers with one digit only[15] — to those listed in Table 1 on pp. 42-43, *sub* 4 should be added also the 12 sexagesimal numbers, all written according to the model "*X šu-ši*," without lower order digits. The use of these numbers will become less and less frequent in the subsequent inscriptions (cf. Table 1).

In effect, all the numbers surpassing *60* are devoid of digits of an order lower to the *šuššu* or to the *me* (except *641*, which is a *Distanzangabe*), while out of a total of 33 numbers larger than *9*, only 7 have units: 3 times *25*; once each *16, 23, 42* and *641* (*16* and *23* are numbers making up the "total" of two lists of names, cf. p. 190).

The most frequent numbers are: 6 times: 1 *šu-ši* (= *60*, once has the variant 2 *šu-ši*); 5 times: 2 *šu-ši* (= *120*, including when appearing as a variant); 4 times: *1*; 3 times: *5, 25* and 20 *lim* (= *20,000*).

The booty most of the time is not quantified, and, if we except one instance (in which some gods are mentioned), the type of goods taken are not even specified. When numbers are given, then they concern objects that will subsequently be donated to the Assyrian gods.[16] Tribute is quantified 3 times out of 9 (imposed tribute; only once does the text speak of tribute received).

---

11 Its poor condition does not allow a clear idea of the category to which this text belongs. Tadmor (ARINH, 17 and n. 16) thinks it could represent an early attempt at editing a royal inscription, in which the royal scribes experimented with some of the chronistic conventions.

12 Cf. A.0.77.1 in RIMA 1, 180ff (= IAK, XXI, 1, 110ff and ARI 1, 526ff). The extant numbers are (in ascending order): *3, 8, 9, 51* (var. *41*), 3 *šu-ši* (= *180*), 4 ŠÁR (= *14,400*), plus two *Distanzangaben* in the passage on the building activities: 2 *šu-ši* 39 (= *159*) and 9 *šu-ši* 40 (= *580*).

13 Cf. below pp. 150-151 (esp. n. 171 there).

14 1 R, 9-16 = AKA, 27-108 (RIMA 2, 7ff, A.0.87.1; ARI 2, 3ff). The 49 does not include dates, the fraction in III 100 and the expression *kibrat* 4-*i* appearing in the titulary (cf. the remarks at p. 41 n. 91). The various copies of the inscription show only one variant (p. 46 no. 4), concerning one copy only (cf. n. in RIMA 2, 19).

15 Note that for "number" I mean the sum of a given category of objects and for "digit," or "digits," the sign or signs forming it. Thus *641* will be the number, while "6," "4" and "1" will be its digits, 3 in all. Any statement on the digits present in a given number will make reference exclusively to the original writing: for instance a number written 1 *me* 20, "120," will be considered as formed by 2 digits, "1" and "2," the zero being not represented in the cuneiform system. 2 *šu-ši* will be considered as formed by a single digit, "2." In fact, a number written 1,10 (= 70) was conceived, at least in some cases, as consisting of two digits "1": see the episode of the "turning" of the number *70*, thus becoming *11*, at Marduk's will, a thing possible only if we conceive of both digits as "1" (cf. below, p. 140). The sexagesimal system provided that "60" and "1" had to be written in the same way, with a vertical stroke, and only to avoid a possible uncertainty in the former case it was (usually) specified as 1 *šu-ši*. It might be hypothesized that, on practical grounds, the number *120* of the above cited example was conceived as "2 × 60" (thus as formed by the digits "2" and "6"), yet the *šuššu* has the technical value of an exponent.

16 The three quantified cases of booty appear in RIMA 2, 15:28ff, 19:102ff and 20:23ff (here only the gods are quantified), and in all three cases it is reported that this booty was then donated to various Assyrian

TIGLATH-PILESER I

QUANTIFIED OBJECTS[17]		N	S	Q	max	comments
People		13	2	**6**	*20,000*	(all "round")
(Assyrian troops)		7	-	-		
Kings		1	1	**5**	*60*	
Cities		6	3	**3**	*25*	
Chariots	(enemy)	-	-	**2**	*120*	(booty)
	(Assyrian)	6	-	**1**	*30*	
Booty		7	4	**3**		(tot. **5** quantif.)
Tribute	received	-	-	**1**		(tot. **2** quantif.)
	imposed	5	1	**2**		(tot. **3** quantif.)

N = not quantified;[18] S = the quantity is specified in generic terms ("without limit," "many," "all," "the rest," etc.); Q = numerically quantified.

A later inscription of Tiglath-pileser I, rather fragmentary, presents a "summary" of the enterprises of the ruler in chronological order.[19] The synthetic narration (and the poor condition of the text) do not permit the undertaking of a thorough analysis of the digits appearing there, but rather limit us to pointing out some significant facts. The first 6 paragraphs of the section containing the military narrative[20] include events already dealt with in the above annals, and report comprehensively 8 numbers, one of which represents a "new" datum (previously omitted), the deportation (?) of *2,000* prisoners,[21] while among the other 7 there are two variations (cf. nn. 5 and 6 at p. 46). A remarkable fact is that each paragraph, in spite of its brevity, has at least one number.[22]

---

gods (cf. subsequent lines; in the first and third case the quantities are specified again, cf. chapter 4 §3 on this argument). There are also two groups of booty represented solely by the enemy chariots (RIMA 2, 17:3f and 21:94, in both cases 2 *šu-ši*). This kind of booty has to be formally separated from the previous one, since sometimes, within a single episode, it refers both to the booty (of chariots) made "on the battlefield" and to the booty captured inside the enemy cities (cf. e.g. RIMA 2, 21:94 and 22:5ff).

17 A limited number of objects is taken into account, since it is rather difficult to group them within homogeneous categories (cf. above p. 8 n. 44). The entries "booty" and "tribute" (for the distinction cf. pp. 11-12) report the number of mentions of booty or tribute quantified with one or *more* numerical data, followed by the total number of quantifications appearing therewith.

18 The numbers reported in this column (and partially those in col. S) should be considered *approximations*. As might be easily understood, this kind of accounting can be very problematic: it is not simply a matter of counting how many times a given term appears in the text. For instance, the same objects may be cited several times within a single passage (and in this case they are *not* counted again), or may appear in contexts that do not allow in any way their quantification (in this case too they are *not* counted).

19 Weidner, AfO 18, 359ff, Tf. XXX (A.0.87.2, RIMA 2, 31ff; ARI 2, 64ff; cf. EAK 1, 114ff), other copies in 3 R, 5 nos. 2 and 5; KAH 2, 160 a-b. It is a *summary inscription* reconstructed from several fragmentary tablets, on the whole much shorter than the previously mentioned annals.

20 Lines 18-36 of the edition in RIMA 2. The narrative of the campaigns continues (lines 37ff), but the text, from this point on, is very lacunal. The scribe has subdivided it into paragraphs by horizontal lines.

21 The passage referring to the Qumānu in the annals did not speak of prisoners, but only related a collision with *20,000* of their troops, a digit taken up also by the later text under discussion (cf. RIMA 2, 34:34 with 24:82ff).

22 If excluding the paragraphs containing the introduction and the concluding ones with the *building inscription*, we would record one or two quantifications for each paragraph (it is not necessary to stress how the quantifications taken up are among the most significant ones):
lines 18-20 **1** number (*12,000* soldiers)
21-22 **1** number (*4,000*, term not specif.)

A still later inscription, reconstructed by Weidner from several fragments,[23] presents in a very brief fashion the events covered by the previous ones (lines 18-23, with only 2 numbers), while the narration of the later campaigns includes only 4 numbers, all multiplicatives.[24] Quite frequent, however, are such expressions as *ana lā mīna, ma'dūte*, etc., employed also for objects that often were numerically quantified in the above cited annals (cities, prisoners, etc.).[25] Some measurements ("improvements," cf. below at p. 153) appear in the passage referring to the building enterprises, rather detailed and placed, as usual, in appendix to the inscription.

(4) The most complete version of **Ashur-bel-kala's annals** (actually rather fragmentary) preserves only two numbers, both within the paragraph concerning the hunting deeds.[26]

The so-called "broken obelisk," probably belonging to this reign, reports 11 numbers — too few, however, to draw any general indication.[27]

(5) The **annals of Ashur-dan II** also are fragmentary:[28] only 4 numbers have been preserved, which, in this case too, appear within the hunt paragraph. The booty, even if in at least 4 cases its composition is specified, is never numerically quantified (in one case the expression *ana lā māni* is used).

(6) The **annals of Adad-nerari II** contain cardinal digits only in the paragraph concerning the hunting deeds,[29] if we except the depth of a canal.[30] Besides these, it is "counted" how many times the king marched to the land of Nairi ("*4th* time"), to the land of Ḫanigalbat ("for the *2nd/3rd/4th/5th/6th/5th* (?) time") and to the city of Kummu ("*2nd* time").[31] Thus, even though

---

23-24 **1** number (*25* gods)
25-27 **1** number (*30* kings)
28-29 **2** numbers (*1* day and *6* cities)
30-36 **2** numbers (*2,000* prisoners and *20,000* soldiers).
The subsequent lines are very fragmentary. Note that each paragraph is marked off by horizontal lines.

23 Weidner, AfO 18, 347ff, Tf. XXVI-XXIX (for other copies cf. KAH 2, 63, 66, 69, 71, 71a and 73) = RIMA 2, 38ff (A.0.87.4; ARI 2, 87ff). Cf. also the inscriptions from Nineveh (A.0.87.10-11), that relate a very similar report on the military activities. The inscription RIM A.0.87.3 is older (it repeats, more synthetically, part of the narrative of the text under discussion) but does not have many quantifications. The other texts of this reign are either very fragmentary or very short (A.0.87.2, 5-9, 12-14).

24 X-*šu*, cf. lines 15, 34 and 50. On the crossings of the Euphrates ("28 times, twice in one year," line 34) cf. below, p. 89.

25 Appear: *ana siḫirtišu* (3 times), DAGAL (twice), *gabbu, ana lā mīna, ma'ttu* and *ma'dūte* (each once).

26 Weidner, AfO 6, 80-93 = RIMA 2, 89ff (A.0.89.2), ARI 2, 210ff. The other versions of his annals (A.0.89.1 and 3-6) are nothing more than fragments. The hunting reports will be discussed at pp. 143ff.

27 AKA, 128-149 = RIMA 2, 99ff (A.0.89.7), ARI 2, 227ff; it is written in the form of annals. King (AKA, 128 n. 1 and 131f n. 4) thought that, although it had to be ascribed to a successor of Tiglath-pileser I, the deeds (hunting or others) are those of the latter. Weidner, AfO 6, 92ff, however, strongly supported the attribution to Ashur-bel-kala — see Brinkman, PKB, 383ff and EAK, 1, 135 and 138ff on the matter. In the passage concerning the hunting some numbers appear missing: the scribe left some blank spaces, probably intended to be filled in later with the numbers of killed animals; cf. p. 144.

28 Weidner, AfO 3, 151-161 = RIMA 2, 131ff (A.0.98.1), ARI 2, 359ff. Another version of the annals is even more badly preserved: cf. Weidner, AfO 22, 76f = RIMA 2, 136ff (A.0.98.2). Neither this nor other inscriptions of this reign, however, contain quantifications not appearing in the annals A.0.98.1.

29 KAH 2, 84 (RIMA 2, 145ff [A.0.99.2]; Seidmann, MAOG 9/3, 5-35; ARI 2, 411ff). An earlier version is very fragmentary and includes only two (?) numerals, referring to cities, cf. KAH 2, 83 + KAH 1, 24 (RIMA 2, 142ff [A.0.99.1]; Seidmann, MAOG 9/3, 36-41; ARI 2, 389ff). A later version is too fragmentary to draw any conclusion: only a few lines have survived, cf. RIMA 2, 156ff (A.0.99.4; this applies also to the fragments no. 99.3 and 5).

30 "*9* cubits," at line 65, within the section concerning the building activities.

31 All these ordinal numbers are written according to the model *X-te-šú* (cf. ARI 2, p. 26 n. 120), the verb employed is always *alāku*. A series of 5 subsequent paragraphs begins with the indication of the eponymate (*ina li-me* ᵐPN) followed by an identical expression X-*te-šú a-na mat Ḫa-ni-gal-bat lu a-lik* and by the more or less detailed report of a campaign to Ḫanigalbat (cf. lines 42, 45, 49, 61 and 62, from the *2nd* to the *6th* time — as usual, the scribe has separated the various paragraphs with horizontal lines).

the inscription reports 3 mentions of booty, 9 of tribute received and 3 of tribute imposed, never are the commodities quantified (and often they are not even specified). Frequently, on the contrary, non-numerical expressions are used in relation to booty or to people or cities.[32]

(7) The **annals of Tukulti-Ninurta II** contain[33] several quantifications (84: but some digits are not readable) and also some detailed tribute lists. The round numbers are frequent: out of 50 numbers higher than *9*, only 6 have the unit (*11, 18, 32, 2,702*, 2 times *14*). There are 15 numbers higher than *99* completely readable: 12 are round (that is devoid of digits of order lower than *100*). Only 3 numbers[34] use sexagesimal digits: 2 times 1 *šu-ši* and *470*, written 4 *me* 1,10.

The numbers most frequently appearing are: 8 times: *2* and *10* (once doubtful); 7 times: *20* and *30*; 6 times: *3* and *1*; 5 times: *200* and *100* (once as 1 *me* [...]).

Noteworthy is the presence of low numbers (the highest is *2,702*). Also very numerous are the low digits ("1" and "2"). This is due to some quantifications relating to lengths of time ("the *4th* day the city of Pīru ... I conquered")[35] but especially to the presence of 12 quantified tribute lists, some of which are rather detailed.

TUKULTI-NINURTA II

QUANTIFIED OBJECTS[36]		N	S	Q	max	R	E
People	killed	1	2	-			
	deported	1	-	-			
Cities		1	-	**2**	*30*	-	-
Horses (Assyrian)				**1**	*2,702*	-	1
Time expressions				**4**			
Booty		6?	2	-			
Tribute		1?	-	**12**		11 (+ 1)	(1?)

Columns **R** and **E** report the number of "round" and "exact" quantifications used with each object. In this case, the entries booty and tribute report the *total* number of quantifications appearing there.

Further on (line 98), the same expression introduces the "definitive" conquest of the region, an episode that represents again, according to the text, the "*5th* time" — an oversight for "*7th*" (cf. ARI 2, n. 377), digit possibly confirmed by the (very) fragmentary passage of RIMA 2, 158:10′.

32 The repertory includes DAGAL, *ma'attu*, DUGUD; repeatedly appearing also *ana siḫirtišu* and *ana pāṭ gimrišu* to indicate the conquest of the "totality" of a certain region, or of the "totality" of its cities. In a fragmentary passage it is specified *ša* KI.LÁ-*šunu lā aṣbat* ("whose weight I did not determine," line 72) in relation, as far as it appears, to some metal objects obtained as booty (in turn defined, as a whole, DAGAL, "large"). A couple of imposed tributes are defined GUN *u tāmarta udannin* ("strengthened," 90 and 93).

33 Schramm, BiOr 27, 147-160, *pl.* I-V (RIMA 2, 169ff, A.0.100.5; ARI 2, 462ff). This is the only annalistic inscription of this ruler. From the accounting are excluded the fractions at lines 28f.

34 With possibly one more appearing as 1 [*šu-ši*].

35 Line 122 (*ina* UD 4.KÁM). Other examples at lines 51 (*ina* UD 3.KÁM), 37 and 43 (both *ina* 2-*e* u_4-*me*).

36 Cf. nn. 17f above. People are distinguished according to their fate (killed or deported — those received as tribute are excluded, cf. n. 40 below — and are accounted at the proper entry). Under "tribute not quantified" are taken into consideration only those that specify at least one object (thus, e.g., "a tribute of horses" is, while "the tribute of X I received" is not). In the second part of the table, instead, are distinguished some objects coming *both* from tribute and from booty, and the number of quantifications referring to each entry is reported.

Tribute quantifications:	Q	max	R	E
Sheep	7 (+1)	*1,200*	7 (+1)	-
Oxen	6 (+2)	≥ *100*	1	(1?)
Silver	6	*10 t.*		
Tin	5 (+1)	*32 t.*		
Gold	5	*20 m.*		
Asses	3	*30*		
Iron	1 (+1)	*2 t.*		
Bronze	1	*130 t.*		
Camels	1	*30*		
(other)	24 (+3)	*150*	3	-

The inscription has a particular structure, from which emerges a lengthy passage in a particular narrative style, called by some authors "show of strength."[37] It consists of the detailed report of an itinerary covered, almost peacefully, by the Assyrian army, of which are reported also some marginal episodes.[38] Out of the 13 tribute lists, 12 include quantifications,[39] all referring to very limited areas, if not to a single town. For this reason the numbers there appearing are never particularly relevant ("high"), even if the differences between one case and another are remarkable. Oxen and sheep are almost always quantified with round numbers: sheep *200* (5 times), *500, 1,200,* [*X*] *me*; oxen *30* (3 times), *50* (2 times), *100, 1 me* [+ X?], [*X*] *me 40*. Only once are people quantified (even when taking into account the other inscriptions of this ruler): *2* girls received as tribute.[40]

The preceding section (cf. n. 37) includes the report of four campaigns,[41] all devoid of numbers in relation to booty, even if it is mentioned in 7 (?) cases.

(8) Also **Ashurnasirpal II**'s "annals" carved on the Ninurta temple pavement slabs at Kalah[42] have many round numbers: out of 84 that are higher

The parentheses with such indications as (+ 1) etc. are explained by the presence of some doubtful cases originating from the lacunae in the text.

37 The inscription can be subdivided in the following way: (lines according to RIMA 2) after a lacuna (the beginning is missing) lines 1-40 military achievements; lines 41-121 "itinerary," devoid of any military engagements, except in its appendix, 121-127; lines 128-135 summary of the royal deeds; lines 132-147 building enterprises, curses, date. Thus, at least within the part that has been preserved, the passage referring to the itinerary is clearly predominant. This kind of narrative makes its appearance with the annals of Adad-nerari II, where it occupies only lines 105-119 (RIMA 2, 153f) and does not include any number.

38 Even information on the supplies for the army is reported, including particulars on springs or wells — in one instance some wells are quantified: 4 *me* 70 PÚ.MEŠ*(burāte) uḫ-tap-pi*, "*470* wells I destroyed (= exhausted);" Schramm, BiOr 27, pl. II and p. 150:43 (RIMA 2, 173).

39 These tribute lists will be more fully discussed at pp. 97ff. At the end of the "itinerary" passage there appears the only military engagement of the campaign, which yielded booty, not quantified (lines 122f — cf. also at line 50, where probably with *šal-la-su-nu ma-ʾa-ta a-sa-la* only some prisoners are meant).

40 They are placed at the end of a tribute list: 2 MUNUS.NIN.MEŠ-*šú iš-tu nu-ud-ni-ši-na ma-ʾ-di na-*[*mur-tu ša* ...]*a-a-ia mat La-qa-a-ia* "2 sisters of his with their rich dowry, trib[ute of Ḫamat]āya of Laqû;" line 101 = Rs. 19 of Schramm's edition.

41 Lines 1-10 contain the synthetic report of three campaigns (in three different years?), all of them concluded by a short booty list, while lines 11-40 relate the campaign of the 4th year, which yielded 3 or 4 (cf. ll. 35f) groups of booty. The long passage with the itinerary is datable to the 5th year (cf. Grayson in CAH 3/1, p. 252 n. 64).

42 AKA, 254-387 = Le Gac, *Asn.*, 3-125 for the copies; edition in RIMA 2, 191-223 (A.0.101.1; transl. ARI 2, 533ff) — actually not authentic "annals," cf. below. The accountings include 1-*en* ... *ul ezib* of col. I 108. The variants given by the "Kurḫ monolith" or by the "Nimrud monolith," on the contrary, are ignored.

than *9*, only 3 have units (*172, 326, 332*); while out of 63 higher than *99* only 16 have digits of order lower than a hundred (one case is in doubt since the numeral appears as 1 *lim* 4 *me* [...]). The only number higher than *1,000* not having the form "*X lim* (*X me*)" is 1 *lim* 4 *me* 1 *šu-ši* (= *1,460*; and perhaps the doubtful case just mentioned). The presence of these round numbers, together with the fact that the lower digits ("1" and "2") are very frequently used, gives the impression that there have been many cases of rounding-up.

There are 8 numbers combining the sexagesimal and the decimal systems.[43]

It is to be noted that the volume of numbers falling between *100* and *1,000* and of those equal to or higher than *1,000* has increased as compared with Tukulti-Ninurta II's annals (cf. Table 1 at the end of the chapter). While there are no extremely high numbers (the highest is *10,000*, referring to a tribute of oxen), the quantifications on the order of the hundreds and of the thousands are abundant.

The quantifications of people (enemies, soldiers or civilians) are far the most frequent:

ASHURNASIRPAL II

QUANTIFIED OBJECTS[44]		N	S	Q	max	R	E
PEOPLE	killed	8	10	**26**	*6,500*	17	3
	captured	1	4	**8**	*3,000*	3	-
	prisoners	20	7	**9**	*3,000*	9	-
	(other)	-	-	**1**	*6,000*	1	-
	(Assyrian)	16	1	**-**			
Cities		25	1	**19**	*250*	1	-
Multiplicatives				**2**	*3*		
Time expr. (days)				**5**	*6*		
Booty		12	6	**-**			
Tribute	received	34	1	**4**	(tot. **25**)	16	-
	imposed	2	2	**1**	(tot. **1**)	-	-

The other annalistic inscriptions of this king are not taken into account because they are excessively fragmentary or incomplete (RIMA 2, A.0.101.4-8, 14-16, 20-22) or are very short and anyway duplicate sections of the "annals" (101.9-14). The "Nimrud monolith" (101.17) repeats the first two columns of the "annals" (cf. here below n. 49), while that of Kurḫ (101.19) contains only the text of the 5th campaign, even if more detailed. The "white obelisk" (101.1) does not contain any quantification (actually its attribution is not certain).

43 Besides *1,460*: 5 *me* 1,20 (= *580*), 4 *me* 1,10 (= *470*), 4 *me* 1 *šu-ši* (= *460*), 2 *me* 1 *šu-ši* (= *260*), 1 *me* 1,10,2 (= *172*, number of 4 digits), 1,10 (*70*), 1 *šu-ši* (*60*). In spite of this, the number *120* is twice written 1 *me* 20 (AKA, 345:132 and 387:136). The use of sexagesimal digits appears to be declining, at this period, within the *corpus* of the royal inscriptions: compare e.g. the 3 *me* of AKA, 342:122 (Ashurnasirpal II) with the 5 *šu-ši* of AKA, 81:31 (Tiglath-pileser I) — but cf. also n. 53 here below.

44 In the table, under "people" distinction is made between "captured" (*ṣabāṭu*, usually referring to ERÍN) and "prisoners," that is those who have been deported (*šalālu, nasāḫu, waṣû* Š), be they troops (ERÍN) or "civilians" (this group includes the prisoners, *šallatu*, burned, GÍBIL, which therefore do *not* appear amongst the "killed"). The reason for this distinction is that sometimes, in this inscription, within a single episode either the killed, the captured or the deported are quantified (cf. one instance at p. 9). The hostages (*līṭu*) are not taken into account, never being quantified within the *corpus*. Under the entry "(Assyrian)" are reported the references to foreign contingents incorporated into the Assyrian army.

The prisoners are always given in round numbers: deported: *300, 500, 1,200, 2,000, 2,500, 3,000* (twice); burned: *200, 3,000*. The only three exact numbers in the text are reserved for contingents of enemy soldiers killed: *332* (*dâku*; col. I 112); *326* (*napāṣu* D; II 36); *172* (*dâku*; II 41). The numbers appearing in the tribute list, when higher than *9*, are always round ones: *10, 20* (twice), *100* (7 times), *200, 250, 300, 460, 1,000* (3 times), *2,000, 3,000, 5,000, 10,000* (including oxen and sheep).[45] Comprehensively only four tributes are quantified (three consecutively),[46] notwithstanding the frequent references to tribute received, which mostly appear in some passages in the style of "show of strength" of which we spoke in connection with Tukulti-Ninurta II.[47]

In this case it is possible to examine the distribution of the numbers through the various campaigns. But first it is necessary to point out that these "annals" have a particular structure: the first two columns consist of a long introduction and an annalistic passage including the campaigns of the accession to the 5th regnal year, dated by eponyms, followed by the text (almost independent) of a *display inscription*. The third column includes another annalistic passage, relating some campaigns of which the first and the last ones are dated to the 6th and the 18th year, and is concluded by one further *display inscription* (different from the previous one).[48] Obviously, we are being confronted with a "compilation" of heterogeneous material.[49] The numbers are distributed thus:[50]

(year)	passage	lines	NUMBERS	(year)	passage	lines	NUMBERS
Introd.	I 1-42	42	-	*D.I.*	II 125b-135	$10\frac{1}{2}$	**1**
(1)	I 43-68	26	**3**	(6)	III 1-26a	$25\frac{1}{2}$	**4**
	I 69-98	30	**1**		III 26b-50a	24	**14**
(2)	I 99-II 23a	$42\frac{1}{2}$	**11**		III 50b-56a	6	**2**
(3)	II 23b-33a	10	**5**		III 56b-92a	36	**15**
	II 33b-49a	16	**8**	(18)	III 92b-113a	21	**7**
(4)	II 49b-86a	37	**9**	*D.I.*	III 113b-136	$22\frac{1}{2}$	**2**
(5)	II 86b-125a	39	**28**				

The quantifications are proportionally distributed, in accordance to the length of each section. Yet the various sections tend to quantify certain objects instead of others. Thus the quantifications of tribute appear only in the text of two campaigns, that of the 5th year and that preceding the 18th

45 Note that, in the table, the "round" numbers (**R**) include only those based on the models "X00" (*100, 200*, etc.), "X,000" and "X,X00" (*1,200, 2,800*, etc.). Consequently, the cities could never be quantified only by "round" numbers, since the highest of these is *250* and only 5 of them reach or surpass *100*.

46 Plus an imposed tribute of *1* mina of silver. One passage appears at col. II 120ff and the other three at col. III 63-76. In the above table, because of the low number of quantified tributes, I make no distinction between the various commodities received, indicating only the total of quantifications. The silver is, however, the most frequently quantified term (5 times in total, that is, every time).

47 Col. II 91ff; III 1ff, 56ff.

48 On their structure cf. EAK 2, 18ff; Grayson, BiOr 33, 138f and Paley, *Ashurnaṣirpal*, 145ff.

49 According to Tadmor (*Fs Finkelstein*, 210) the scribe recopied the text of the "Nimrud monolith," with the campaigns of the first 5 years, including its summary (represented by the first *display inscription*), then adding some other annalistic material and a second conclusion. Thus, the "Nimrud monolith" is not examined in these pages because it is very similar to the first two columns of these "annals."

50 *D.I.* = "Display Inscription." The text for 12 times reports some time indications (eponymate or day/month), and therefore I have subdivided the annalistic sections of the text according to these indications. Only certain campaigns can be dated (those introduced by the name of the eponym), and in this case the year of reign appears in parentheses. The dating and the number of the campaigns of the third column represent a quite complicated problem: Brinkman thinks there are at least 4 between those of the 6th and of the 18th year (PKB, 394) — cf. also EAK 2, 29ff and Grayson, BiOr 33, 138ff.

year (actually the latter has *only* quantifications of tribute). The text of the campaigns of the 2nd and 3rd year contain many quantifications of cities (6 in all) and people (10 times killed, 3 times captured or deported), including the sole 3 "exact" numbers (cf. above; besides these also *1,460*). The campaign following that of the 6th year contains 4 quantifications concerning a hunting expedition.[51]

(9) **Annals of Shalmaneser III: Edition "A"** — the "monolith inscription" is relatively rich in numbers, most of which are round.[52] Subdividing the numbers according to their composition and representing a number of vertical strokes or of *Winkelhaken* from 1 to 9 with "X" and "X0" respectively, we get the following table:

no.	composition	no.	composition
16	units (X)	11	thousands (X *lim*)
10	tens (X0)	4	thousands and hundreds (X *lim* X *me*)
2	tens and units (*14* and *12*)	5	tens of thousands (X0 *lim*)
2	sexantine and tens (*90* and *70*)	1	tens and thousands (*10+4 lim*)
22	hundreds (X *me*)	1	tens, thousands and hundreds (*10+4 lim 6 me*)

Besides these, there is a number written 5 *šu-ši*.[53]

Out of 60 numbers surpassing *9* only 2 have the units: *14* cities (1st year) and *12* kings (6th year). No number higher than *100* includes digits of lower order (tens or units). This noteworthy preponderance of round numbers (highlighted by the absence of exact numbers) is partly due to the presence of some tribute lists (2nd year; cf. below, p. 99), but also to the quantification of a series of military contingents that gave battle to the Assyrian king at Qarqar (6th year; cf. p. 103), all these being given in round numbers. Most of the quantifications, in effect, are concentrated in the campaigns of the 2nd and the 6th year:

year	passage	lines	NUMBERS	year	passage	lines	NUMBERS
a.a.	I 14b-29a	15	**4**	*3*	II 30b-69a	39	**4**
1	I 29b-II 13a	38	**4**	*4*	II 69b-78a	9	**1**
2	II 13b-30a	17	**42**	*6*	II 78b-102	24	**21**

Here too the low digits ("1," "2" and "3") are more frequent.

Tribute (received or imposed) is often quantified while war booty never is.[54] Within the tribute, the livestock (oxen and sheep) are always given in round numbers:

---

[51] Cf. below, pp. 146f for Ashurnasirpal's hunting exploits.

[52] F.E. Peiser in KB 1, 150-175 or A. Amiaud - V. Scheil, *Salm.*, 4-43 (= ARAB 1, 594ff), cuneiform text in 3 R, 7-8 and Rasmussen, *Salm.*, pl. Iff; cf. the collations of Craig, *Monolith Inscr.*, 31f. Years are identified through the eponyms.

The subdivision of the various editions of Shalmaneser III's annals ("recensions" A-F) follows EAK 2, 70ff.

A tablet published in Mahmud - Black, *Sumer* 44, 137ff will not be taken into consideration here, since it contains a report of the first two campaigns similar to that of the "monolith" — yet showing, at the same time, some deviations (including some variants in the numbers: cf. p. 48 nos. 21f). Other duplicates of the "monolith" (cf. EAK 2, 71) do not show deviations as far as the numbers are involved.

[53] Note how "*300*" is written 5 *šu-ši* (slaughtered *mun-daḫ-ṣi*) in 3 R, 7 col. I 34 (= KB 1, 156:34) and 3 *me* elsewhere (5 times, referring to objects received or imposed as tribute). From the table have also been excluded the two partially readable numbers mentioned in n. 92 below.

[54] For further remarks, cf. p. 99.

SHALMANESER III — Edition "A"

QUANTIFIED OBJECTS[55]	N	S	Q	max	R	E
Soldiers killed	1	5	**5**	*14,000*	4	-
Cities	7	-	**1**	*6*	-	-
Cities of the neighborhood	1	-	**3**	*200*	2	-
Prisoners (*šallatu*)	4	-	**1**	*14,600*	1	-
Booty (generic)	2	2	**-**			
Booty (specif.)	2	4	**1**	*100*	(people)	
Tribute receiv.	10	-	**7**	tot.: 34	18	-
Tribute imposed	1	-	**4**	tot.: 10	3	-
In the tribute:	N	S	Q	max	R	E
Silver	7		**8**	*100 t.*	1	-
Gold	7		**4**	*3 t.*		
Sheep	8	1	**4**	*5,000*	4	-
Oxen	9	1	**4**	*500*	4	-
Asses			**3**	*30*		
Iron			**3**	*300 t.*	2	-
Copper	2		**3**	*300 t.*	1	-
Dromedaries			**2**	*7*		
Tin	2					
Chariots	3	3				
Cavalry	2	1				
Horses	7	2				
(other)			**13**	*1,000*	8	-

(10) **Edition "B"** — the second edition (up to the 9th campaign)[56] is extremely sparing with numbers: only 3, all of them very high (*3,000, 17,500* and *40,400*, which is a "total," cf. p. 94). The inscription concentrates on the campaigns to Babylonia (8th and 9th), which are devoid of numbers. Besides these, only a few events pertaining to the first 4 years are reported, thus ignoring the years 5-7.

Comparing edition "A" with "B," only one quantification appears in both texts: in the latter it has dropped, passing from *3,400* to *3,000* (cf. p. 49 no. 23). A "new" datum makes its appearance in the text of the 4th campaign (*17,500* soldiers).

55 In many cases the booty is indicated with *šallassu(nu) ašlula*: these occurrences have been placed under "prisoners." Alternatively, sometimes it is specified *šallatu* + *būšu* + *makkūru*, and these appear under "booty - generic," while the instances in which some object is specified (most of the time chariots or units of cavalry-*pētḫallu*) appear under "booty - specif." What has been brought (*târu* D) by Aḫuni (col. II 74f) is here considered "booty."

56 Michel, WO 2, 408-415 and WO 4, 29-35 (ARAB 1, 615ff and Delitzsch, BA 6, 133-144); Most editions in copy are incomplete, cf. however Th. Pinches, "The Bronze Gates discovered by Mr. Rassam at Balawat," *Transactions of the Society of Biblical Archaeology* 7 (1882), 83-118, fig. = Pinches, *Balawat*, 1ff or Rasmussen, *Salm.*, pl. XI-XIV, to which add Unger, *Balawat*, 16ff and *id.*, MDAIA 45 (especially 90ff).

(11) **Edition "C"** — Has only 25 numbers, 6 of which are greater than a thousand.[57] Within the first six campaigns appear 3 quantifications comparable to those of edition "A" (one has "increased," passing from *14,000* to *25,000*), plus there are two "new" numbers: one already appearing in "B" (here it has passed from *17,500* to *22,000*),[58] and one appearing within the report of the 5th campaign, given for the first time.[59]

There are several exact numbers: *89, 97, 373, 399, 2,002* and *5,542* (the first four written employing the sexagesimal system). The average number of digits is higher than in previous editions (cf. p. 43 bottom). The (relatively) low percentage of round numbers is due, partly, to the leaving out of the tribute lists appearing in edition "A." Booty and tribute, which are mentioned quite often, are never quantified (only in three cases they are indicated with *ana lā māni*):

	N	S	Q
Booty:			
šallatu (+ *šalālu*)	5	1	-
šallatu and/or NÍG.GA etc. (generic)	2	1	-
specifies at least one object	2	1	-
Tribute received:			
mandattu (generic)	6	-	-
specifies at least one object	4	-	-

The quantifications of cities are the most frequent:

SHALMANESER III — Edition "C"

QUANTIFIED OBJECTS	N	S	Q	max	R	E
Soldiers killed	8	1	**2**	*25,000*	2	-
Prisoners (ERÍN)	-	-	**1**	*22,000*	1	-
Cities	4	-	**5**	*100*	1	-
Cities of the neighborhood	5	-	**3**	*100*	2	-
Euphrates crossings	5	-	**3**	*10-šú*		

(12) **Edition "D"** — Is a duplicate of "C" as concerns the 3rd-15th campaigns (the differences are on the level of writing), but adds a report of the 18th, containing 4 quantifications, with a round and an exact number: *16, 16,000,* plus a booty of *1,121* chariots and 4 *me* 1,10 [= *470*] *pētḫallu*.[60]

---

57 Cameron, *Sumer* 6, 6-26, or Michel, WO 1, 454-475; cuneiform text: cf. the photographs in *Sumer* 6, pl. I-II, WO 1, Tf. 22-24 or in JCS 5 (1951), pl. I-II. Other texts with passages of this edition (listed in EAK 2, 73ff) confirm the digits of the Cameron tablet (cf. below, p. 49 n. 21).

58 For this quantification, cf. below pp. 93ff.

59 According to Olmstead, JAOS 41, 362 (cf. *Historiography*, 22, 24 and 27), a later campaign "was in the later editions of the annals moved forward to fill the gap in the year 855" = 5th *palû*.

60 Delitzsch, BA 6, 144-151 (ARAB 1, 640ff); copies: Layard, ICC, 12-16 and 46f or Rasmussen, *Salm.*, XVff; the fragment 3 R, 5 no. 6 (Michel, WO 1, 265-268; ARAB 1, 672) contains only the text of the 18th campaign.
For *pētḫallu* = "horse" (in this epoch) cf. AHw, 858a.

(13) **Edition "E"** — Comes up to the 20th campaign.[61] The first 16 show only minor deviations with respect to the previous editions, except the 15th, which has been shortened (two numbers are also missing). At the 18th campaign a number varies: *16,020* instead of *16,000* (cf. p. 50 no. 27). The quantifications appearing within the reports of the "new" campaigns (17th-20th) as well as the "total" booty of the first 20 years given at the end of the inscription are "exact."[62] This applies also to the numerals referring to the Assyrian forces appearing at the very bottom of the inscription, "*2,001* chariots (GIŠGIGIR) and *5,242* horses (*pētḫallu*)," showing some differences from those of the previous edition (cf. p. 50 nos. 29-30).

(14) **Edition "F"** — The latest edition, given by the "black obelisk,"[63] to leave place for the latest events shows a much shortened narrative for the first 20 campaigns, and this caused the dropping of many quantifications, while one of those surviving varies: 20 *lim* 5 *me* (= *20,500*) instead of 20 *lim* 5 *lim* (= *25,000*; cf. p. 49 no. 25). Most of the numbers appearing in the new narrative refer to the Euphrates crossings, which are "counted." For this reason too, the percentage of high numbers is not remarkable.[64] Of the new campaigns, the 21st, 22nd and 23rd are given in a very synthetic fashion, while the subsequent ones are much more detailed. Overall, they include 12 numbers, 6 of which refer to cities — cf. particularly the campaign of the 31st year: 5 numbers, all of cities — while one single booty (or tribute) is quantified (at the 18th *palû*,[65] already mentioned by the earlier editions):

61 Safar, *Sumer* 7, 3-21 = Michel, WO 2, 27-45. Text: *Sumer* 7, pl. I-III and WO 2, Tf. 2-4 (photographs). A fragmentary duplicate of the final section of the inscription is published in Michel, WO 1, 389ff, Tf. 18-20.

62 See especially 1 *me lim* 80 *lim* 4 *lim* 7 *me* 55, a number as much high as exact, in Safar, *Sumer* 7, 13:38 (the passage is quoted at p. 93). The campaigns 17 and 19 do not mention military events: the digits appearing there concern the animals killed by the king. His sporting activities are used to cover the lack of military activity (cf. below p. 148-149).

63 Michel, WO 2, 137-157 and 221-233 (ARAB 1, 553ff); text in ICC, 87-98 or Rasmussen, *Salm.*, XXIIIff (cf. also Layard's drawings reproduced in Börker-Klähn, *Bildstelen*, Tf. 152a-d). Another text belonging to this edition is included on a fragmentary statue from Nimrud (Læssøe, *Iraq* 21, 150ff, cf. also EAK 2, 80). The accounting includes the "*11* cities" of line 52 (the number is ignored by Michel, but cf. below, p. 126 n. 51). The fractions written *per extenso* (*šá-nu-te*) at lines 77 and 174 are not included.

64 Also for the suppression of the "totals" reported at the end of edition "E." Altogether, only 2 numbers reach one thousand: *1,121* ("exact") and *20,500* (lower than the earlier editions), which refers to the enemies killed during the battle of Qarqar (6th campaign).

65 Note that throughout the work the term *palû*,in relation to Shalmaneser III, will be translated either as "campaign" or, more properly, as "year of reign" (actually, campaigns and years coincide). Thus, the chronological framework furnished by his later annals will be followed, even if it is probably incorrect, being inconsistent with that given by the eponym-lists: cf. ch. 5 §7.

SHALMANESER III — Edition "F"

QUANTIFIED OBJECTS	N	S	Q	max	R	E
Soldiers killed	6	-	**1**	*20,500*	-	-
Prisoners	-	1	-			
Kings	3	-	**4**	*27*		
Cities	10	2	**5**	*250*	-	-
Cities of the neighborhood	4	-	**4**	*100*	1	-
Euphrates crossings	10		**8**	*22-šú*		
Crossings of other rivers	4		-			
Booty (*šallatu*)	7	-	-			
Booty (generic)	2	-	-			
Booty (specif.)	5	1	**1**	*1,121*	-	1
Tribute received	15	1	-			
Tribute imposed	1	-	-			

(15) Also the "monolith inscription" of **Shamshi-Adad V**, with his **annals**,[66] contains several numbers referring to cities destroyed, conquered, etc.: out of a total of 16, 9 refer to cities.[67] These numbers sometimes represent a *record*, since *1,200* cities, or even *500* or *447*, have never appeared in the earlier Assyrian royal inscriptions.

Here too tribute and booty are quantified only in two cases:[68]

66 L. Abel in KB 1, 174-187 (ARAB 1, 713ff); text in 1 R, 29-31 (archaic signs) and 32-34 (transcription in ordinary signs, with one oversight, cf. p. 50 n. 29), dupl.: Reade - Walker, AfO 28, 114ff. At col. IV 13 we have 3 LIM EN?.MEŠ *a-di* UN.MEŠ-*šú-nu* NÍG.GA-*šú-nu šá-šu-šú-nu iš-tú ki-rib* URU *šú-a-tú al-qa-šú-nu-ti*, "*3,000* (?) together with their people, their properties, their things, from that city I took away." The sign may be EN ("*3,000* functionaries," cf. KB 1, 184: n. 1; ARAB reads "*3* chieftains") or even URU (thus in the transcr. in ordinary characters of 1 R, 34), in this case the translation would not make much sense. The number *3,000*, however, is clearly readable.

67 Only twice is reference made to cities (plural) without quantifying them: in 1 R, 30 and 33 col. II 28 (cf. KB 1, 178) and at col. III 9 (KB 1, 180; cf. AfO 28, 115:19′), where however they are qualified as *ma-ʾ-du-ti*. Besides these, two cities are cited at col. IV 14ff, being conquered together with *200* towns of the neighborhood. See for instance as in 1 R, 30 and 33 col. II 52-59 (KB 1, 180 = ARAB 1, 718) the cities are quantified, but not the booty (*a-na la ma-ni*) taken nor the enemies killed (*ma-at-tu*). Note that already the last edition of Shalmaneser III's annals, given by the "black obelisk," shows, within the text of the 31st *palû* (the last one), a very high frequency of quantifications of cities (5 out of 6 — cf. ARAB 1, 588).

68 The *120 pētḫallu* of 1 R 30 and 33 (= KB 1, 180) and 1 *me* GIŠ.GIGIR.MEŠ-*šú* 2 *me pet-ḫal-lu-šú*, "*100* of his chariots and *200* of his horses" in 1 R, 31 and 34:44f (KB 1, 186 = ARAB 1, 726, in both cases battle booty). Of other booty it is specified only that the camels had *2* humps. Tribute, as a rule, is composed only of horses, which are never quantified: *ma-da-tú* ANŠE.KUR.RA.MEŠ LAL-*at ni-ri ša ... am-ḫur*, "the tribute of horses broken to the yoke of ... I received" (col. II 1f) or similarly (this applies also for the sole instance of tribute imposed, col. III 44-64). Only in III 41f is the composition of booty (called "tribute," *ma-da-ta*) specified (passage mentioned above, p. 12 n. 67).

SHAMSHI-ADAD V						
QUANTIFIED OBJECTS	N	S	Q	max	R	E
Soldiers killed	-	1	**7**	*13,000*	5	1
Prisoners	-	-	**3**	*3,000*	3	-
Cities conquered	2	1	**4**	*447*	1	1
Cities of the neighborhood	-	-	**5**	*1,200*	5	-
Booty (generic)	3	-	**-**			
Booty (specif.)	2	3	**3**	*3,000*	3	-
Tribute received	5	-	**-**			
Tribute imposed	1	-	**-**			

The round numbers are quite frequent: only two numbers have 3 digits: 4 *me* 47 and 1 *lim* 1,10 (= *1,070*). The most frequent number is *200*, which appears 4 times.[69]

The Ashur stela, duplicating the "monolith," but continuing with the 5th and the 6th campaign, also seems to prefer to quantify the cities (3 times out of 5 readable numbers).[70]

(16) The annalistic texts of **Tiglath-pileser III** are very fragmentary, yet they preserve a good number of digits, and some conclusions can be drawn. His annals[71] have many high numbers: out of a total of 60, 47 reach or surpass one hundred and 18 reach or surpass one thousand.[72] Frequent are the "exact" ones: out of 56 numbers surpassing *9*, 16 have units, and out of 47 surpassing *99*, at least 22 have tens and/or units. "Round" numbers, however, are also frequent: one may note a clear prevalence of "exact" numbers within the text of the 8th *palû* and of the "round" ones within the preceding sections. It is

69 In ed. "A" of Shamaneser III's annals, *2* and *200* are the most frequent numbers (6 times each); for the remaining editions ("B" - "F"), however, there are no numbers appearing significantly more often than others (i.e. no number occurs more than twice — for the exception of the *12* Syrian kings, see below, p. 134) and, therefore, this fact has not been taken into account.

70 Partly published by Weidner, AfO 9, 89ff — the inscription is rather fragmentary.

71 Tadmor, *Tigl. III*, 40ff; copies: Pl. I-XXVIII; Rost, *Tigl. III*, Pl. I-XIX, XXI-XXIII and XXVIII; 3 R, 10 no. 1 A-B and 9 no. 1; ICC, pl. 19b, 29b, 34, 45b, 50-52, 65-69 and 71-73a. On the problems concerning these annals cf. Tadmor, *Tigl. III*, 27ff; *id.*, *Intr. Remarks* and EAK 2, 125-131. There exist several parallel versions, all very fragmentary and incomplete. The "statistical" remarks here expressed refer to the whole of the texts belonging to all versions indiscriminately. To outline each of them separately would be problematic and perhaps non-productive: of some editions only a few lines have been preserved, while there are just a few overlappings (at the 8th, 9th and 13th year). Moreover, the various versions are more or less contemporary, the text being carved on the walls of various rooms of Tiglath-pileser's palace at Kalah.

Tadmor's Ann. 18 and 24 represent two parallel texts, partially duplicating, which report several digits, mostly referring to prisoners. Now, since where the digit has been preserved, the name of the land involved is missing from at least one of the two fragments, it is impossible to state whether they refer to the same deportations or not. In the former case we might be facing some "variation." For the purposes of these accounts, however, all the numbers have been considered as referring to different quantifications.

72 The total of 60 includes all the at least partly readable numbers (there are certainly two more, completely missing, at least judging from the copies). In some cases it is not certain whether certain numbers are "round" or "exact," because of the very fragmentary context, nor can their magnitude be ascertained, since the exponent (*lim* or *me*) is doubtful, as in the case of some items of the tribute of Raḫianu (Tadmor, *Tigl. III* pl. XIX:5'f — Ann. 21; *id.*, *Intr. Remarks*, p. 20 fig. 1; cf. also ICC, $45b_2$). Note that the quantity of gold is, in this case, 3 GUN (cf. also Weippert, ZDPV 89, 35), the numeral being represented by three horizontal strokes (cf. here above p. 4 n. 18).

noticeable that for tribute and booty "round" numbers are almost always employed, while for prisoners/deported, very frequently quantified (some 30 times), the "exact" ones are preferred.

	TIGLATH-PILESER III					
QUANTIFIED OBJECTS	N	S	Q	max	R	E
Soldiers killed	3	1	-			
Prisoners (deported)	1?	-	**13**?	*83,000*	4	5
Prisoners (settled)	2	-	**18**?	*6,208*	7	9
Cities conquered	7?	1	**2**	*591*	-	1
Cities of the neighborhood	9?	-	-			
Booty (*šallatu*)	9?	-	-			
Booty (chariots-horses)	1	-	-			
Booty (specif.)	2	5	**4**	*19,000*	6	1
Tribute rec. (specif.)	2?	1	**2**	*3,000*?	5?	-

Note that the prisoners 3 times out of 13 are simply listed together with the booty (and in this case round digits are used).[73]

The most frequent digit in this text is "5."

(17) The Khorsabad **annals of Sargon**[74] also contain several "exact" numbers: out of 64 numbers greater than *9*, some 20 have the units; out of 43 greater than *99*, from 21 to 25 have tens or units (depending on the variants or on the lacunae).

---

73 *12,000, 8,650* and *900*. The sole deportation not quantified appears in Tadmor, *Tigl. III*, Ann. 25:6′ (line 95 in Rost's edition; doubtful case). Note that the four quantified lists of booty include 9 numbers in all (including one instance in which only the prisoners are counted), while the two lists of tribute include, it seems, 8 numerals.

The table includes some approximations, since, due to the fragmentary nature of many passages, it is not always clear whether a given numeral refers to people or not, whether it is "round" or not, etc. The table takes into consideration only the cities subject to conquest, destruction, etc. (thus excluding those where deportees have been settled) and does not compute, amongst the "soldiers killed," the 8 occurrences of the expression *dīktašunu adūk* ("I defeated them"), in one instance accompanied by *ma'attu*.

74 Lie, *Sargon* (ARAB 2, 2ff; a new edition is given in Fuchs, *Sargon*, 82ff and 313ff); text in Botta, 65-65 bis and 70-92 (or Winckler, *Sargon*, nos. 1-52, 55 and 58-60); cf. also Thureau-Dangin, RA 24. The copies belong to two (or three) different versions. For the purpose of simplicity, I have followed the "composite" edition of Lie, which integrates the inscription of "room II" with that of "room V" (the latter being considered too fragmentary and incomplete to deserve a separate edition). In this instance I am considering the two versions as duplicates, which is not the case. The various versions of Sargon's Khorsabad annals, however, do not contain much significantly different information (cf. the remarks of Renger, *Introd.* [10 and 58]).

The numbers restored by Lie on the basis of the "display inscription," or of the "letter to the God," have not been taken into account since in at least nine cases they differed from those of the annals (cf. chapter 3, nos. 44f, 47, 48 and 49-53) — with the exception of the two numbers *7* of Lie, *Sargon*, 68:457f that may be safely restored (cf. p. 133). Note that lines 170-82 of Lie's edition include two numerals (cf. Fuchs, *Sargon*, 118ff and 322f).

The prisms from Nineveh (cf. Tadmor, JCS 12) and from Kalah (Gadd, *Iraq* 16) are too fragmentary to be considered here.

		Sargon				
QUANTIFIED OBJECTS	N	S	Q	max	R	E
Soldiers killed	5	2	-			
Prisoners (deported)	8	1?	**12**	*90,580*	3	9
Prisoners (settled)	9	-	-			
Cities (main)	19	-	**11**	*30*		
Cities (fortresses)	1	-	**6**	*22*		
Cities of the neighborhood	4	1	**7**	*140*	-	1
Booty (not specif.)	9	3	-			
Booty (specif.)	7	2	**4**	*100,225*	2	4
Tribute (not specif.)	8	2	-			
Tribute (specif.)	4	-	**1**	*4,609*	-	1
Tribute imposed	1	1	**1**	*2,000*	1	-

Most of the "round" numbers (actually 14) are related to military contingents (either Assyrian or enemy),[75] whose size was probably pre-arranged in a (at least conventionally) round measure.[76]

As can be seen in the table, the quantifications of cities and deportees are very frequent. For the former rather low numbers are used, contrary to what happens with the deportees, who are never less than *2,400*.

The first half of the annals (roughly up to the 9th year) shows a different narrative "style" from the following part, at least as far as the quantifications are concerned.[77] This is underscored by those referring to cities and booty: within the second part, when some cities are named *per extenso*, their total number is almost always explicitly given (cf. p. 190), while in the first part this never happens. Thus, within the second part the composition of the captured booty is always specified, while it is simply indicated as *maršītu* or NÍG.GA or *šallatu* in the first part. See the following table, reporting the number of quantifications (in parentheses the cases *not* quantified numerically):

Cities			Booty		
palê	not named	named	*palê*	not specif.	specif.
1-9	(4) 14	(9) -	*1-9*	(11) -	(2) 3
10-14	(7) 2	(3) 8	*10-14*	(1) -	(7) 1

75 E.g. the "*20,000* bowmen" and the "*10,000* shield-bearers" of Lie, *Sargon,* 72:11 (cf. below p. 56 no. 60).

76 Cf. p. 82 n. 156.

77 This actually appears to concern the whole narrative style, very terse and prosaic in the first part, but more loose from (about) the 10th year on (this section of the text is very lacunal), where the narration becomes quite rich with "decorative" elements, and frequently includes marginal episodes. This is only partly explainable by the greater amount of text given to the later years (244 lines preserved *versus* 164 of the earlier *palê*).

Also the ratio between round and exact numbers varies between the two sections (note that the 9th camp. includes one exact and three round numbers):

	Tot.	R	E
1-9	45	6 (+$^?$)	10
10-14	49	13	5

The text of each campaign is sufficiently ample to allow an analysis of the distribution of the quantifications.[78] As can be noted, the eighth campaign is highlighted by the presence of several quantifications:[79]

		- n. quantifications -						- n. quantifications -			
year	no. of lines	tot	≥100	R	E	year	no. of lines	tot	≥100	R	E
1	11*	**2**	-	-	**	*8*	$38\frac{1}{2}$*	**21**	10	1	5
2	$7\frac{1}{2}$*	**1**	1	-	1	*9*	39*	**6**	4	3	1
3	10	**-**				*10*	30	**2**	1	1	-
4	4	**1**	1	-	1	*11*	20	**-**			
5	$6\frac{1}{2}$	**3**	2	2	-	*12*	$105\frac{1}{2}$*	**26**	6	3	3
6	$21\frac{1}{2}$*	**2**	-			*13 f.*	$88\frac{1}{2}$*	**21**	13	9	2
7	26*	**9**	$4^?$	$2^?$	2 (+)	*b.a.*	43*	**3**	1	-	1

Line counting follows Lie's edition[80] — with *b.a.* the building activities are meant — * the text includes broad lacunae — ** a number is only partly readable.[81]

(18) The latest edition of the **annals of Sennacherib**[82] contains only 17 numbers, thus distributed:

camp.	no. lines	numbers	camp.	no. lines	numbers
1	45	**7**	*6*	22	**-**
2	54	**1**	*7*	44	**2**
3	96	**4**	*8*	107	**1**
4	25	**-**	*bldng.*		
5	38	**1**	*activ.*	48	**1**

78 A thing not possible with the annals of his predecessors (cf. p. 33 for Shalmaneser III).

79 To those shown in the table should be added also some not (anymore) readable quantifications once appearing in the fragmentary passage of Lie, *Sargon,* 26:154ff. The stress given to this campaign is confirmed also:
• by the presence of many "exact" numbers;
• by the presence of the highest number: *100,225* (even if there is only one other number greater than *1,000*: the *6,170* deported from Muṣaṣir); a number that is the corruption of an originally much lower one (cf. p. 54 no. 52);
• by the presence of two quantified booty lists (Lie, *Sargon,* 26:154ff and 28:158ff: booty carried off from Muṣaṣir). Elsewhere, only one group of booty is quantified (cf. 62:6f), besides the horses of a tribute received (*maḫāru* — cf. 30:193f — yet the other animals are not quantified) and one tribute imposed (*kânu* D — cf. 46:284f).

80 Only the (at least) partially preserved lines of each year are counted, including lines 170-182 (for which see Fuchs, *Sargon*, 118ff), while also Fuchs' lines 322-326 have been added (12th *palû*).

81 Perhaps it is exact: cf. Botta, 70:9′ (= Winckler, *Sargon*, no. 2; Lie, *Sargon,* 6:22).

82 Edition in Ling-Israel, *Fs Artzi*, 220-242 (Luckenbill, *Senn.*, 23-47 and 128-131; ARAB 2, 233ff and 423ff). Text: Luckenbill, *Senn.*, 163ff (Oriental Institute prism), 1 R, 37-42 ("Taylor" pr.) and Ling-Israel, *Fs Artzi*, pl. IV-XVI (photographs of the "Jerusalem prism"). Cf. also 1 R, 43f, 3 R, 14 and VS 1, no. 77 (cf. BAL^2, 65f and Reade, JCS 27, 193f).
The other annalistic inscriptions of Sennacherib repeat the text of the "Taylor prism," except for leaving out some of the last campaigns (cf. below p. 119 n. 19), or present the events of only one or two campaigns (the "cylinder Smith," with 1 campaign, and the "cylinder Bellino," 2 campaigns).

Though there are few numbers present, there is a high frequency of non-numerical quantifications,[83] as appears also from the following table (*sub* S):

SENNACHERIB

QUANTIFIED OBJECTS	N	S	Q	max	R	E
Prisoners (deported)	4	1	**2**	*208,000*	1	1
Prisoners (settled)	3	-	-			
Cities (main)	12	-	**4**	*75*		
Cities of the neighborhood	1	3	**2**	*420*	-	-
Booty (not specif.)	6	-	-			
Booty (specif.)	4	4	-			
Tribute (not specif.)	-	1	-			
Tribute (specif.)	-	1	**1**	*800*	1	-
Tribute imposed	4	1	**1**	*20*		

(19) The Nineveh prisms of **Esarhaddon**[84] contain only 19 numbers, with the "exact" ones quite frequent (out of 15 numbers greater than *9*, 7 have the units).

The quantifications are distributed thus, following the subdivision of the text into "episodes" as done by Borger (the other episodes do not contain quantifications):

episode	2:	2 numbers	episode	14:	6 numbers
"	6:	1 "	"	17:	3 "
"	9:	1 "	"	21:	3 "
"	13:	1 "	"	22:	2 "

The quantifications most of the time concern foreign rulers (3 times), distances (3 times) and imposed tribute (2 groups of tribute for a total of 6 quantifications; in both cases only the augment over the previous tribute is indicated). Rare or absent are quantifications typical of the annals of his predecessors: enemies killed, prisoners or deportees, cities (quantified only once). Non-numerical quantifications are quite frequent.[85]

83 These too are concentrated within the first part of the inscription, which, as in the case of Sargon's annals, is more crude in narrating the events and therefore also more inclined to furnish "data" or, anyway, the tangible results of the actions. Within the text of campaigns 1-3 (195 lines) appear the expressions *nību* (6 times), *kabittu* (5 t.), *mala bašû, ana lā mīna, gimru* and *šadlu* (each once), while in the text of campaigns 4-8 (236 lines) appear only *nību* and *gimru*, each once.

84 Borger, *Asarh.*, §26-27: *Nin. A* (*Klasse* A1-19 = ARAB 2, 499ff). Cf. also Heidel, *Sumer* 12 (duplicate). They are not annals, the material not being subdivided into years or campaigns (only in the fragmentary prisms *Nin. D* and *E* is it subdivided into *palû*; cf. Borger, *Asarh.*, p. 38f). Principal editions in copy: Thompson, PEA, pl. 1-13; 3 R, 15f; ICC, 54-58 (photos of another exemplar in Scheil, *Prisme S*, Pl. 1ff).

85 *gimru* (5 times), *kabittu* (3 times), *ana muʾdê* (twice), *nību* and *šadlu* (each once) appear, spreading over all the various "episodes."

(20) **Ashurbanipal: "Prism B"** — His earliest sufficiently complete annalistic inscription that we have at our disposal.[86]

It has only 17 numbers (including those reconstructible from parallel passages from other prisms). Certain campaigns are completely devoid of numerals: the 1st, the 3rd, the 4th and the 7th (and, if the report of the fourth one is very short, that of the seventh is quite long). The low numbers prevail: if we except the *2,500* talents of the *2* obelisks from Thebes (for which cf. p. 124), the highest number is *85*. Rare also are quantifications of enemies killed, prisoners or cities, while non-numerical quantifications are quite frequent.[87]

(21) **"Prism C"** — Adds a list of names of Syrian kings, with a numerical total, and modifies the introduction (two numbers thus are missing). A "new" datum appears in the 5th campaign: "*30* horses I imposed on him (the king of the Manneans) as supplementary tribute (*eli ma-da-at-te-šú maḫ-ri-te ú-rad-di*)." The building operations narrated in the appendix (not present in "B") contain some quantifications.[88]

(22) **"Prism F"** — Presents a different arrangement of the campaigns (the 8 campaigns of "B" are concentrated into 4):[89] the 2nd (= 3rd of "B") and the 4th (= 7th of "B") are here also devoid of quantifications (the 1st and the 4th of "B" have "disappeared"). Of the new campaigns, the 5th (in Elam) does not have quantifications while the 6th (Elam and sack of Susa) has some.

NOTES ON TABLE 1

THE NUMBERS IN THE ANNALISTIC INSCRIPTIONS

In the following table are synthesized many of the data reported in the preceding pages. However, some inscriptions have been excluded: those containing a too low number of quantifications (annals of Adad-nerari II and of Esarhaddon and successors) or those preserved in fragmentary or incomplete condition (annals of Ashur-bel-kala, edition "B" of Shalmaneser III's

86 Piepkorn, *Asb.*, 28-93 (= ARAB 2, 841ff; also in Streck, *Asb.*, 92-139). The text has been reconstructed from several fragments: cf. Piepkorn, *Asb.*, 23ff; other fragments added by Thompson, *Iraq* 7, 98f and 103ff and Knudsen, *Iraq* 29, 49ff — some others mentioned in HKL 1, 385f and 2, 223; cf. also Cogan, JCS 32, 149f and Cogan - Tadmor, JCS 40.

87 *75* cities are quantified in IV 6 and *8* in III 37 (a number reconstructed on the basis of "prism C"). There are non-numerical quantifications of several different types: *mala* (appears 8 times, incl. 2 as *mala bašû*), *kabittu* (7 times), *ma'du* (4 times, incl. *ma'assu*), *ina lā mīni, kālu* (each 3 times), *ša nība lā išû, gimru, ana siḫirtišu* (each twice), *napḫaru* and *rapšu*.

88 Streck, *Asb.*, 138ff + Bauer, IWA, 13ff (= ARAB 2, 847ff). Copies: 3 R, 27 (cf. HKL 1, 485); Bauer, IWA, Tf. 5-13; for other fragm. cf. Wiseman, *Iraq* 13, 24ff; Knudsen, *Iraq* 29, 57ff (ND 4378B etc.); Millard, *Iraq* 30, 105; M. Cogan, "Ashurbanipal Prisms K and C: Further Restorations," JCS 41 (1989), 96-99. A newer edition, together with an important new fragment, appears in Freedman, *St. Louis*, 46ff (note that texts P and Q of his catalogue turn out to belong to prism "K").

On the Syrian kings cf. p. 150.

89 Cf. Aynard, *Prisme*, 28-65. On the sequence of campaigns in the historical inscriptions of Ashurbanipal cf. *ibid.*, 91ff and Cogan - Tadmor, Or 46, 82f and 85. On the chronology of the campaigns cf. Grayson, ZA 70.

annals). Also excluded is edition "D" of Shalmaneser III, since it is very similar to edition "C."[90]

The table aims at showing the number of quantifications present, their composition (in terms of digits), their magnitude, etc. The data appearing represent:

1. The number of numerical quantifications present in the inscription,[91] or safely reconstructible from duplicates. The "safety" concerns the presence of the number, and not its exact value. Thus for instance the (rather fragmentary) annals of Tukulti-Ninurta II contain 84 numbers (identifiable, plus perhaps some others), out of which only 75 are clearly and completely readable. The number reported in the table is, however, "84."[92]

2. The numbers written using the sexagesimal system (including also, for example, 4 *me* 1,10 = *470*).

3. The number of the digits making up the numbers (cf. n. 15 at p. 23).

4-5. The quantity of "round" numbers (numbers higher than *100* devoid of digits on the order of tens or units) and of "exact" ones (*idem*, with tens and/or units); the sexagesimal numbers, quite common in Tiglath-pileser I's annals, are excluded — it is difficult to establish if, say, 5 *šu-ši* (= *300*) should be considered "round" or not.

6. The absolute value of the quantifications (how many numbers reach or surpass *100, 1,000, 10,000* and *100,000*).

7. The recurrence of each digit (cf. the just mentioned n. 15). For Tiglath-pileser I, considering that his annals contain several sexagesimal numbers, I report, in parentheses, also the same data excluding those numbers.[93]

In consideration of the variations, the numbers are sometimes separated by a hyphen, thus 8-9 will means "there are 8 or 9 numbers of this kind, according to which variant is considered." The parentheses or other mathematical symbols refer to partially readable numbers, thus 8 (+1) will be "8 numbers plus one probable"; 8 (+) "at least 8 numbers," etc.

---

90 Generally speaking, the reconstructions based on the inscriptions that are not listed in the notes to the chapter (i.e. those originating from diverging versions of the annals or other kinds of inscriptions) have not been taken into consideration, since the numbers are quite often subject to variations (cf. chapter 3 §1).

91 Naturally, the following have been excluded from the accounts: numerical ideograms (such as ^{d}15 = Ishtar), dates appearing in the text or in the colophon (such as number of campaign/year and day), the expression *kibrat* 4-*i* (which mostly appears within the titulary; cf. n. 4 on p. 22), and the fractions. On the other hand, I have included such expressions as "I subdued them to one (1-*en*) authority" (cf. e.g. RIMA 2, 25:46) or "in a second day," *ina* 2-*e* u_4-*me* (cf. RIMA 2, 173:37), *ina* UD 2 KÁM, etc. given that the number is written in digits and not in letters (thus, the number "1" written *iš-ten* in 1 R, 30 and 33 II 52 = KB 1, 180 — Shamshi-Adad V — will not be considered).

92 Thus, for ed. "A" of Shalmaneser III, in the total are included the [... X? +] 2 ANŠE ÚŠ.MEŠ *e-ri-ni* of II 25 (cf. EAK 2, 72) and the last number referring to the military contingents of Qarqar, [... X? +] 1 *lim* (cf. p. 50 n. 243), although the numbers are excluded from other "statistics" (points 2-6.), having only the digits "2" and "1" computed at point 7.

93 It is obvious that using the sexagesimal system allows the use of lower digits, for instance writing the number "70" as 1,10: two digits "1" instead of a "7," and this could give a false impression of the frequency of digits "1" and "2."

TABLE 1 NUMBERS IN THE

Inscription:	T-P I	T-N II	Asn II
Parameters			
1. Total Numbers:	48	84	110
2. Sexagesimal system:	12	3 (+1)	8
3. Number of Digits, 1:	38	65 (+3)	88
2:	7	8 (+2)	16
3:	1	2	5
4 or more:	-	-	1
4. Round Numbers, comp.:	9	13	45 (+1)
X (*me*) *lim* :	7	-	17
X *lim* X *me* :	1	1	4 (+1)
X *me* :	1	11 (+1)	24
5. Exact Numbers, comp.:	1	1	4 (+1)
lim + 10/1 :	-	1	1 (+1)
me + 1 :	1	-	3
6. Size of numbers:			
100 to 999	8-9	15 (+2)	39
1,000 to 9,999	5	2 (+)	22
10,000 to 99,999	3	-	1
100,000 and over	-	-	-
7. Digits: 1 :	15-16 (10)	28 (+2)	45
2 :	16-17 (12)	25	34-35
3 :	5 (4)	16	17
4 :	5 (5)	8	8-10
5 :	9 (8)	5	15-18
6 :	5 (5)	2	5-6
7 :	- (-)	2	3-4
8 :	1 (1)	3	7
9 :	- (-)	-	1

ANNALISTIC INSCRIPTIONS

Shalm III						
Ed. A	Ed. C	Ed. F	Sh-Ad V	T-P III	Sargon	Senn
78	25	26	27	60	97	17
3	4	2	1	3	3	1
63	10	8	17	24	56	9
10	10	15	8	20 (+)	28 (+)	6
1	2	2	2	11 (+)	10 (+)	2
-	3	1	-	5	5	-
45	7	2	17	22 (+)	19	3
17	4	-	6	9	9 (+1)	1
6	-	1	3	3	5-6 (+1)	-
22	3	1	8	10 (+)	4-5	2
-	4	1	2	15 (+)	18-19	1
-	2	1	1	6	15-16	1
-	2	-	1	9 (+)	3	-
23	5	3	11	29	14 (+)	3
15	2	1	9	11	21 (+)	-
7	3	1	1	7	5	-
-	1	-	-	-	1	2
29	17	15	11	24 (+ 1)	29-30	6
23	14	15	11	15 (+2)	40	6
16	5	1	9	13	17	6
4	1	5	2	7	19	4
9	4	4	3	27	18-20	2
3	-	2	1	14	14-15	1
3	1	2	2	1	8	-
1	1	2	-	4-5 (+1)	6-7	2
-	4	2	-	5-6	5	-

3. VARIATIONS AND INFLATIONS

1. LIST OF VARIATIONS

First we list the cases in which a single quantification, reported by more than one inscription, shows diverging numbers.[1] Whenever it is possible to establish a chronological succession between the texts, the varying numbers are placed in different lines, otherwise they are simply separated by a bar. Each case is numbered, in a rough chronological order. Some uncertain cases are listed as well, the number (of the variation) being enclosed in parentheses.[2]

Adad-nerari I

(1) Thickness in bricks (SIG$_4$) - Wall of the "new city" (URU GIBIL) in Ashur[3]

10 KAH 1, 3 Vs. 41 (RIMA 1, A.0.76.10; IAK, XX, 6; ARI 1, 423)

14 KAH 1, 4 Rs. 4 (RIMA 1, A.0.76.13: 41; IAK, XX, 9; ARI 1, 434)

Shalmaneser I

2. Cities (URU) destroyed and burned (*naqāru* + *šarāpu*)

41 / 51 KAH 1, 13 col. I 37 (and cf. p. 74; RIMA 1, A.0.77.1:37, cf. n.; IAK, XXI, 1 I 37 and n. 4; ARI 1, 527 and n. 167)

---

1 In the list, if publications in copy exist, they are given first (semi-colon or "and" distinguishes different texts), while edition(s) in transliteration and translation are put in parentheses.

Note that, for the sake of simplicity, the quantified terms are usually indicated without the determinative of plurality MEŠ, which appears almost always.

2 For example, sometimes the quantification could refer to two different events, or it could represent a "total" (and it is not possible to determine with certainty if this is the case). The list will not include the "variations" bearing incomprehensible quantifications, the obvious consequences of an oversight (a good example is the measure of the perimeter of Dur-Sharruken as given by cylinder "L$_2$" of Sargon, having a superfluous *Winkelhaken* — see Lyon, *Sargon*, 10:65 and n. 25). Certain "variations" originating from an error in the copies or in the transliterations (therefore of "modern" origin) are excluded as well: cf. e.g. nn. 24 and 29 here below or p. 9 n. 47 above, or, still about the perimeter of Dur-Sharruken, the oversights reported in Lyon, *Sargon*, 17 nn. 13-14, and originating from Oppert's copies. In many cases it is difficult to establish whether the variation is owed to the "ancient" or to the "modern scribe," since the original inscriptions have been lost, and only the copies are available (cf. for example notes 7, 12, 35 and 40 below) — some of these "variations" have been included in the list, with the number enclosed in parentheses.

3 Cf. also the 10 SIG$_4$, thickness of the wall of the "inner city" of Ashur (*du-ur* URU *lìb-bi*-URU) in Edzard, *Sumer* 20, 50:13 (RIMA 1, A.0.76.1; ARI 1, 440).

Tukulti-Ninurta I

(3) Digging related to the "new palace" of Ashur, in SAR (surface measure)[4]

50 Weidner, *Tn*, Tf. 5a I 5 (RIMA 1, A.0.78.1 IV 49; ARI 1, 696)

80 Andrae, FWA, 165:25 (RIMA 1, A.0.78.3; Weidner, *Tn*, no. 3; ARI 1, 708); KAH 2, 58:76 (RIMA 1, A.0.78.5; Weidner, *Tn*, no. 5; ARI 1, 717)

Tiglath-pileser I

4. Copper kettles (*ruq-qi* URUDU) brought out (*waṣû* Š) from Murattaš

60 / 120 (1 var. 2 *šu-ši*); AKA, 59:103 and n. 9 (RIMA 2, 19; ARI 2, 2 and n. 52)

(5) Mušku warriors (year of accession)[5]

20,000 men (LÚ), fought with (*šanānu* Gt) - AKA, 35:62 and 36:74 (RIMA 2, 14; ARI 2, 12)

12,000 troops (ERÍN.ḪÁ) "conquered" (*kašādu*) - KAH 2, 71a:1′ (RIMA 2, 42; Weidner, AfO 18, 349:18; ARI 2, 92); Weidner, AfO 18, Tf. XXX:18 (p. 360; RIMA 2, 33; ARI 27, 66)

(6) Kings (LUGAL) of Nairi (cf. p. 92-93)

23 assemble their forces (*kaṣāru* Št) - AKA, 67:83 (RIMA 2, 21; ARI 2, 30)

30 have submitted (*kanāšu* Š) - KAH 2, 68:13 (RIMA 2, 37; Weidner, AfO 18, 343:12; ARI 2, 80); AfO 18, Tf. XXX:26 (p. 360; RIMA 2, 34; ARI 2, 69; cf. AKA, 118:9)

(7) Cities (URU) conquered (*kašādu*)[6]

6 AKA, 74:59 (RIMA 2, 23; ARI 2, 34); AKA, 118:12 (RIMA 2, 34; ARI 2, 70)

17 Millard, *Iraq* 32, pl. XXXIV: (122630) 6′ (p. 168; RIMA 2, 59; ARI 2, 143)

Ashurnasirpal II

8. Soldiers (ERÍN) killed (*maqātu* Š) near the city of Kinabu (2nd year)

600 / 800 AKA, 291:107 = Le Gac, *Asn.*, 40:107 and n. 6 (RIMA 2, 201; ARI 2, 549 and n. 519)

---

[4] = *mūšaru*, or "garden plot," equivalent to 144 sq. cubits (RlA 7, 479; cf. also Weidner, *Tn*, p. 5 n.).

[5] The attribution of the two numbers to the same event is disputable, as far as the inscriptions with the number *12,000* relate only a general and non-chronological summary of the events (cf. also the fragmentary passage of RIMA 2, 53:21).

[6] The first version pertains to the towns conquered "at the foot of Mount Bešri," the second appears in a very fragmentary passage, and therefore it is legitimate to doubt that it refers to the same event.

9. Soldiers (ERÍN) impaled (*zaqāpu*); Dirru, 5th year[7]

5<00> AKA, 336: n. 14 (RIMA 2, 251:80 - "Nimrud monolith")

700 AKA, 234:27 and n. 1 (RIMA 2, 260:76; ARI 2, 638 and nn. 711f - "Kurḫ monolith"); AKA, 336:109 (RIMA 2, 210; ARI 2, 571 - "annals")

10. Cities (URU) conquered (*kašādu*); Dirru, 5th year[8]

40 AKA, 235:29 = Le Gac, *Asn.*, 145 (RIMA 2, 260:78; ARI 2, 639)

50 AKA, 337:111 and n. 7 = Le Gac, *Asn.*, 83 (RIMA 2, 210; ARI 2, 572 and n. 576); ICC, 84: (E) 3 ("annals"). For the Nimrud monolith cf. RIMA 2, 251:86.

11. Soldicrs (ERÍN) captured (*ṣabātu*); Dirru, 5th year[8]

40 cf. above

50 cf. above (AKA, n. 12; ICC, line 5)

Tribute received (*mahāru*) from the nobles of Amme-ba'lī of Bīt-Zamāni[9]

	(K)	(A)
12. Talents of bronze[10] (GUN ZABAR)	***200 + 300***	***100***
13. Bronze casseroles (UTÚL ZABAR)	***1,000***	***100***
14. Bronze receptacles (*kap-pi* ZABAR)	***2,000***	***3,000***

(K) AKA, 238:39 and n. 2 = Le Gac, *Asn.*, 147 (RIMA 2, 261:88; ARI 2, 641 and nn. 715f - "Kurḫ monolith") — (A) AKA, 342:122 = Le Gac, *Asn.*, 87 (RIMA 2, 211; ARI 2, 574 - "annals"); RIMA 2, 252:114ff ("Nimrud monolith")[11]

15. (Soldiers) captured alive ((ERÍN) TI.LA + *ṣabātu* D) near the city of Ṣibātu, land of Suḫu

20 / 30 AKA, 355:33 and n. 3 = Le Gac, *Asn.*, 97 and n. 7 (RIMA 2, 214; ARI 2, 579 and n. 600)

---

7 The "Nimrud monolith," which has "numerous scribal errors" (Grayson in ARI 2, 615), reports 5 <*me*> ERÍN, "*5* soldiers," thus omitting the *me*. Also the "Kurḫ Monolith" bears 7 <*me*>, which AKA proposes to read, alternately, *420* (= *7* × *60*), but here it is likewise clear that the *me* has been left out by an oversight. Note that Le Gac reads 5 <*me*> on the Kurḫ monolith (*Asn.*, 145:27 and n. 1) and 7 *me* in both the "annals" and the Nimrud monolith (83:109). The second version (*700* soldiers impaled) might be due to the fact that just a little more than one line below the text speaks of (another) *700* soldiers killed (in both cases the term is ERÍN.MEŠ).

On the chronology of the inscriptions of Ashurnasirpal II cf. EAK 2, 18ff and W. de Filippi, *The Royal Inscriptions of Aššur-nāṣir-apli* (*Assur* 1/7), Malibu 1977, 25-47.

8 In these two cases (nos. 10 and 11) the number of both the cities and the soldiers, placed on the same line of text, is *40* in the Kurḫ monolith and *50* in the Nimrud monolith and in the "annals."

9 These three data (nos. 12-14) belong to a relatively long list (reported at p. 128), given in two versions (here only the three varied numbers are represented).

10 In the "Kurḫ monolith" the numeral referring to the bronze has been repeated twice (by mistake?): 2 *me* GUN ZABAR 3 *me* GUN ZABAR.

11 Since the "Nimrud monolith" largely duplicates the first five campaigns of the "annals," it has not been fully published in copy.

(16) Bulls (GU$_4$.AM KALA) killed (*dâku*) during a hunting campaign

40 / 50 Le Gac, *Asn.*, 101:48 and n. 13 (RIMA 2, 215; ARI 2, 581 and n. 607)[12]

17. Troops (ERÍN.ḪÁ) deported (*nasāḫu*) from Bīt-Adini to be brought to Kalah

2,400 / 2,500 Le Gac, *Asn.*, 103:53 and n. 6 (RIMA 2, 216; ARI 2, 583 and n. 609; cf. AKA, 362:53f)

18. Soldiers (LÚ.ERÍN) captured alive (TI.LA + *ṣabātu* D) near the town of Udu, 18th year

580 / 780 Le Gac, *Asn.*, 116:111 and n. 6 (RIMA 2, 220; ARI 2, 587 and n. 630; cf. AKA, 379:111)

(19) Cities (URU) of the neighborhood of Bunasi[13]

20 / 30 Le Gac, *Asn.*, 55:35 and n. 1 (RIMA 2, 204 and n.; ARI 2, 556; cf. AKA, 306)

(20) Cities (URU) of the neighborhood of Ḫudun[13]

20 / 30 AKA, 314:57 = Le Gac, *Asn.*, 63:57 (RIMA 2, 206; ARI 2, 560)

Shalmaneser III[14]

21. Cities of the neighborhood of Sugunia burned (URU^{MEŠ}-*ni ša limētušu ina* IZI *ašrūp*) - Year of accession[15]

4 Mahmud - Black, *Sumer* 44, 140:37 (p. 144)[16]

14 3 R, 7 I 25 (KB 1, 154 - ed. "A")

22. Prisoners (*šallatu*) deported from the cities of Tajâ and Ḫazazu, at the end of the campaign of the 1st year[17]

4,600 Mahmud - Black, *Sumer* 44, 145:42 (p. 142)

14,600 3 R, 7 II 12 (KB 1, 160 - ed. "A")

---

12 AKA does not report the variant (cf. 360:48). It is impossible to verify the issue since both the original of the inscription and the squeezes no longer exist (cf. RIMA 2, 192).

13 In some exemplars of the inscriptions the three "Winkelhaken" representing the numeral *30* are written close together, while in other cases the first is separated from the other two, justifying in this manner a reading *u* 20, "and *20* (cities)" (cf. p. 4). Schramm in EAK 2, 24 proposes to read *30* anyway, and this is confirmed by the parallel passage URU *Ḫu-du-un a-di* 30 URU.MEŠ, "Ḫudun together with *30* cities" (cf. AKA, 314 n. 1 = Le Gac, *Asn.*, 63:57; RIMA 2, 206 n.).

14 The "edition" of each annalistic text of this king is indicated in parentheses ("A," "B," etc.), according to the classification of EAK 2. This will help to fit the variants in their proper chronological context (therefore, contemporary variants are *not* separated by a bar).

15 On the same occasion, a pile (*a-si-tu*) of heads was erected in front of the city gate (3 R, 7), or 3 towers (*di-ma-te*) of heads according to *Sumer* 44 (line 36). This is a "variation" too, even if in the first case no numeral is used.

16 The number *4* is placed at the beginning of the line. Now, the following line (*Sumer* 44, 144:38) begins with the sign *šá*, which in cuneiform writing is identical with the number *4*: it is therefore possible that the scribe got confused and took the "number" from the following line. However, this is quite unlikely, since this inscription, most probably, is *older* than edition "A" (cf. p. 30 n. 52 end). Note also that the transformation of the number *4* into *14* can only with difficulty be due to a graphical misunderstanding (cf. the remarks of Na'aman, *Iraq* 46, 117 n. 15 for similar problems arising from time-span statements).

17 In the *Sumer* 44 inscription the prisoners are ascribed to the cities of Tajâ and Ḫazazu, while the version of 3 R, 7 ascribes them to the cities of Tajâ, Ḫazazu, Nulia and Butāmu. The context and the

23. Soldiers of Arramu killed (*maqātu* Š), 3rd year[18]

3,400 (*mundaḫṣu*) 3 R, 8:49 (KB 1, 166 - ed. "A")

3,000 (ERÍN) Pinches, *Balawat*, p. 3 III 1 = Rasmussen, *Salm.*, XI (Michel, WO 2, 414 - ed. "B")

13,500 (*mundaḫṣu*) Hulin, *Iraq* 25, 54:41

24. Soldiers of Aḫuni deported (*nasāḫu*), 4th year[19]

17,500 (ERÍN.ḪÁ) Pinches, *Balawat*, p. 3 III 5 = Rasmussen, *Salm.*, XII (Michel, WO 2, 414 - ed. "B")

22,000 Cameron, *Sumer* 6, 12 II 8 = Michel, WO 1, 462 (ERÍN - ed. "C"); Safar, *Sumer* 7, II, 4 = Michel, WO 2, 30 (ERÍN.ḪÁ - ed. "E").

25. Soldiers (ERÍN^{MEŠ} *ti-du-ki*) killed (*maqātu* Š) at Qarqar, 6th year[20]

14,000 3 R, 8:97 (KB 1, 172 - ed. "A")

25,000 Cameron, *Sumer* 6, 13:30 = Michel, WO 1, 464 ("C");[21] ICC, 46:7 (Delitzsch, BA 6, 146:73 - "D"); Safar, *Sumer* 7, 8:24 = Michel, WO 2, 32 ("E")

29,000 KAH 1, 30:16 (Michel, WO 1, 57 - statue)[22]

20,500 ICC, 90:66 (Michel, WO 2, 148 - "F")

26. Cities (URU) conquered (KUR) together with Aštamaku, 11th year

86 (cf. Michel, WO 4, 36:18)[23]

89 Cameron, *Sumer* 6, 15 III 2 = Michel, WO 1, 466 ("C"); Safar, *Sumer* 7, 9:57 = Michel, WO 2, 34 ("E"); ICC, 91:88 (Michel, WO 2, 150 - "F")

99 ICC, 15:36 and 47:24 (Delitzsch, BA 6, 147:92 - "D")

---

singular similarity between the numbers, however, make it clear that this must be considered a "variation" (one may suggest that the "10,000" has been added to balance the inclusion of two more cities).

Within the same context, "numerous warriors" (GAZ.MEŠ-*šú-nu* ḪÁ) have been killed according to the earlier inscription, which become *2,800* in the later version (2 *lim* 8 *me* GAZ.MEŠ[-*šú-nu*] — in both cases *a-duk*, "I killed").

18 Cf. below p. 95 n. 202. To this event is possibly also connected a quantification appearing in a rather fragmentary literary composition (apparently a heroic epic), mentioning *18,000* enemies (LÚ.KÚR) killed (*mâtu* Št) with the battle weapons (STT 1, 43:47 = SAA 3, 17 r. 15; older ed. in Lambert, AnSt 11, 152). The text possibly relates the events of the 3rd *palû* of Shalmaneser III — actually neither the king's name nor Arramu's nor the year of reign appear on the document: for the problems in dating the narrated events cf. *ibid.*, 154f and J. Reade, "Shalmaneser or Ashurnasirpal in Ararat?," SAAB 3 (1989), 93-97.

19 On this quantification cf. below pp. 94ff. Both passages are reported in Sader, *États araméens*, 70ff.

20 On this variation cf. also the similar scheme in Michel, WO 2, 41 n. 10. The version given by the black obelisk (*20,500*) is possibly the result of an oversight or a dittography (cf. the remarks of S. Timm, *Die Dynastie Omri. Quellen und Untersuchungen zur Geschichte Israels im 9. Jahrhundert vor Christus*, Göttingen 1982, p. 187 n. 7 and 11) — however, the passage from 𒌋𒌋𒐊𒇷 (20+5 *lim*) to 𒌋𒌋𒇷𒐊𒈨 (20 *lim* 5 *me*) is not so immediate. It is also rather unlikely that the *25,000*, in turn, derives from an erroneous reading of *14,000* ("Verlesung" — S. Timm, *loc. cit.*).

21 A fragment belonging to this "edition," as published in Boissier, RT 25, 83f (line 19), reads [...] 5 *me* ERÍN^{MEŠ} *ti-du-ki-šu*, which thus would represent an original [*20*],*500*. However, judging from the photograph (*ibid.*, 82) it would rather seem that a reading [...] 5 [*l*]*im* is preferable, since the small lacuna between the *5* and the *me* allows enough space for a *Winkelhaken*.

22 Statue containing a "Bauinschrift" of uncertain date, but in any case subsequent to the 26th campaign (cf. EAK 2, 83). Here the term employed is *alīli mundaḫṣi* "powerful fighters."

23 Text engraved on a bronze band found at Balawāt, contemporary with ed. "B" — for a reproduction of the band, the 13th, cf. L.W. King, *The Bronze Reliefs from the Gates of Shalmaneser, king of Assyria B.C. 860-825*, London 1915. All the passages appear also in Sader, *États araméens*, 195ff (however, she gives *89* also for the texts of ed. "D," without justifying this reading).

27. Soldiers (ERÍN^MEŠ *ti-du-ki*) of Ḫazael of Damascus killed (*maqātu* Š), 18th year[24]

16,000	3 R, 5, no. 6:9 (Michel, WO 1, 265 - ed. "D"); ICC, 13, "on Bull No.2": 48 (Delitzsch, BA 6, 151 - "D"); Kinnier-Wilson, *Iraq* 24, pl. XXXIV:23 (p. 94, here *mundaḫṣu* - "E")
16,020	Safar, *Sumer* 7, 11:51 = Michel, WO 2, 38 ("E")[25]

28. Crossing of the Euphrates - 19th year

17-šú	Safar, *Sumer* 7, 12:16 = Michel WO 2, 40 (ed. "E")[26]
20 <*-šú*>	Kinnier-Wilson, *Iraq* 24, pl. XXXV:30 (p. 94 - "E")[27]
18-šú	ICC, 92:99 (Michel, WO 2, 154 - "F")

Chariots and cavalry of the Assyrian army ("total")

	(C)	(E1)	(E2)
(29) Chariots (^GIŠGIGIR)	***2,002***	***2,001***	***2,001***
(30) Cavalry (*pētḫallu*)	***5,542***	***5,242***	***5,241***

(C) Cameron, *Sumer* 6, 18:47f = Michel, WO 1, 474 (ed. "C") (E1) Safar, *Sumer* 7, 14:last line (after the colophon!) = Michel, WO 2, 44 (ed. "E") (E2) Michel, WO 1, 391 l.r. II 1 (cf. photo Tf. 20 - ed. "E")

(31) Kings (LUGAL) of Tabal, 22nd year

20 / 24	Læssøe, *Iraq* 21, 154:27 and *ibid.*, 155:12 give *20* — ICC, 92:105 gives *24* (all ed. "F")[28]

Shamshi-Adad V[29]

32. Cities (URU) of the neighborhood of Nimittišarri destroyed, demolished, burned (*napālu, naqāru, šarāpu*)

255	Weidner, AfO 9, 93:35 (p. 92 - stela from Ashur)
[*2*]***56***	KAH 2, 142:5 (Weidner, AfO 9, 102 - letter to the God, fragm. passage)

---

24 The passages pertinent to this variation are collected in Sader, *États araméens*, 231ff.

Another apparent variation is relative to the forces of Ḫazael captured (*ekēmu*) on the same occasion. Most of the texts have *1,121* chariots (^GIŠGIGIR) and *470* cavalry (*pētḫallu*) (cf. following lines - edd. "D" / "E"). "Bull No. 2," as published by Layard in ICC, 13 (lines 50f - ed. "D") has 𒁹𒌋𒁹𒁹𒌍𒁹 which is sometimes read *1,131* (cf. edition in Delitzsch, BA 6, 151 and ARAB 1, 663). However, since Rasmussen, *Salm.*, pl. XXXI:51f (and p. 56) has *1,121*, it would seem that Layard confused a horizontal wedge with a "Winkelhaken." Note how also the number *470*, relating to the *pētḫallu*, is written in an imprecise way: 𒐉𒁹𒁺[*sic*] (= *4,1,70*).

25 Safar's translation at p. 19 erroneously gives "*1,602*" (cf. Michel, WO 2, 43 n. 10).

26 Note that the text of the 19th *palû* is almost identical with that of the 17th (copied, with a possible confusion between *palû* and crossing? — cf. below, p. 148).

Cf. below, p. 137 (and n. 97 there) for another discrepancy in the counting of the Euphrates crossings.

27 The text states that *ina* 19 BALA.MEŠ-*ia* 20 ÍD.BURANUN *e-bir*, while in the following line the entry for the 20th campaign starts with *ina* 21 BALA.MEŠ-*ia* 20-*šú* ÍD.BURANUN *e-bir*; obviously the scribe confused the crossings with the *palê* (cf. the table at p. 137).

28 Michel doubts this reading (cf. WO 2, 154:105 and n. zz). However, Rasmussen, *Salm.*, pl. XXVIII (and p. 70) agrees with Layard's copies, having *24* (cf. also the reproduction in Börker-Klähn, *Bildstelen*, Tf. 152b: lower register, line 10).

For another possible variation in the number of kings, cf. below, p. 134 n. 89 (*12* kings / *15* cities).

29 Another "variation" originates from an inconsistency between the two copies of the monolith given

Adad-nerari III

Tribute paid by Mari' of Damascus[30]

	(R)	(S)	(N)
(33) Silver (KÙ.BABBAR)	***2,000*** t	***1,000*** t	***2,300*** t
(34) Gold (GUŠKIN)	-	≥ ***100*** t	***20*** t
Copper (URUDU)	*1,000* t	-	-
Bronze (ZABAR)	-	-	*3,000* t
(35) Iron (AN.BAR)	***2,000*** t	***60*** (***100***?) t	***5,000*** t
Garments (*lubulti birme*)	*3,000*	-	*n.q.*

(R) Page, *Iraq* 30, *pl.* 39 and p. 142:6ff = Cazelles, CRAIBL 1969, 109 (Tadmor, *Iraq* 35, 143; Page, VT 19 - stela of Rimaḥ) (S) Unger, PKOM 2, 10:20 (Tadmor, *Iraq* 35, 145 - stela of Saba'a) (N) 1 R, 35, no. 1:10f (Tadmor, *Iraq* 35, 148 - "Nimrud slab," it is posterior to the other two inscriptions)[31]

Tiglath-pileser III[32]

36. People (LÚ) deported (?) from Buda (and) Duna

588 / 589 Tadmor, *Tigl. III*, pl. IX:6 (Ann. 13) and pl. II:5 (Ann. 2). Older publications: ICC, 50a = Rost, *Tigl. III*, Pl. XV (p. 24:146) and ICC, 69b:5 = Rost, Pl. III respectively. Cf. 3 R, 9 no. 3:46.

(37) Tribute received from Metenna of Tyre[33]

	(Summ. 9)	(Summ. 7)
Gold (GUŠKIN)	***50*** t	***150*** t
Silver (KÙ.BABBAR)	*2,000* t	...

(Summ. 9) Wiseman, *Iraq* 18/26, pl. XXIII rev. 26 (p. 124 and 126; Tadmor, *Tigl. III*, Summ. 9) (Summ. 7) 2 R, 67:66 = Rost, *Tigl. III*, pl. XXXVII:last line (Tadmor, *Tigl. III*, Summ. 7 rev. 16′)

---

by Rawlinson: the *120 pēthallu* of 1 R, 30:33 (reproducing the original archaic characters) are copied as *140* in 1 R, 33:33 (version in standard characters); cf. KB 1, 181:33 and pertinent n.

30 A hyphen indicates that the term is ignored, *n.q.* that it appears but is not quantified. Cf. Page, *Iraq* 30, 144 nos. 6-7 for a similar scheme (taken up also by Pettinato, *Semiramide*, 297f).
The three passages are collected in Sader, *États araméens*, 238ff. On the identity of Mari' cf. Pitard, *Ancient Damascus*, 165ff.

31 According to Page (*Iraq* 30, 147 and 149) it seems probable that we are dealing with two (or three) different events (Cazelles, CRAIBL 1969, 110 agrees). Other authors, however, are inclined to believe a single event is described — cf. recently Tadmor, *Iraq* 35, 144 and Shea, JCS 30, 107. For an overview of the problem of the dating of the Syrian campaigns of Adad-nerari III see Millard - Tadmor, *Iraq* 35, 61ff and Pitard, *Ancient Damascus*, 160ff. Considering that the numbers tend to grow, they could be, at least in the case of the Nimrud slab, "totals."

32 The annalistic fragments published as ICC, 72b-73a (end of each line) and 29b (= Rost, *Tigl. III*, Pl. XVIII a and b) contain some errors which might let us think of some other "variation" (e.g. compare 6 *me* 55 *šal*-[...] of ICC, 72:8 with 6 *me* 56 *šal-lat* URU *Sa*-[...] of ICC, 29:4f (Rost's Pl. XVIIIb gives the latter as 6 *me* 50 !) — cf., however, the recent re-edition in Tadmor, *Tigl. III* (as Ann. 18 and 24).

33 As in the case of Mari' of Damascus (above, n. 31), one might doubt that a single tribute is described, since the pertinent passages appear in two *summary inscriptions* in which the events are not dated in any manner. Moreover, both passages are rather fragmentary, and Summ. 9 does not even record the name of the tribute-bearer. Note that most of Tiglath-pileser III's *summary inscriptions* are more or less contemporary, being composed shortly after the 17th *palû* (cf. Tadmor, *Tigl. III*, 118 and 154). Cf. also *ibid.*, 171 n. 16′ and *supplementary study D*.

Sargon II[34]

38. People (UN) counted as booty (*šallatiš manû*) in Samaria

27,280	Botta, 122:17 (display inscr. from room VII);[35] Gadd, *Iraq* 16, XLVI iv 31 (p. 179 - prism from Nimrud)
27,290	Botta, 145a:last line = Winckler, *Sargon*, no. 64 (p. 100:24 - display inscr. from room X)

39. Chariots (^{GIŠ}GIGIR) annexed (*kaṣāru*) to the royal army on the same occasion

200	Prism from Nimrud: iv 33 (cf. above)
50	display inscr. (cf. above, incl. Botta, 122:17 and 138a:1); Botta, 70:2′ = Winckler, *Sargon*, no. 2 (Lie, *Sargon*, 4:15 - annals)

Booty from Karkemish, after its conquest (5th year)[36]

	(N)	(XIV)
40. Gold (GUŠKIN)	***11*** t ***30*** m	***1***[***1***] t
41. Silver (KÙ.BABBAR)	***2,100*** t ***24*** m	[***2,1***]***00*** t

(N) ICC 34:21 = Winckler, *Sargon*, Tf. 48 (vol. 1, p. 172 - Nimrud inscription) (XIV) Botta, 158:last line and 159a:first line = Winckler, *Sargon*, nos. 58 and 59 (p. 86:42f and Fuchs, *Sargon*, 94:72b-c - slab from room XIV at Khorsabad)

42. Forces collected (*kaṣāru*) and added (*redû* D) to the royal army - Karkemish (5th year)

	(N)	(A)
Chariots (^{GIŠ}GIGIR)	...	*50*
Cavalry (*pētḫallu*)	***500***	***200***
Infantry (^{LÚ}*zūk šēpē*)	...	*3,000*

(N) Gadd, *Iraq* 16, XLV iv 21 (p. 179 - prism from Nimrud, fragm.) (A) Botta, 72:10′ and 159a:2′ = Winckler, *Sargon*, no. 4 and no. 59 (Lie, *Sargon*, 12:75 - annals)

---

34 Reference to the recent work by A. Fuchs on Sargon's inscriptions from Khorsabad is made only if solicited by the discussion or in the case where older editions are uncertain. It will be easy to check each passage in his edition since the references to older publications are consistently indicated.
For other variations appearing in two fragmentary duplicates of Sargon's inscriptions, see here below n. 40 and p. 133 nn. 83 and 86.

35 Variant not credited by Winckler, but cf. the recent edition by Fuchs, *Sargon*, 197. The duplicate from room VIII, according to Botta, 138a:1 apparently reads *24*?*,290* (or *24* [ ] *280* according to Fuchs, *loc. cit.*). The prism from Nimrud has [2]*7,280*, while in the annals from Khorsabad the number is broken away.
On the conquest of Samaria, connected with variations 38 and 39, cf., recently, Becking, *Samaria*, especially 25ff and 39ff, where he argues that the conquest of Samaria is *not* described in the annals (Botta, pl. 70, mentioning *50* chariots, cf. below variation 39).

36 Inscription (N), which has no narrative of military events, reports the depositing of the booty (*ki-šit-ti*) of Pisiris king of Karkemish in the "juniper palace" of Kalah. The version (XIV) reports the looting (*šalālu*) of Pisiris' treasure in Karkemish during the 5th campaign. The first number is readable as *10* + [...] talents, the second as [...] *me* talents (no minas are given in either case) — a number of talents of bronze were also recorded, but the ciphers have been lost. Note that the inscription from Nimrud is probably older, and that the booty is there explicitly defined as "enormous," *rabītu* (GAL-*ti*) (translation in ARAB 2, 138).

43. Fortresses of Ullusunu the Mannean

12 (*ḫalṣu*) Weidner, AfO 14, 43:16 (p. 46 - fragm. of a prism from Ashur, "6th" year)[37]

22 (*birāte*) Botta, 74:10′ and 119:11′ = Winckler, *Sargon*, no. 6 and 35 (Lie, *Sargon*, 16:101 - annals, 7th year); Botta, 146a:3, 123:last line and 95:10 = Winckler, *Sargon*, no. 66 (p. 104:39 - display inscr.)

44-45. Cities of Mitatti of Zikirtu conquered (*kašādu*), 8th year

12 URU *dannūti* + *84* URU *ša seḫrišunu* — Thureau-Dangin, TCL 3, pl. V:89 (p. 16 - letter to the God)[38]

3 URU *dannūti* + *24* URU *ša limētišunu* — Botta, 76:1′ = Winckler, *Sargon*, no. 9 (Lie, *Sargon*, 22:131 - annals)

23 URU BAD$_4$ *ša limētišunu* — Botta, 146a:last but one and 94:7 (cf. also 124:7) = Winckler, *Sargon*, no. 66 (p. 106:47 - display inscr.)

46. Nobles[39] of the family of Ursâ, 8th year[40]

260 Thureau-Dangin, TCL 3, pl. VII:138 (p. 24 - "letter"); Botta, 76:4′ = Winckler, *Sargon*, no. 9 (Lie, *Sargon*, 24:134 - annals)

250 Botta, 146a:6 (cf. also 94:3 and 135b:6) = Winckler, *Sargon*, no. 66 (p. 104:42 - display inscr.)

47. Distance between mount Uauš and mount Zimur in *bēru* (KASKAL.GÍD), 8th year

6 Thureau-Dangin, TCL 3, pl. VII:145 (p. 24 - "letter")

5 Botta, 76:5′ = Winckler, *Sargon*, no. 9 (Lie, *Sargon*, 24:135 - annals)

---

37 Corresponds to the 7th year of the later annals, see Tadmor, JCS 12, 22 and 93ff and Ford, JCS 22, 84.

38 The *12* main cities are mentioned by name (actually, 13 names are given: cf. also p. 190 no. 26).

39 According to the "letter to the God" (TCL 3) the captured were "his nobles, the officials his governors (and) his cavalrymen" (NUMUN *šarru-ti-šu* LÚ *šu-ut* SAG^{MEŠ} LÚ EN.NAM^{MEŠ}-*šu* LÚ *ša pet-ḫal-lì-šu*). The annals relate the number to "his nobles (and) the cavalry" (NUMUN *šarru-ti-šu* LÚ *pét-ḫal-lì-šu*), while the display inscr. speaks only of *250* NUMUN *šárru/šarru-ti-šú*.

40 According to his display inscriptions, Sargon conquered *55* fortified cities of Ursâ the Urartian (thus the inscription from room X, Botta, 146a:7 = Winckler, *Sargon*, no. 66 / p. 104:43). The fragmentary passage appearing on the display inscr. from room IV apparently has [5]*6* cities, cf. Botta, 94:3, variant credited also by Renger, *Sargon*, 291 and Fuchs, *Sargon*, 204. However, both the original and the squeezes have disappeared, and therefore could not have been collated. This quantification is ignored by the other inscriptions.

48. Cities (URU) of the Bīt-Sangibūti area, 8th year

21 URU^{MEŠ}*-ni dannū[ti]* ... ***146*** URU^{MEŠ}*-ni ša limētišunu*	Thureau-Dangin, TCL 3, pl. XIII:268 (p. 42 - "letter")[41]
21 URU^{MEŠ}*-ni dannūti adi* ***140*** URU^{MEŠ}*[-ni ša] limētišunu*	Botta, 76:10′ = Winckler, *Sargon*, no. 9 (Lie, *Sargon*, 24:140 - annals)

Booty from Muṣaṣir, 8th year[42]

	(TCL 3)	(A)	(D)
49. People (UN)	***6,110***	***6,170***	***20,170***
50. Mules (^{ANŠE}*ku-dini*)	***12***	} ***692***	
Asses (ANŠE)	***380***		
51. Oxen (GU$_4$)[43]	***525***	***92[0]***	
52. Sheep (UDU)	***1,235 / 1,285***	***100,225***	
Gold (GUŠKIN)	[*X* t *X*] m	*34* t *18* m	
53. Silver (KÙ.BABBAR)	***167*** t ***2 1/2*** m	***160*** t ***2 1/2*** m	

(TCL 3) Thureau-Dangin, TCL 3, pl. XVII:349ff and pl. XXI:424 (p. 52 and 66 - letter to the God) (A) Thureau-Dangin, RA 24, 79:11′ff (Lie, *Sargon*, 26:154f - annals) (D) Botta, 148:3 = Winckler, *Sargon*, no. 69 (p. 112:75 - display inscr.)[44]

54. Silver bowls (*kappi* KÙ.BABBAR *ṣuppūti*, "massive"), from the booty of Muṣaṣir

54	Thureau-Dangin, TCL 3, pl. XVIII:360 (p. 54 - "letter")
55	Winckler, *Sargon*, Tf. 45, fragm. B:21′ (Thureau-Dangin, TCL 3, p. 78:34 - prism "B")

---

41 The *21* main cities are mentioned *per extenso* here too, and 29 lines separate them from the numeral of the suburbs.

42 Similar table in Millard, *Fs Tadmor*, 216. These variations will be discussed below, p. 70 n. 102, p. 73 and p. 112. In the "letter to the God" the numbers of the prisoners and of the animals are repeated at the end of the inscription (with a variation). The list of the palace treasure, beginning with gold and silver, continues with several other objects, and is followed by the list of the treasures from the temples (both fully quantified: cf. p. 129). Also, the annals, a few lines after, speak (again) about [*X*] talents *3* minas of gold and *162* talents [*X*] minas of silver, but the passage is very fragmentary, yet it refers to the same incident (Thureau-Dangin, RA 24, 80:2′ = Lie, *Sargon*, 28:158). The version on Nineveh's "prism B" is even more fragmentary, and is excluded from this table since it preserves only a few numbers (cf. Thureau-Dangin, TCL 3, 76ff, and cf. next variation). The list appeared also in the prism from Ashur, of which little survives (cf. Weidner, AfO 14, 43 col. C).

The "display inscriptions," on the contrary, do not specify anything other than people: 20 *lim* 1 *me* 70 UN.MEŠ *a-di mar-ši-ti-šú-nu* ^{d}*Ḫal-di-a* ^{d}*Ba-ag-bar-tum* DINGIR.MEŠ-*šú a-di* NÍG.GA-*šú-nu ma-ʾ-at-ti šal-la-ti-iš am-nu* ("*20,170* people with their goods, Ḫaldia (and) Bagbartu, their gods, with their abundant properties I counted as booty"). The "increase" of people could represent a compensation for the missing quantification of the animals.

43 The annals have 9 *me* 20 [... GU$_4$].MEŠ, the lacuna allowing some smaller digits to be written (cf. also Fuchs, *Sargon*, 114 and 320). Perhaps *92[5]* is to be restored.

44 The display inscr. from room VII, according to Botta's copies (pl. 126:3), reads 20 *lim* 1 *me* 80? [...], which Fuchs, *Sargon*, 215 corrects to *20,170* (while Renger, *Sargon*, 292 reads *20,178*).

(55) Cities captured together with the fortress of Ḫubaḫna (9th year)[45]

25	Gadd, *Iraq* 16, pl. XLVI:48 (p. 177 - prism "D")
24	Botta, 116b:6′ = Winckler, *Sargon*, no. 37 (Fuchs, *Sargon*, 120:175 and n. there - annals)

56. Horses (ANŠE.KUR.RA) received (*maḫāru*) from the Medes, 9th year[46]

8,609 (?)	Gadd, *Iraq* 16, pl. XLVI:56 (p. 177 - prism "D")
4,609	Botta, 80:11′ = Winckler, *Sargon*, no. 13 (Lie, *Sargon*, 30:193 - annals)

57. Chiefs (*nasikāte*) of the land Yadburu, bringing tribute to the king (12th year)[47]

5 / 6	Botta, 92b:9′ and 66b:2 (= Winckler, no. 21 and 33; Fuchs, *Sargon*, 151:299 - annals from "room II" and from "room V" resp.)

Booty sacked (*šalālu*) by the Assyrian troops and received (*maḫāru*) in the king's camp, southern Babylonia, 13th year[48]

	(45)	(46)	(55)
People (UN)	...	...	*90,580*
58. Horses (ANŠE.KUR.RA)	***2,500***	*(X +)* ***500***	***2,080***
59. Mules (ANŠE.KUNGA)	***710***	***610***	***700***
Camels (ANŠE.A.AB.[BA])	...	...	*6,054*
Sheep (UDU.NÍTA)	*(X +) 40*	...	...

(45) Botta, 110:6′f = Winckler, *Sargon*, no. 45 — (46) *ibid.*, 116a:6′f = no. 46 — (55) *ibid.*, 163:8′ = no. 55 (cf. Lie, *Sargon*, 62:6 and n. 3; Fuchs, *Sargon*, 164)

45 The passage appearing in the annals from "room V" is very fragmentary, and the identification of these *24* cities with those of the Kalah prism is therefore highly debatable.

46 The tribute refers to a wide group: "tribute from Ullusunu the Mannean, Daltā the Ellipian and Bēl-aplu-iddina of Allabria, from *45* city prefects of the land of the mighty Medes, *4,609* horses, mules, oxen, sheep without number (*ana lā māni*) I received (*maḫāru*)" (lines 191-94 — the numeral refers to the horses only, while mules, oxen *and* sheep were "without number").

47 The passage appearing in the annals from "room II" having *5* is very fragmentary, and has been ignored by Lie (cf. *Sargon*, 52:2). Note that the *6* chiefs are named *per extenso* by the annals from "room V" (the version from "room II" is too broken to allow an understanding of whether it had five names or six).

48 "The numerous (*gapšāte*) troops of Aššur for *3* days carried off spoils without number (*šallat ana la nībi*)." The tribute (or booty) appears in some fragments of the annals, belonging to at least two different versions, whose chronological order is unknown. It is suspected that Botta, pl. 110 and 116a (nos. 45 and 46 of Winckler) are two copies of the same slab (in fact they are almost identical: cf. Lie, *Sargon*, 62: n. *).

Van der Spek, JEOL 25, 61, reads "6 ME 54" camels: this seems due to the fact that the sign LIM is written differently from those ascribed to the people and to the horses (but it is not so unusual: cf. the same sign at line 14′ of the same text — 1 *lim* ANŠE.KUR.RA.MEŠ). A quantification of oxen was possibly inserted between those of camels and sheep, but has been completely broken away.

60. Forces of Bīt-Yakin annexed (*kaṣāru*) to the Assyrian army, 13th year

	(A)	(D)
Chariots (^{GIŠ}GIGIR)	*150*	*150*
Cavalry (*pētḫallu*)	*1,500*	*1,500*
Bowmen (ERÍN ^{GIŠ}PAN)	*20,000*	*20,000*
Shield-bearers (*nāš kabābi*)	***10,000***	***1,000***

(A) Botta, 106:11 = Winckler, *Sargon*, no. 49 (Lie, *Sargon*, 72 - annals); Gadd, *Iraq* 16, pl. XLV:8ff (p. 179 - prism from Nimrud) (D) Botta, 151b:9 (and cf. 128:2) = Winckler, *Sargon*, no. 72:117 (p. 118 - display inscr.)

61. Chariots and cavalry selected (*kaṣāru*) among the people of Hamath and annexed (*redû*) to the royal forces (2nd year)

	(D)	(CS)
Chariots (^{GIŠ}GIGIR)	***200***	***300***
Cavalry (*pētḫallu*)	*600*	*600*

(D) Botta, 145b:11 = Winckler, *Sargon*, no. 65:35 (p. 102 - display inscr.); Lambert, *Ladders*, 125:1 (CS) Winckler, *Sargon*, Tf. 47 l.s. = VS 1, p. 68 (Winckler, *Sargon*, 178:57 - "Cyprus stela"); Renger, *Sargon*, 768:1 ("Hamath stela") - the fragmentary passage in Thompson, *Iraq* 7, 87:19 has no digit readable.

62. Weight of the *8* lions (UR.MAḪ) of shining bronze (URUDU *namri*) set up as protecting deities in the palace of Dur-Sharruken, in talents

1 ŠÁR GÍŠ+U 6 (var. 8) UŠ 50 (var. 40 / 60$^{?}$) (TA.)A.AN
= *1* × *3,600* + *6* / *8* × *60* + *40* / *50* / *60*$^{?}$ (*4,610*, var. *4,730* or *4,600* / *4,620*)

4,610 is given by most texts.[49] The other versions appear in some copies of threshold inscription no. 4.[50]

---

[49] Annals from room V (Botta, 105 = Winckler, *Sargon*, no. 51:1′; Lie, *Sargon*, 76:1; Fuchs, *Sargon*, 183:434); display inscr. from room XIV (*ibid.*, 160b / no. 61:10′ = Weissbach, ZDMG 72, 182:37; Fuchs, *Sargon*, 79:37); bull inscr. (Lyon, *Sargon*, 16 and 44:70; Fuchs, *Sargon*, 69); threshold inscr. no. 2 (Botta, 2:31′ and 5:30′ = Winckler, *Sargon*, Tf. 37 / p. 140:31f; Fuchs, *Sargon*, 253:31); display inscriptions (Botta, 103:5; cf. also 130:18 and 134a:5; Fuchs, *Sargon*, 239:162) — the display inscr. from room X, according to Botta's copies (152b:6) has *40* instead of *50*, but this version is not credited by Winckler (*Sargon*, no. 76:162, cf. edition at p. 130) nor by Fuchs (*loc. cit.*).

[50] Cf. edition in BAL[2], 62 or Fuchs, *Sargon*, 268 (line 109). The copies from doors *g*, *j* and *U* (Botta, pl. 16ter:109; 17:100f and cf. 12ter:86 — version followed by Winckler, *Sargon*, Tf. 39 / p. 154:109) give *4,610*. The copy from door *E* (*ibid.*, 4b:29′; rather fragmentary) as well as, possibly, the copy from door *M* (very fragmentary: cf. *ibid.*, 7ter:92) give the variant with *40*. The copy from door *l* apparently reads 1 ŠÁR GÍŠ+U 6 NI 60 (or 6 UŠ 40/50, if we admit that Botta's copies are confused at this point, NI making no sense: cf. *ibid.*, 19^{quater}:95). The version 1 ŠÁR GÍŠ+U 8 UŠ 50 (= *4,730*) is given by the copy from door *k* (*ibid.*, 18ter:96). In this case, possibly the scribe(s) confused this numeral with that of the *8* lions (in some copies it appears exactly one line above).

Precious metals presented (*qiāšu*) to the Gods of Sumer and Akkad throughout the first three years

	(X)	(VII)	(CS)
63. Gold (GUŠKIN)	***154*** *t 26 m* ***10*** *s*	***164*** *t 26 m [X] s*	*[X] t 26 m* ***6*** *s*
64. Silver (KÙ.BABBAR)	*[1],**604** t 20 m*	***1,804*** *t 20 m*	***1,804*** *t 20 m*

(X) Botta, 152a:9 = Winckler, *Sargon*, no. 74:141 (p. 124 - display inscr. room X); Weidner, AfO 14, 49:2 (p. 51 - prism from Nineveh) (VII) Botta, 129:8 (display inscr. room VII)[51] (CS) VS 1, p. 69 = Winckler, *Sargon*, Tf. 47 r.s. (p. 178:6f - "Cyprus stela")

Sennacherib[52]

Cities of Babylonia, 1st campaign

	(1)	(2)	(3)	(4)
65. Cities (URU-*šu dannūti*)	***88***	***89***	***89***	***75***
66. Suburbs (URU TUR *ša limētišunu*)	***820***	***820***	***620***	***420***

(1) Smith, *First Camp.*, 41:50 (Luckenbill, *Senn.*, 54 - account of the first campaign)[53] (2) ICC, 63:11 (Luckenbill, *Senn.*, 56 - "Bellino" cylinder, 2 campaigns) (3) cf. Bezold in KB 2, 84:34f and n. 14 ("Rassam" cylinder and dupl., 3 campaigns)[54] (4) AD, 298 I 37f ("cylinder C," actually a prism, with 4 campaigns); CT 26, *pl.* 2:43ff = Heidel, *Sumer* 9, col. I 43ff (prism "Heidel-King" - 5 campaigns + "eponym campaigns"); Luckenbill, *Senn.*, 164:36f (p. 25 and add Ling-Israel, *Fs Artzi*, 222:32f - prisms with 8 camp.). On these variations cf. also BAL^2, 70.

---

51 Fuchs, *Sargon*, 231 confirms this version after collating the originals (it was ignored by Winckler, while Renger, *Sargon*, 293 credited only the variant *164*, not *1,804*). Note that the Nineveh prism is older than the *display inscriptions*.

52 For the chronology and classification of the main inscriptions of Sennacherib cf. BAL^2, 64f; Reade, JCS 27; Levine, HIII. The texts are here simply grouped together according to the numerals they bear.

Eckart Frahm very kindly sent me in advance a provisional version of his doctoral thesis on Sennacherib's inscriptions (*Einleitung in die Sanherib-Inschriften*). His readings, based on fresh collations, have been used to correct some numerals: in such cases the (erroneous) digits given by older editions (including copies) are referred to in the footnotes. Reference to unpublished texts that he takes into consideration is not made, except when they give new data or variant readings.

A fragmentary text published in Na'aman, BASOR 214, 26:19 (cf. n. 17) speaks about the "*7th* time" (*ina* 7-*šú*), in reference, it seems, to the "troops of Amurru." A passage of the annals (Luckenbill, *Senn.*, 170 and 30:59 and *passim*) recalls that "the kings of Amurru, all of them, many gifts, as their onerous tribute, for *4* times (*a-di* 4-*šú*) before me they brought." However, it seems far from certain that the two figures make reference to the same event, due to the fragmentary state of the first text, in which the "*7th* time" would seem related with the act of "breaking like a pot" (cf. translation in BASOR 214, 28). Recently, G. Galil suggested the attribution of the text to Sargon ("Conflicts between Assyrian Vassals," SAAB 6 [1992], 61f).

53 This text mentions (most of) the *88* main cities: cf. below, p. 191 nos. 33-36 and n. 12 there.

54 Luckenbill considers this a duplicate of the "Bellino" cyl. as far as it concerns the first two campaigns, and does not note the variants in the numbers. According to Bezold, *loc. cit.*, at least one of the smaller fragments has *820* (fragment B 4) — E. Frahm's collations, however, confirm that all the documents belonging to this prism edition have *620*.

Booty of the 1st campaign

	(1)	(2/3)	(4)
People (UN)	*208,000*	*208,000*	*208,000*
Horses (ANŠE.KUR.RA) and Mules (ANŠE.KUNGA)	*7,200*	*7,200*	*ša lā nību*
Asses (ANŠE)	*11,073*	*11,073*	
67. Camels (^{ANŠE}GAM.MAL)	***5,230***	***5,230***	
68. Oxen (GU$_4$)	***80,050***	***80,100***	
69. Sheep (U$_8$.UDU.ḪÁ)	***800,100***	***800,600***	

References as above, except: (1) line 60 (2/3) "Bellino" line 16[55] - "Rassam" cf. KB 2, 85:48ff and nn. 33ff.[56] (4) "C" lines 55ff - "Heidel-King" 64ff - 8 camp. prisms (Ling-Israel) 45-47. Cf. also BAL^2, 70.

On some duplicates of (3), cylinder "Rassam," the following variants appear[57]

	80-7-19,1	79-7-8,302
67. Camels	***5,230***	***5,233***
68. Oxen	***80,100***	***200,100***
69. Sheep	***600,600***	***800,600***

70. Cities (URU) of the province of Ukku, 5th campaign

35	Heidel, *Sumer* 9, col. IV 89 (prism "Heidel-King" - 5 camp.)
33	Luckenbill, *Senn.*, 176:28 (p. 37 and Ling-Israel, *Fs Artzi*, 231:23 - prisms with 8 campaigns)

Building of the "palace without rival" (É.GAL *šānīna lā išû*) in Nineveh

In the following table are related all[58] the quantifications concerning the palace, even if some of them do not show any variation, for easy comparison. They have the characteristic of "totals," as dimensions of the building(s), number of colossi, etc., can change owing to modifications to the architectural plans.[59]

55 The number of sheep, given by Layard's copies as *800,500*, has been corrected according to the original copy of Bellino (published in Borger *et al.*, *Die Welt des alten Orients*, Göttingen 1975, 52f). Bezold (KB 2, 85 n. 38) read *900,500* for the "Bellino" (2) and *800,600* for the texts under (3).

56 According to E. Frahm's collations, the number of asses is 11 *me* 73 in all exemplars (Bezold, KB 2, 85 n. 34 gives *11,173* for a duplicate of cylinder "Rassam").

57 Both cylinders are unpublished (of 80-7-19,1, *alias* cylinder "Rassam," only the concluding section has been published), and have been taken into consideration only as variant-sources (in KB 2 they are marked B 2 and B 4 and in Luckenbill, *Senn.*, C 1 and C 3 respectively). This table, therefore, is based exclusively on E. Frahm's forthcoming new edition of these texts (cf. n. 52 above), which he will mark, in turn, T 3 and T 4 respectively. His notes also allow us to infer that three exemplars of this cylinder edition have *5,230*, four have *80,100* (another fragm. text may have *200,100* too) and five have *800,600*. Thus, these variants are obviously originating from miscopyings.

58 To add that in Luckenbill, Senn., 121:53, text (7), it is stated that the bull colossi and the alabaster statues were carved out of one (1-*en*) stone.

59 The building reports, in most of cases, were buried as foundation deposits, *before* the work was actually carried out, or during the initial stages, and therefore they only outline the construction plans (this also explains the absence of certain information), even if these are referred to in the past tense (see Porter, IPP, 43 and 57). In many cases, it is not clear where the Sennacherib texts here considered were actually buried, if in the foundations or in the walls — however, even some early texts, as cylinder "Rassam" and some of its duplicates (here no. 3), were buried in the walls of the palace (cf. e.g. J.E. Reade, "Foundation Records from the South-West Palace, Nineveh", ARRIM 4 (1986), 33).

References (in a roughly chronological order): (1) Smith, *First Camp.*, lines 63ff (Luckenbill, *Senn.*, 94ff) (2) ICC, 63f lines 34ff (Luckenbill, *Senn.*, 94f and 99f - "Bellino" cyl.). (3) Evetts, ZA 3, lines 61ff (KB 2, 86 and Luckenbill, *Senn.*, 102 - "Rassam" cyl.); cf. KAH 2, 120 (4) "cylinder C"[60] (5a) "cylinder D"[61] (5b) CT 26, pl. 18ff. col. V 23ff (p. 21ff; Luckenbill, *Senn.*, 103ff); Heidel, *Sumer* 9, 152ff col. V 53ff; cf. Knudsen, *Iraq* 29, pl. XXII (ND 5416) (6) older colossi: ICC, 59-62;[62] Meissner - Rost, BS, Tf. 6f (7) later colossi: 3 R, 13; ICC, 38-42; Meissner - Rost, BS, Tf. 8;[63] Borger, ARRIM 6, 10 (8) Thompson, *Iraq* 7, fig. 3f (p. 89ff.)[64] The inscriptions relate the data in the same order, save for the absence of some information.[65]

(71) Dimensions of the former palace (É.GAL *maḫrītu*)

(1)	***30* × *10*** GAR [= *360* × *120* cubits]
(2) (3) (4) (5a)	***360* × *80* × *134* × *95* *ina*** 1.KÙŠ (cubits)[66]
(5b) (6) (7)	***360* × *95* *ina*** 1.KÙŠ
(8)	***330*** ... (fragm. passage)[67]

Blocks of limestone (NA$_4$ *pīli* GAL^{MEŠ}) on the Tebiltu river

(5b)	*4*, in (an area of) ½ GÁN ("field")

For the sequence of the building of this palace cf. Russell, *Sennacherib's Palace*, 78-93 (where also the written reports are discussed).

For most of the mentioned length measures cf. p. 16 n. 98 (and n. 103 there for the *bēru*). For the GAR, or NINDA (= *akalu*, or *nindanu*?), equivalent to 12 cubits, cf. RlA 7, 463b; Postgate, FNALD, 71; AHw, 26b; CAD A/1, 245a. Among the surface measures, the GÁN (*eqlu*, "field") corresponds to *ca.* 3,600 sq.m., while the PI (*pānu*, "seah"), which is both a capacity and a surface measure (equivalent to 6 *sūtu*), is *ca.* 2,700 sq.m. (RlA 7, 483 bottom and 487a). For the terminology (*tipku, tamlû, temmēnu*, etc.), besides the dictionaries, cf. also Turner, *Iraq* 32, 70f.

60 Of the building section of prism K 1674 ("cylinder C") no copy, edition or translation exists (cf. also BAL^2, 65). G. Smith in AD translated only the campaign narrative, while in *Senn.*, 140ff he edited the "Bellino cylinder," here indicated as (2), noting the variants given by "cylinder C" — among these, *176* (instead of *162*) and *443* (instead of *383*) for the dimensions of the platform, cf. below no. 72 (*ibid.*, p. 151). Luckenbill in his edition of Sennacherib's inscriptions totally ignored this document, taking into consideration, for the prism edition with 4 campaigns, only the fragments published in CT 26, pl. 39 and KAH 2, 121, though simply reporting in the footnotes the variants from the "Rassam," having 3 campaigns. Of these fragments, K 4492 obv. does not preserve any numeral in the passage concerning the dimensions of the platform.

The data has been drawn from E. Frahm's forthcoming work on Sennacherib (cf. n. 52 above), including a complete edition of the building sections of texts (4) and (5a), which he marked as T 10 and T 11.

61 Only some fragments of this prism edition have been published (for K 1675, identified as Smith's "cylinder D," cf. Meissner - Rost, BS, Tf. 3f). It contains 5 campaigns and is only slightly older than the texts indicated as (5b) — cf. BAL^2, 65. Also in this case I follow E. Frahm's preliminary edition.

62 Cf. also ICC, 38ff, notes. This text is probably contemporaneous with "cylinder D" (5a) *ca.* 696 BC (the military account ends with the 5th campaign).

63 Luckenbill, *Senn.*, 117-125 takes into consideration only these later texts (yet ignoring the variants recorded in ICC, 38ff and belonging to an older bull colossus). 3 R, 12f (identified as George Smith's "Bull 4") contains an annalistic account running through the 6th campaign. For a general overview of Sennacherib's bull inscriptions cf. Galter - Levine - Reade, ARRIM 4. Note that the building reports appearing on the various colossi show different arrangements.

64 Only two fragments of this prism have been published (it will be edited by E. Frahm). It included 8 campaigns (as many as the "Taylor" prism).

65 In texts (2) and (3) the reference to the size of Nineveh ("its site I enlarged") is placed at the very end of the building report. Of texts marked as (6), the fragment published in Meissner - Rost, BS, Tf. 6 differs in relating the dimensions of the new palace before the height of the terrace. Text (8) is very fragmentary, but probably it once showed many deviations (cf. n. 68 below). For the "bull inscriptions" cf. n. 63 above.

66 According to E. Frahm one unpublished fragment belonging to the class here marked as (3) reads 1 *me* 30 [...], and another one 1 *me* 1,34 (= *194*) instead of *134* (clearly, a mistake of the ancient scribe).

67 Cf. Thompson, *Iraq* 7, 89 (the number 3 *me* 30 is clearly readable).

(72) Dimensions of the platform (*temmēnu*)[68]

(1) ***60*** (1 UŠ) × ***34*** (GAR) [= *720* × *408* cubits]

(2) ***700* × *162* × *217* × *386*** *ina aslum* GAL-*ti* (great cubits), *tamlû* + *malû* D

(3) ***700* × *176* × *268* × *383*** (var. ***400***)[69] × ***386*** *ina* (1.KÙŠ) *aslum* GAL-*ti*

(4) (5a)[70] ***3'40* × *288*** *ina* 1.KÙŠ (cubits) added (*redû* D) to the measure of the former terrace (*ṣēr mešiḫti tamlî maḫrī*), a total (*napḫar*) of ***700* × *176* × *268* × *443* × *386*** *ina aslum* GAL-*ti* [For the resulting measurements cf. no. 74 below][71]

(5b) ***340* × *289*** *ina* 1.KÙŠ (cubits) added to the measure of the former terrace

(6) ***340* × *289*** *ina* 1.KÙŠ (cubits) added to the measure of the former terrace

(7) ***454*** (var. ***554***)[72] × ***289*** *ina* 1.KÙŠ ground reclaimed and added to the measure of the former palace (*ṣēr mešiḫti* É.GAL *maḫrīti*)

Inscribes a stela with his name (and buries it) below the terrace at a depth of

(2) (3) [4][73] *160 tibki* (courses of bricks) *tamlî kiribšu*

(73) Height of the terrace (*rēšu*)

(1) ***160*** (2 UŠ 40')[74] *tibki*

(2) (3) (*160*, cf. above)[75] + *20* added (*ṣēr maḫrī* + *waṣāpu* D) = ***180*** *tibki*

(4)[76] [*160*, cf. above] + [*30*] added = ***190*** *tibki*

(5a) (5b) (6) (7) (8) ***190*** *tibki* (*elû* D)

---

68 An unpublished fragment (BM 127903+), belonging to the prism class here marked as (8), apparently gives the dimensions of the former platform (?), no. (72), as 3 *me* [...] and [...] *me*? 80? cubits. It relates also, immediately after, the dimension of the new palace, no. (74), 9 [*me*?] and [...] *me* 60 cub[its], and the height of the terrace, no. (73), [1 *me* 1,]30 (= *190*?) *tibki* (information drawn from E. Frahm's forthcoming edition).

69 The text published by O. Schroeder as KAH 2, 120 — actually a conflation of two texts — has the variants *386* and *400* (line 78 and n.). E. Frahm when collating the texts read ⸢3⸣ ⸤*me*] 8⸢3⸣ (= either *383* or *386*) and [4⸣ *me* (= either *400* or *700*) respectively. Still according to his collations, the number *4* (in the 4 *me* version) is only partly readable. He suggests that the sign 4, read *limmu*, could derive from a misreading of the sign 7 (of 7 *me*), read *imin*, appearing just one line above. We may also note that the *386* version probably derives from a similar oversight (either on the part of the modern — Schroeder — or the ancient scribe), since the following dimension, again *386*, is recorded exactly one line below.

70 The reading 2 *me* 40 in the copies of "cylinder D" (5a) given by Meissner - Rost, BS, Tf. 3:20 should be corrected to 3 *me* 40 according to Frahm's collations.

71 These prisms, (4) and (5a), specify both the "addition" and the resulting "total" mensuration (I follow Frahm's preliminary new edition). Two fragments belonging to version (5a) and showing these numerals have been published in H. Winckler [review of C. Bezold, *Catalogue* ...], OLZ 1 (1898), 77 and F.E. Peiser, "Aus Rom," OLZ 7 (1904), 39.

72 ICC, 38:16 gives 4 *me* 54, while Meissner - Rost, BS, Tf. 8:6′ has 5 *me* 54.

73 "Cylinder C" (4) is fragmentary at this point.

74 Text editions have 2 UŠ 50, to be corrected to 40 according to E. Frahm's collations.

75 Texts (2), (3) and (4), after relating the burying of the stela, *160 tibki* deep in the terrace, specify that at a later time (*arkānu*) *20 tibki* were added, thus bringing the height of the terrace to *180 tibki*.
The later texts do not mention the stela at all.

(74) Dimensions of the new palace

(2) (3)	its area (*tarpašû*) enlarged (*rabû* Š); the former measure enlarged (*ṣēr mešiḫti* É.GAL *maḫrīti* + *redû* D)
(5b) (6)	***700* × *440*** *ina aslum* GAL-*ti* (great cubits);[77] its site (*šubat-sa*) enlarged (*rabû* Š)
(7)	***914* × *440*** *ina aslum* GAL-*ti*; its site enlarged

(75) Colossi

(1)	*8* (?) lions (UR.MAḪ) of ŠÁR ŠÁR ŠÁR GÍŠ+U GUN ZABAR (= *11,400* t of bronze); ***2*** pillars (*timme*) of ŠÁR 4 × GÍŠ+U GUN *pitiq siparri* (= *6,000* t of copper); ***2*** pillars (*timme*) of cedar (GIŠ.ERIN); ***er-bet*** (= *4*) sheep of the mountains (UDU *šad-di* ^{d}LAMMA);
(4) (5a)[78]	***12*** lions of bronze; ***2*** pillars of bronze; ***4*** pillars of cedar; ***10*** cows (^{f}ÁB.ZA.ZA-*ate*) of copper;[79] ***10*** cows of alabaster (^{NA_4}GIŠ.NU$_{11}$.GAL); ***12*** cows of *pitiq* GU.AN.NA (unkn. metal); ***2*** pillars of ebony (GIŠ.ESI); ***12*** sheep of the mountains of copper; ***2*** sheep of the mountains of alabaster; 1 UŠ 12 (= ***72***) sheep of the mountains and cows of white limestone (^{NA_4}*pīli peṣê*)
(5b) (6) (7)	***12*** lions (UR.MAḪ); ***12*** bulls (^{d}ALAD.^{d}LAMMA); ***22*** cows (^{f}ÁB.ZA.ZA-*ate*); (all made of bronze, *2* other bulls were silver-coated, *zaḫalû*)[80]

Dimensions of Nineveh

(2) (3)	Its site (*šubat-su*) enlarged (*rabû* Š)	
(4) (5a) (5b)	former dimens. (*šubat limētišu*)	*9,300 ina* 1.KÙŠ
	added (*ṣēr mešiḫte maḫrīti* + *redû* D)	*12,515*
	overall perimeter	*21,815 ina aslum* GAL-*ti*

---

76 "Cylinder C" (4) is rather fragmentary at this point, but, according to E. Frahm's reconstruction, possibly it specified that a stela had been buried at a depth of [*160 tib*]*ki* and that later on [*30 tib*]*ki* were added to the height of the terrace, thus obtaining *190 tibki* (this is the only number actually appearing in the text).

77 These meaurements do not result from the sum *360* × *95* (former palace, no. 71) plus *340* × *289* (meas. added to the platform, no. 72), giving *700* and *384*.
For the bull inscriptions (6), cf. Meissner - Rost, BS, Tf. 6:3′f and ICC, 62:27 (the latter having 7 [*me*]). ICC 39, n. 4 credits a version 9 *me*, to be corrected to 7 *me* according to J.M. Russell, *Sennacherib's Palace without Rival*, Diss., Univ. of Pennsylvania 1985, 24 (coll. E. Leichty).

78 Data drawn from E. Frahm's forthcoming new edition. For a fragment of "cylinder D" (5a) including information on these colossi cf. Meissner - Rost, BS, Tf. 4:5ff.

79 *Apsasāte* indicates "sphinxes" rather than cows (cf. Turner, *Iraq* 32, 75 n. 49). Most of the surviving colossi from Sennacherib's palace are stone bulls or lions, but cf. e.g. Galter - Levine - Reade, ARRIM 4, 31 for a couple of such "sphinxes."

80 Borger, ARRIM 6, 10:26′ mentions another *12* bronze lions.

(76) Dimensions of the main wall

(4) (5a) (5b)	Thickness (*kabāru* D)	*40* SIG$_4$ (bricks)
	Height (*rēšu* + *elû* D)	***180*** ***tibki***
(8)	Thickness	*40* SIG$_4$
Height of the battlement	(*naburru* + *têru* D)	*39* SIG$_4$
Height (wall)	***200*** (3 UŠ 20) *tibki* SIG$_4$	

(77) Number of gates (KÁ.GAL)[81]

(4) (5a)	***14***	in the East (^{d}UTU.È)	*8*
		in the North (*ša meḫret* IM *iltāni*)	*3*
		in the West (*ša meḫret* IM *amurri*)	*3*
(5b)	***15***	in the East	*7*
		in the North	*3*
		in the West	*5*
(8)	***18***	in the ⌜East]	[*8*]
		in the North and West	*10*

Depth of the foundations (*ušše*) of the external wall[82]

(4) (5a) (5b) (8) *45* GAR

Width of the external moat (S) *100* great cubits[83]

(78) Width of the royal road (S) ***62*** great cubits
(S) (8) ***52*** great cubits[84]

(79) Land in the plain (A.ŠÀ *tamerti elēn ali*) subdivided in plots of (*pilku* + *palāku*)

(1) (2) (3) (4) (5a)[85]	***2*** PI
(3b)	***4*** PI.TA.A.AN[86]

Length of the canal (^{ÍD}*ḫarru*)

(1)[87] (2) (3) *1½* KASKAL.GÍD (*bēru*)

81 The texts give the name of each gate: cf. the discussion in Thompson, *Iraq* 7, 92f; CT 26, p. 18f; Reade, RA 72, 50ff and, recently, B. Pongratz-Leisten, *Ina šulmi īrub* (BaF 16), Mainz 1994, 211-216.
The text STT 2, 372 also lists the names of the gates of Nineveh (cf. Reiner, JNES 26, 197f).
For text (5a) see *Cat. Suppl.*, XIX-XXV.

82 Also reported by an unpub. prism fragment (BM 127932+) related to (8) (to be edited by E. Frahm).

83 (S) indicates a stela from Nineveh, published as 1 R, 7 F (lines 17 and 21 = Luckenbill, *Senn.*, 153:18 and 22). The short text also recalls that "the site of Nineveh, my capital, I enlarged" (*ibid.*, 14f).

84 This measure appears on a prism fragment (BM 127932+) related to fragm. no. (8) (according to E. Frahm's forthcoming work) and in a duplicate of the Nineveh stela mentioned in the previous note: cf. C. Bezold, "Ein Duplicat zu I R 7, VIII, F," ZA 4 (1889), 284-290.

85 According to E. Frahm's collations, all texts have 2 PI. Published editions of cylinder "Rassam" (e.g. the copies of Evetts, ZA 3, 318:86) have 4 PI.

86 "Four PI each." Tablet K 3752, presumedly written one year later than (3), cf. Meissner-Rost, BS, Tf. 5 II 21 (collated by E. Frahm).

87 An early text, duplicate of (1), according to Smith, *First Camp.*, line 90, which used it to restore the broken passage in (1), and to E. Sollberger's collation mentioned in Reade, RA 72, 61 n. 36, has *3 bēru*. E. Frahm's recent collation, however, does not confirm this reading: he thinks the quantification is either *1* or *2 bēru*.

(80) Fields (ŠE.NUMUN) watered in the surroundings of the city[88]

[(4)] (5a)	***500***
(5b)	***1,000***
(8)	***600***

Esarhaddon

Tribute (*mandattu*) added (*redû* D) to that of Ḫazael, father of Yauta'

	(D)	(B/C)	(A)
Minas of gold	*10*	*10*	*10*
81. Choice gems ($NA_4{}^{MEŠ}$ *bi-ru-ti/e*)	***100***	***1,000***	***1,000***
Camels	*50*	*50*	*50*
82. Bags of spices (KUŠ *kun-zi* ŠIM.ḪÁ)	***100***	***1,000***	***100***

Cf. Borger, *Asarh.*, p. 54:20f[89] (D) Thompson, *Iraq* 7, p. 95 and fig. 5 (*Nin.* D) (B/C) 1 R, 46 III 22ff; Harper, *Hebr.* 4, 21:9′f (*Nin.* B and C) (A) Thompson, PEA, pl. 7:20f; Scheil, *Prisme S*, 20:18ff (*Nin.* A)

83. Distance from Bāzu, in double-hours (*bēru*)[90]

140 KASKAL.GÍD 1 R, 46 III 27 and Heidel, *Sumer* 12, 20:11 (Borger, *Asarh.*, 57 - *Nin.* B)

120 KASKAL.GÍD Thompson, PEA, pl. 8 IV 55 (Borger, *Asarh.*, 56:55 - *Nin.* A)

84. *Distanzangabe* in years (between Adad-nerari I and Esarhaddon)

580 / 586 9 UŠ 40 A.AN MU.AN.NA: KAH 2, 126 c:12 (Borger, *Asarh.*, §2 III 33)[91] 9 UŠ 46 MU.AN.NA: KAH 2, 125 col. I 24 (Borger, *Asarh.*, §3)

This measure probably appeared also on the later colossi (7), but both ICC, 42:42 and 3 R, 13 slab 4:35 are lacunal (cf. Luckenbill, *Senn.*, 124:42).

88 Texts (4) (5a) according to E. Frahm's forthcoming edition. Still according to him, text (8) has 6 *me* (Thompson, *Iraq* 7, fig. 3 C:5 has 5 *me*).

89 Text classified according to Borger, *Asarh.* Note that the dating of these prisms is very uncertain (the order seems to be *D B C A*, cf. Porter, IPP, 174). A fragment gives *10* minas of gold, *1,000* gems and *50* [camels] (Winckler, ZA 2, Tf. 2 = Borger, *Asarh.*, §72 Rs. 6) — cf. also the fragment MacGinnis, SAAB 6, no. 2 (pp. 4 and 16, the photo is illegible to me). Note also that the 1 LIM leather bags in (B) are exactly below the 1 LIM choice gems, in the following line.

90 The land of Bāzu has probably to be located on the western shore of the Persian Gulf: cf. E.A. Knauf, *Ismael*, Wiesbaden 1985, p. 55 n. 267; M. Liverani [Review of D.T. Potts, *The Arabian Gulf in Antiquity*, Oxford 1990], *Or.* 61 (1992), 160f; Ephᶜal, *Ancient Arabs*, 130-137.

91 The variant *580* may be considered a mistake, since it appears in one text only: cf. J. Boese - G. Wilhelm, "Aššur-dān I, Ninurta-apil-Ekur und die mittelassyrische Chronologie," *Wiener Zeitschrift für die Kunde des Morgenlandes* 71 (1979), 33f.

Ashurbanipal

(85) The passage containing the *topos* of abundance appearing in prisms "B" and "D" relates that "the grain grew *5* cubits tall in its furrow, the ear was *5/6* of a cubit long."[92] The later prism "A," giving a shorter description of this *aurea aetas*, also has *5* and *5/6*, but shows also the variants *4* and *2/3*.[93]

(86-87) Market prices in Assyria

In the older prisms, the above mentioned passage is followed by the list of market prices. According to the earliest version, given by prisms "B" and "D," each of the following commodities was bought for *1* shekel of silver throughout the land:[94]

10 / 12 *emāru* of barley
1 / 2 *emāru* of wine
(1) PA (= *2* *sūtu*) of oil
1 talent of wool

The later prism "K" confirms *12* for the barley, but records ***3*** *emāru* of wine.[95]

88. Elamite nobles (*qin-ni-šú* NUMUN *bit* AD-*šú*) and princes (NUN) captured:[96]

17 nobles + ***86*** princes	K 2825 l. 7 (3 R, 37 no. 5 = Streck, *Asb.*, 206);
17 + ***88***	K 1609 + K 4699 Rs. 4 (CT 35, 47 = Bauer, IWA, 46);
85	Prism "B" (and "D") VII 63 (Piepkorn, *Asb.*, 78; copies: cf. SKT 3, 47 VII 11 and 53:32); "F" III 21f (Aynard, *Prisme*, 75 and 42); "A" IV 24 (5 R, 4 and 3 R, 20:87 = Streck, *Asb.*, 34). The first group, that of the nobles, is mentioned but not quantified.

---

92 5 (1.)KÙŠ *še-am iš-qu ina* AB.SÍN-*ni-šu e-ri-ik šu-bul-tu* 5/6 (1.)KÙŠ, Piepkorn, *Asb.*, 28:29f; Thompson, *Iraq* 7, fig. 6:23. Cf. also Parpola, LASEA 2, 104f, discussing the passage as appearing in prism "B".

93 According to Streck, *Asb.*, 6:46f (cf. nn. c, f and 6 there). Of the two published cuneiform copies, 3 R, 17 gives *5* (var. *4*) and *2/3* ("prism A," see also G. Smith, *History of Ashurbanipal*, London 1871, p. 8), while 5 R, 1 (prism "Rassam") has *5* and *5/6*. Note that Piepkorn, in his edition of prism "B" (cf. previous note), listing also the variants relative to prism "A," does not confirm the variants *4* and *2/3* (nor are they noted by ARAB 2, 769).

94 Cf. Millard, *Iraq* 30, p. 111 for the correct translation. The passage is edited in Piepkorn, *Asb.*, 30:36 and nn. 9f (for the variants *10* and *1* cf. Thompson, *Iraq* 7, fig. 6 and 98:28; Bauer, IWA, Tf. 13 I 3′). Note that, according to Piepkorn, *Asb.*, 31 n. 10, the variant "2 *emāru* of wine" appears only on the fragment K 13730, which actually reads, as E. Weissert informed me, "[...] *2* homers of barley (and) *2* [... homers of wine]," the lacuna allowing the digit "3" to be written.

The ANŠE = *emāru*, "ass(load)," as a capacity measure corresponds to *ca.* 160 litres. The PA represents a special writing for 2 *sūtu*: one neo-Assyrian *sūtu* corresponds to *ca.* 8 litres (cf. RlA 7, 501ff and Postgate, FNALD, 67f).

95 The fragment ND 5524, published in Knudsen, *Iraq* 29, pl. XXIII (line 9′), has *3* homers for the wine. It joins ND 4306 (+) 4378C+ (*ibid.*, p. 65), which, in turn, indirectly join the prism "K" fragment published as Mahmud - Black, *Sumer* 44, p. 152 No. 4 (cf. E. Weissert, "On the Recently Discovered Neo-Assyrian Royal Inscriptions from Nimrud," NABU 1990, p. 103). A 5-sided prism fragment (similar to prism "T") published in Millard, *Iraq* 30, pl. XXV:19′ / p. 111:36, also has *3* homers of wine. For the barley, *12* homers are apparently recorded on the unpublished fragment A 8104 (cf. Piepkorn, *Asb.*, 31 n. 11), which recently R. Borger joined to A 8011, a prism "K" fragment (cf. Cogan - Tadmor, Or 50, 237 on the latter). No fragment of prism "C" recording the passage has survived (*contra ibid.*, p. 235 n. 15, ascribing A 8104 to prism " C"). I thank E. Weissert (Jerusalem) for his help in connection with this passage.

96 At first a group of *60* nobles (*zēr šarri*) escaping from Elam after Teumann's usurpation took refuge in Assyria (Piepkorn, *Asb.*, 60:83; Bauer, IWA, 93:4; 63:4 [79-7-8,176] and 87 Vs. last line). Afterwards, the war with Elam led to the capture of Teumann and the remaining nobles (GAL.MEŠ) — cf. A.K. Grayson in CAH 3/2, 147ff — prism "B" dates the first episode to the 6th campaign and the second to the 8th.

The variants are placed on different lines even though their chronological relation is often uncertain. The "K" collection tablets, often fragmentary, are difficult to date: cf. S.S. Ahmed, *Southern Mesopotamia in the Time of Ashurbanipal*, The Hague 1968, 168ff for some of them, and cf. Grayson, ZA 70, 245 for the prisms.

(89) Length of time during which Elam has been devastated

In prisms "F" and "A," in connection with the campaign which brought about the sack of Susa and the restoration of the statue of the goddess Nanna to Uruk (cf. no. 90 below), Ashurbanipal affirms that "(for) 60 *bēru* of ground, at the command of Ashur and Ishtar, who had sent me, I entered the land of Elam and advanced victoriously" (Streck, *Asb.*, 50:123ff; Aynard, *Prisme*, 52:65ff). Later on, the destruction of the land is reported in the following terms:

ma-lak ITI 25 $U_4^{MEŠ}$ *na-gi-e mat* ELAM.KI *ú-šaḫ-rib* MUN *ú* ZÀ.ḪI.LI.SAR *ú-sap-pi-ḫa* EDIN-*uš-šu-un*, "(for) a journey of a month (and) *25* days in the province of Elam I laid waste, scattering salt and cress upon them," (thus prism "A," 5 R, 6:77ff = Streck, *Asb.*, 56; dupl. in 3 R, 22:111ff; prism "F" shows a slightly different phraseology: cf. Aynard, *Prisme*, 83 and 56:55f).

Prism "T," three years older than prism "A," employs the same wording, but quantifies the extent of the destruction in terms of distance, instead of time:

60 KASKAL.GÍD *qaq-qa-ru qé-reb mat* ELAM.KI *ú-šaḫ-rib* MUN *ú* ZÀ.ḪI.LI.SAR *ú-sap-pi-ḫa* EDIN-[*uš*-]*šu-un* "(for) *60 bēru* of ground into Elam I laid waste," etc. etc. (Thompson, PEA, pl. 17 / p. 34 V 5ff; dupl. Bauer, IWA, Tf. 26 III 8).

Thus, in prism "T" we have some sort of "fusion" between the two quantifications. However, within the context of a general summary of the conquests, inserted before the campaign narrative, it also includes the reference to a similar journey of a month (and) *20* days: "from the upper sea to the lower sea, to which the kings my fathers had pressed forward, so did I too, (for) a journey of a month (and) *20* days (*ma-lak* ITI 20 $U_4^{MEŠ}$), amid sea and dry land" (Thompson, PEA, pl. 16 / p. 34 IV 19ff).[97] Now, since prism "T" contains only the narrative of this campaign to Elam, besides some building accounts, it is possible to consider the last mentioned passage as referring to the Elam episode.

Therefore, two possible variants may be taken into consideration: (a) between a month and *25* days and *60 bēru*,[98] (b) between a month and *25* days and a month and *20* days.

---

97 Cf. also the fragments K 4521 (Bauer, IWA, Tf. 41 / p. 58:5) and K 7596 (CT 35, 22:13 = Bauer, IWA, 96), which reads *a-lak* ITI 20 KÁM (the sign for "day" after the number *20* has been omitted, cf. *ibid.*, n. 7).

98 Note that a text of Esarhaddon (Borger, *Asarh.*, §76; mentioned above, p. 17 n. 104) gives 2 *bērus* as the usual day's march of the Assyrian army (cf. also RlA 7, 467a) — however, it also suggests that the army was advancing over very difficult terrain (cf. e.g. Vs. 17f, Rs. 4, 5 and 14).

90. Number of years which the goddess Nanna has been in Elam[99]

1,635 1 *lim* 6 *me* 30 A.AN 5 MU.AN.NA.MEŠ; Prism "A" VI 107 (5 R, 6 = Streck, *Asb.*, 58; cf. 3 R, 23:9); prism "T" V 9 (Thompson, PEA, pl. 17 / p. 35); K 2664 III 12 (Bauer, IWA, Tf. 26 / p. 34); K 2631 etc. obv. 16 (3 R, 38 no. 1 = Streck, *Asb.*, 180; here 2 × *nēr* 7 UŠ 15 MU.AN.NA.MEŠ);

1,630 [1] *lim* 6 MEŠ 30 A.AN MU.AN.NA.[MEŠ] in the prism "F" fragment Thompson, *Iraq* 7, fig. 17 (cf. p. 106) no. 32:5; prism "F" in the Louvre gives 1 *lim* 6 *me* 30 A.[AN ...], cf. Aynard, *Prisme*, 83 and 58 V 72; another prism "F" fragment gives the variant ***1,830*** (mentioned by Freedman, *St. Louis*, 135 no. 42 II 17′);

1,535 1 *lim* 5 *me* 30 A.AN 5 MU.AN.NA.MEŠ; K 3101[a] + (3 R, 35: no. 1, l. 22 = Streck, *Asb.*, 220); K 4455 (cf. Bauer, IWA, 61: n. 6); K 2628:4 (Bauer, IWA, Tf. 22 / p. 43); cf. also the fragments K 2638:11 (Bauer, IWA, Tf. 24 / p. 62) giving 1 [*li*]*m* 5 *me* 30 A.AN [...] and YOS 9, 77 (last line), giving ⌞1⌟ *lim* 5 *me* 30 A.AN [...]

---

99 Both Streck (p. 58 n. 4 and 221 n. 2) and Nassouhi (MAOG 3, 34f) thought that *1,535* was a scribal error, but further texts subsequently confirmed this variant (see also Streck, CCCXXXVII n. 1 and CCCXLI n. 2).
These figures are obviously exaggerated (cf. Brinkman, *Prelude to Empire*, 103).

2. OVERVIEW OF THE VARIATIONS[100]

The variations refer to:

1-3 Predecessors of Tiglath-pileser I
4-7 Tiglath-pileser I
8-20 Ashurnasirpal II
21-31 Shalmaneser III
32-37 between Shalmaneser III and Sargon
38-64 Sargon
65-80 Sennacherib
81-84 Esarhaddon
85-90 Ashurbanipal

It is inadvisable to draw conclusions from the fact that there are more variations in certain reigns than others, since this could be due to reasons such as:

- number and length of the inscriptions of each reign;
- number (frequency) of quantifications;
- recurrence of the information.

Thus, Shalmaneser III's and Sargon's inscriptions exhibit a lot of variations, but they also form a consistent bulk of texts, crowded with numbers and frequently relating the same events (and quantifications). On the contrary, the inscriptions of Tiglath-pileser III only in rare cases overlap. On the other hand, certain inscriptions show many variations, for instance the "Kurḫ monolith" of Ashurnasirpal II seven times gives numbers differing from those of the "annals" (no. 9-14), while the various copies of these same "annals" contain five variants in the numbers, four of which appear within the concluding section (nos. 8, 15-18; for nos. 19-20 cf. n. 13). Also several numbers reported by Sargon's "letter to the god" show many differences in regard to his Khorsabad annals (nos. 44-45 and 47-53) and to his *display inscriptions* (nos. 44-46 and 49).

A) *Type of variation*

Whenever it is possible to establish a chronological succession between the variants, these are separated by the symbol > and are followed by a **+** if the value of the digit has increased and by a **-** if it has decreased. Other symbols used: **a** digit added; **s** digit suppressed; **<+>** the exponent increases (e.g. 1 *me* > 1 *lim*); **<->** decreases.

A digit is suppressed or added:	21.	*4 > 14*	a
	22.	*4,600 > 14,600*	a
	23.	*3,400 > 3,000*	s
	27.	*16,000 > 16,020*	a
	45.	[1,20+4 =] *84 > 24*	s
	48.	*146 > 140*	s
	49.	*6,110 > 6,170* [= 6.1.1,10]	a
	53.	*167 > 160*	s

100 The doubtful cases are not taken into account: when the variation may refer to two different events (e.g. nos. 33ff — but nos. 5 and 6 are included); the variants relative to minas or shekels in the case of booty expressed in talents (nos. 40f and 63); those relative to building accounts (cf. n. 59 above), and to the Euphrates crossings (no. 28). No. 62 is ignored since it implies several variations on the level of digits — most likely due to simple oversights — which would result in a too complicated classification. Also ignored are the passages *14,000 > 20,500 / 29,000* (no. 25); *86 > 89* (no. 26); *88 > 75* (no. 65) and *820 > 420* (no. 66), since in each of these cases an intermediate version may easily explain the issue.

59. *700 / 710*
67. *5,230 / 5,233*
84. *580 / 586*
90. *1,630 / 1,635*
cf. also nos. 40f

A digit varies:

2.	*41 / 51*	1	50.	*12 + 380 > 692*	+ 3
4.	*60 / 120*	1	51.	*525 > 92[5]*	+ 4
8.	*600 / 800*	2	54.	*54 > 55*	+ 1
9.	*500 > 700*	+ 2	56.	*8,609 > 4,609*	- 4
10.	*40 > 50*	+ 1	57.	*5 / 6*	1
11.	*40 > 50*	+ 1	59.	*610 / 710*	1
12.	*200 > 100*	- 1	61.	*200 / 300*	1
14.	*2,000 > 3,000*	+ 1	63.	*154 > 164*	+ 1
15.	*20 / 30*	1	64.	*1,604 > 1,804*	2
17.	*2,400 / 2,500*	1	65.	*88 > 89*	+ 1
18.	*580 / 780*	2	66.	*820 > 620*	- 2
26.	*86 > 89*	+ 3		*620 > 420*	- 2
	89 / 99	1	69.	*800,100 > 800,600*	+ 5
32.	*255 > 256*	+ 1		*600,600 / 800,600*	2
36.	*588 / 589*	1	70.	*35 > 33*	- 2
38.	*27,280 > 27,290*	+ 1	83.	*140 > 120*	- 2
42.	*500 > 200*	- 3	88.	*85 / 86*	1
43.	*12 > 22*	+ 1		*85 / 88*	3
45.	*24 / 23*	- 1		*86 / 88*	2
46.	*260 > 250*	- 1	90.	*1,535 / 1,635*	1
47.	*6 > 5*	- 1	add also p. 133 nn. 83 and 86		

A digit changes exponent:	13.	*1,000 > 100*	<->
	25.	*25,000 > 20,500*	<->
	60.	*10,000 > 1,000*	<->
	81.	*100 > 1,000*	<+>
	82.	*100 / 1,000*	

A digit varies and changes exponent:	25.	*20,500 > 29,000*	<+> + 4
	39.	*200 > 50*	<+> - 3
	49.	*6,170 / 20,170*	4
	68.	*80,050 > 80,100*	<+> - 4 [50×2=100]
		80,100 / 200,100	<+> - 6 [= $\frac{1}{4}$]

Two digits vary:	25.	*14,000 > 25,000*	+ 1 + 1
	65.	*89 > 75*	- 1 - 4

Two digits are added:	23.	*3,000 > 13,500*	s s

A digit varies and one is added or suppressed:

23.	*3,400 > 13,500*	a + 1	52.	*1,235 / 1,285* [= 1.2.1,20+5]	1
44.	*12 > 3*	a + 1	59.	*600 / 710*	1
45.	[1,20+4 =] *84 > 23*	s - 1	90.	*1,535 / 1,630*	1

A digit varies and one changes exponent:	52.	*1,235 > 100,225*	<++> - 1
One changes exponent and one is suppressed:	(5)	*20,000 > 12,000*	s <+>
	(6)	*23 > 30*	s <+>
	52.	[1.2.1,20+5 =] *1,285 > 100,225*	<++> s
One varies and changes exponent and one is suppressed:	49.	*6,110 > 20,170*	<+> - 4 s
	58.	*2,080 / 2,500*	
Two digits vary and one is suppressed:	24.	*17,500 / 22,000*	+ 1 - 5 s

B) *Relation between the variants*

More than double:

13.	*1,000 > 100*	<->	45.	*84 > 24 > 23*	s -
21.	*4 > 14*	<+>	49.	*6,110 > 6,170 > 20,170*	- <+> s
22.	*4,600 > 14,600*	<+>	60.	*10,000 > 1,000*	<->
23.	*3,400 > 3,000 > 13,500*	a a	68.	*80,050 > 80,100 / 200,100*	<+> - <+> -
25.	*14,000 > 29,000*	+ +	81.	*100 > 1,000*	<+>
39.	*200 > 50*	<+> -	82.	*100 / 1,000*	
42.	*500 > 200*	-			
44.	*12 > 3*	s +			

More than tenfold:	52.	*1,235 / 1,285 > 100,225*	<++> -

C) *Relation between categories of texts*

For a statistical investigation, one needs to keep in mind on one hand that the classification of the inscriptions into well-defined categories presents several problems, and on the other that there should be a place for such factors as the number of inscriptions of each kind (and for each reign), their length, etc. In any event, the most evident cases[101] are:

---

101 Certain entries appear twice since a variation may concern similar and different texts simultaneously. The uncertain cases are excluded: variations **61** and **62** and those of the reign of Ashurbanipal have been ignored since the texts are of uncertain classification. Variants **40-41** have been ignored since they represent only minor alterations.

Variations between copies of the same text:
2, **4**, **8**, **15**, (**16**), **17**, **18**, **36**, **38**, **57**, **58-59**, **63-64**, **67-69** [total 19, if those mentioned at p. 133 nn. 83 and 86 are included].

Variations between similar texts (e.g. subsequent editions of the annals):
21, **22**, **23**, **24**, **25**, **26**, **27**, **28**, (**31**), **42**, **43**, **56**, **61**, **65-66**, **67-69**, **70**, **81-82**, **83**, **84** [total 23].

Variations between texts belonging to a different category:
9, **10**, **11**, **12-14**, **23** (A < S), **25** (A < S), **32** (A < L), **38** (A < S), **39** (A > S), **43** (A < S), **44** (L > A), **45** (L > A > S), **46** (L+A > S), **47** (L > A), **48** (L > A), **49** (L < A < S), **50-52** (all L < A), **53** (L > A), **54** (L < A), **60** (A > S), **63-64** (A < S) [total 26].

A = annalistic text; L = "letter to the god"; S = *Summary* (or *display*) inscr.
> = higher than (the number decreases); < = lesser than (augments).

Variations **10-14** refer to the "annals" and to the "Kurḫ monolith" of Ashurnasirpal II, containing the text of a single campaign, while, as it appears, most of the other variations concern the "letter to the God" and the "annals." These, together with other variations between texts of different types, are common in Sargon's reign, while those between similar types are common in Sennacherib's reign — this is not surprising since the former left us examples of inscriptions of many different categories, while the latter left several series of annalistic texts but no "letters to the God" and only a small number of *summary* texts.

3. ORIGIN OF THE VARIATIONS

"Oversight" — An explanation for a great part of these variations can be easily given in terms of copying errors due to carelessness, and this is probably always true when the variation concerns a digit of low order varying by one unit: cf. no. **2**, **17**, **26**, **32**, **36**, **38**, **46**, **49** and **52**,[102] **54-55**, **59** (*610 / 710*), **62-70**, **83** and **88**. Variations no. **19-20** are certainly due to oversights or copying errors.

An oversight can be the origin of the variation also when an exponent changes (without changes in the digits): cf. exp. no. **13** (*1,000 > 100*), **25** (*25,000 > 20,500*), **60** (*10,000 > 1,000*), **81** and **82** (cf. n. 89). For other possible examples of oversights or scribal errors cf. no. **18** (*580 / 780*), **28** (cf. nn. 26f), **45** ([1,20+4 =] *84 > 24 > 23*) and **90**. Some copying errors explain also certain variants within the building account of Sennacherib's palace at Nineveh (cf. e.g. nn. 66 and 69 above).

In some cases, it is possible that two similar numbers had been confused during the copying of the text, see for instance no. **9** (cf. n. 7) and no. **58** (*2,080* — note the *90,580* above). Cf. also nn. 12, 16 and 20 above and 105 below and nn. 83 and 86 at p. 133.

"Updating" — Some quantifications refer to situations that meanwhile have (or could have) changed: it is the case of the "palace without rival" of Sennacherib (no. **71-80**) or of other building works (no. **1** and **3**). Another

---

102 Concerning the passage *6,110 > 6,170*, note that *70* is written 1,10. For *1,235 / 1,285* cf. here below n. 109.

possible case of "updating" concerns the prices reported by some inscriptions of the reign of Ashurbanipal (cf. no. **86-87**).

Certain quantifications have, on the other hand, the characteristic of "totals," and for this reason their variation may also be considered an "updating": cf. no. **6** (?), **33-35** (cf. n. 31 above), **39** (?), **42** (?) (if one supposes that these last two "totals" have been "updated" to a lower figure, respectively *200 > 50* and *500 > 200*, chariots and cavalry annexed to the Assyrian army).

Nos. **29-30** certainly represent some "totals" referring to the whole reign, and therefore their variation is not surprising. But if we look at the numerals, it seems difficult to believe that the variation is due to an "upgrading," but rather due to an oversight. The same thing could be said for no. **22**, a quantification which refers, in its higher version, to a wider context (cf. note 17 there), but still represented by singularly similar digits.

"Rounding" Possible cases of "rounding" are represented by no. **6**, **23** (*3,400 > 3,000*), **48** (*146 > 140*), **53** (*167 > 160*), **58** (*2,080 / 2,500*), **59** (*610 / 700 / 710*) and **84** (*580 / 586*). Cf. also the variations **40-41**, where the minas have been left out, and above, p. 17 n. 104.

"Specification" — Cf. especially case **27** (*16,000 > 16,020*), but also no. **67** (*5,230 > 5,233*), where digits of low order are added. In this case, a number already formed by several digits is (still further) "specified."

"Regularization" — For one example of "regularization" of a number cf. no. **69**,[103] *800,100 > 800,600 / 600,600* (for the "regularization" of a series of quantifications see p. 127 n. 57 and variations no. **12-14**).

"Inflation" — In some cases the suspicion arises that the variation could be intentional, made with the aim of "improving" the outcome of a campaign, etc. (cf. next paragraph). Actually, as noted also by Millard, *Fs Tadmor*, 220, there are only a few examples of numbers being augmented greatly over the course of time. The most evident cases are:

- no. **22** (prisoners): *4,600 > 14,600* (cf. also **21**);
- no. **23** (soldiers killed): *3,400 > 3,000 > 13,500*;
- no. **24** (soldiers deported): *17,500 > 22,000*;
- no. **25** (soldiers killed): *14,000 > 25,000 > 20,500 / 29,000*;
- nos. **49-53** (deportees and booty; cf. n. 42 end):
  6,110 > 6,170 > 20,170;
  12 + 380 > 692;
  525 > 92[*0*];
  1,235 / 1,285 > 100,225.

One should consider also what is quantified, soldiers killed and deportees, both having a particular ideological significance (cf. next section for discussion).

---

103 In this case the resulting "corrected" number is formed by a repetition (6 *me lim* 6 *me*). In a series of other cases it can be observed that a varying digit is identical with another one contained in the same number (the varying or added digits are in bold type): no. 24 (***22**,000*), 26 (***99***), 51 (*100,2**25***), 67 (*5,23**3***), 70 (*3**3***), 88 (*8**8***). The opposite happens in 29 (*2,002 > 2,001*), 30 (*5,542 > 5,242 > 5,241*), 32 (*255 > 256*), 65 (*88 > 89*). Cf. also no. 68 (*80,050 > 80,100 > 200,100*): the varying numbers are not the "rounding" of the former version, yet they still have digits that connect to each other in some manner (as *2* and *1* do).

In some cases the varying digit surpasses the preceding one of a unit: 5 (***12**,000*), 26 (*8**9***), 32 ([2]*5**6***), 45 (*2**3***), 65 (*8**9***), 88 (*17 + 8**8***; cf. there also *17 + 8**6***). Cf. also 25 (*14,000 > 25,000*; + 1 and + 1). This could show an unintentional "accounting" as the origin of the new version.

"Reduction" — This is the opposite case: there are quantifications that decrease from one inscription to another.[104] The most noteworthy example comes from the reign of Sennacherib, nos. **65-66** (cities and forts: *88 > 89 > 75* and *820 > 620 > 420* respectively).

And, besides these: no. **5** (*20,000 > 12,000* — doubtful case, cf. n. 5 above), **12** (*200 + 300 > 100*), **13** (*1,000 > 100*), **39** (*200 > 50*), **42** (*500 > 200*), **44-45** (*12 > 3*; *84 > 24 > 23*), **53** (*167 > 160*; booty), **56** (?) (*8,609* (?) *> 4,609*), **60** (*10,000 > 1,000*) and **83** (*140 > 120*).

4. "INFLATION" OF NUMBERS: GENERAL REMARKS

As we saw, examples of quantifications being augmented in later inscriptions are not wanting; however, in most of the cases they can be explained as oversights or copying errors, and are equalled by a similar number of diminishings. The most noteworthy cases of "inflation" have just been mentioned. They refer to three specific situations, that is:

(1) the transition between the earliest annalistic tablet of Shalmaneser III (Mahmud - Black, *Sumer* 44) and edition "A" of his annals (nos. **21-22**);

(2) the transition between edition "A" or "B" and edition "C" of the annals of Shalmaneser III (nos. **23-25**, cf. also no. **26**: *86 > 89 / 99* cities);

(3) the booty taken at Muṣaṣir at the conclusion of the 8th campaign of Sargon (nos. **49-53**).

The "inflations" under discussion must therefore be interpreted independently and according to contingent factors (narrative contexts, etc.).

The first group (1) includes two very similar alterations, *4 > 14* and *4,600 > 14,600*. In both cases a digit "10" has been prefixed to a "4," appearing at the beginning of a line. The possibility that this derives from a scribal mistake is therefore quite likely — e.g., assuming that the text was dictated, one might think of a misunderstanding.[105] However, in the second case, where the numeral passes from *4,600* to *14,600*, the later text includes also the quantification of *2,800* warriors killed on the same occasion (earlier given as "numerous"), a numeral showing some affinities with the *4,600* of the prisoners.[106] Therefore, the possibility that the number of killed had been deliberately invented cannot be excluded, and this, in turn, might let us infer that the "10" had been twice added to the "4" according to a similar licentious line of action by a not so imaginative scribe.

As for (2), variation no. **25** is the more prominent case and, obviously, it has been already noted.[107] It is difficult not to think of an intentional act of

---

104 Obviously, one must take into account the fact that a lacuna (a break) could hide a number higher than it actually appears (as suggested by Th. Jacobsen, *The Sumerian King List*, Chicago 1939, 17f), cf. n. 43 above (and nos. 32 and 58).

105 One should hypothesize a misunderstanding between *erba* (4) and *erbēšer* (14; cf. GAG, §69), which is rather unlikely. Apart from this case, no other oversights of this kind occur (not even between "10," *ešer*, and "20," *ešrā*). On the possibility that the texts were dictated, cf. the discussion in Grayson, *Fs Tadmor*, 265f (he suggests that the texts were "auto-dictated"). Cf. also n. 69 on p. 60.

106 *4,600 > 2,800*, - 2 and + 2 (a sort of "compensation"); note also that all digits are even.

107 Cf. e.g. Olmstead, *Historiography*, 7 and 22: he thought that this fact (the variation) shows the higher reliability of the monolith inscription (edition "A") in respect to the other and later inscriptions of the reign (an indication held valid, as a general principle, for all the reigns).

corruption of the data, unless one supposes it represents a "total" or an "upgrading," including the enemies killed by Shalmaneser's army during the three subsequent camapigns against the Syrian coalition.[108] However, this is very unlikely to be the case, since the numerals show a clear affinity: cf. e.g. *14,000* > *25,000*, + 1 for both digits. The discussion will be resumed again below, §12 (see esp. p. 94 for no. **24** and n. 202 there for no. **23**).

In the case of the booty of Muṣaṣir (3) the number relative to the sheep, passing from *1,235* to *100,225*, is the most showy inflation of all the *corpus*. The passage had been carried out through the addition of a "hundred" on the level of thousands: thus, from 1 *lim* 2 *me* 30+5 we get 1 *me lim* 20+5 (leaving out for the moment the variation *30 / 20*, probably due to an oversight).[109] According to Ungnad, ZAW 59, the lack of familiarity of the Assyrian scribes with the decimal system, accustomed as they were to the Babylonian sexagesimal system, led them to read 1 *lim* 2 *me* 30+5 as "*1,002* hundred *35*," or *100,235*, which they transcribed 1 *me lim* 2 *me* 25. All can be explained, therefore, not as an intentional alteration or negligence, but as an attempt at correctly representing the thousands, avoiding any ambiguity. Ungnad from this inferred (*loc. cit.*, p. 201) that in some other cases too, even in inscriptions of other reigns (notably Sennacherib), numbers of the kind *200,150*, *800,100* etc. should be understood as "2,150" or "8,100" and so on.[110] However, as Borger pointed out (BAL², 136), the hypothesis of the lack of familiarity is very unlikely: it must have happened, sometimes, that those scribes had to write high and/or complicated numbers, considering that they were "professionals" and that they, with all probability, had to deal with many kinds of texts and subjects. How would they manage with an economic document if they were not able to resolve this kind of ambiguity? In any case, as a last resort, they could have copied the number "same as sample," without changes. The very list of booty from Muṣaṣir shows a variation (no. 49: prisoners, from *6,110* to *6,170* to *20,170*) not easy to explain in terms of oversight or lack of familiarity with numbers (cf. also n. 42 [end] above). The feasibility of this quantification will be discussed below, pp. 112ff, while the editorial habits asking for the use of such "exact" numbers will be examined at pp. 86ff and pp. 171ff.

Thus, to sum up, the "inflating" of numbers does not seem to have taken place during the phase of the re-edition of pre-existing inscriptions, except in the three instances just referred to. Wherever we may think we are in the presence of numbers exaggerated with the aim of giving importance to the deeds there narrated, we should consider that this operation has been undertaken during the phase of the original editing of the texts. Besides, the Assyrian kings certainly had not a little need of magnifying the deeds of the

108 But his inscriptions quantify the "victory" against the Syrian coalition in one other instance: cf. n. 205 below.

109 Incidentally, note how the "source" for the number *100,225* was not *1,235* (as implied by Ungnad) but rather *1,285* (the inscription repeats some data, cf. above, n. 42), written 1 *lim* 2 *me* 1,20+5, arising out of 1 *lim* 2 *me* 30+5 (*1,235*), perhaps becuse of an oversight (a vertical wedge replaces the first of the three *Winkelhaken*).

110 Olmstead, by contrast, thought that the "true" number hidden behind *200,150* was *150* only: cf. *Historiography*, 8; *History*, 580.

first years of their reigns.[111] The first inscriptions did already show a "correct" version of the events, and further "adjustments," usually, were not needed, at least for the quantifications.

Many variations are actually pertinent to (contemporary) copies of the same text, or even belonged to a single document.[112] Considering that in most of the cases the variation is minimal (it remains in the same order of magnitude) and is almost certainly due to oversights, and that the very frequency of the variations is rather high, we can maintain that, generally speaking, the Assyrian compilers paid much more attention to the rough value of the number, to its magnitude, than to the precision or to the exactness of the digits (this is less true for the Sargonid period). It was not important if the enemies were *580* or *780*, but that they were "so" large a quantity. Also, the function of a "round" or "exact" number is always respected: note how rarely the numbers pass from one "category" to another because of a variation: never does an "exact" number become "round" and only once *vice versa* (no. **27**).

5. CHECKING THE NUMBERS

The suspicion that the numbers presented by the Assyrian royal inscriptions (especially the later ones) are in many cases nothing more than inventions or gross exaggerations is quite common and is often more or less explicitly expressed in several historiographic works, especially by authors outside the circle of specialists in Assyrian history. Now, it is clear that an attitude of generic skepticism regarding these numbers or accepting them *in toto*, because they represent objective "facts," or for other reasons is not productive. The best way to proceed is, I think, to evaluate each quantification individually, before taking into consideration the "general" aspects. I would suggest that this could be done basically in four ways, or in four phases, that have a well defined hierarchic order:

1. checking the data against other sources, or possibly with "direct" proofs (cf. below, §6);
2. evaluation of the probability of the number, in consideration of the quantified object, of the historical context, etc. with the aim of establishing if the number is "possible" (below, §7);
3. evaluation of the inconsistencies within the *corpus* (the inscriptions taken as a whole); "global" evaluation of the quantifications present in the inscription under examination or in the other contemporary inscriptions (below, §8);
4. evaluation of the number in itself ((epi)graphical factors, etc.; cf. §9).

---

111 For the tendency to concentrate in the first year or in the beginning of the reign as many important undertakings as possible, cf. Tadmor, ARINH, 16f and M. Cogan, "Omens and Ideology in the Babylonian Inscription of Esarhaddon," in HHI, p. 87 and n.

112 This is the case of Sargon's "letter to the God" in which two versions of the booty of sheep are given (cf. no. **52**). Cf. also p. 69 above.

6. DIRECT CHECKING

The possibilities of checking the data are scarce, inasmuch as valid alternative sources are not available either to the "modern" reader or to the ancient one. The absence of a true and proper "counter-information" caused the possible Assyrian readers to be unable to reject the numbers given by the inscriptions on the basis of facts, cipher against cipher. I said possible readers, because in a society formed mostly by illiterate people the fruition of written texts could not be so easy. Some of the royal inscriptions under examination were not even made public, since most of the longer texts, especially if carrying building descriptions, were buried as "foundation deposits."[113] The common people, a great number of whom did participate directly in the events narrated by the inscriptions, then, simply did not have access to the inscriptions,[114] the contents of which they could know only orally (and "second hand").[115] The "audience" was, thus, the very Assyrian court, and the circle of palace officers,[116] who were responsible for the production of the texts.[117] They could certainly, in some manner, have "first hand" news and information, but it is not easy to know if and how they shared the "official" view, since no "unofficial," or "semi-official," historical literature existed, as far we can see.

Anyway, as we just saw, the collocation of most of the inscriptions make it clear that they were (theoretically) addressed not to contemporaries, but, according to the assertions of some of them, to the gods and (especially) to

113 "... of all the stone, metal and clay tablets with building inscriptions from the Isin-Larsa to the Achaemenid period, most were probably, and some were certainly, building deposits" (Ellis, FD, 104, for the prisms and cylinders, cf. 108ff). Cf. also the remarks of Oppenheim, *Propaganda and Communication*, 118f.

114 Note that the access to the royal palace was strictly limited — most of the court *literati* and even the crown prince, even if living at a short distance from the palace, had to communicate with the king through written letters or dispatches (cf. the remarks of Parpola, LASEA 2, XX and Tadmor, AAC, 206).

115 The possibility that some royal inscriptions were read publicly has been suggested by Oppenheim, JNES 19, 143 and *Propaganda and Communication*, 124f (cf. Grayson, Or 49, 157f and 167; Porter, IPP, 112ff), but is based on indirect elements and, in any case, would apply to the so-called "letters to the God." As regards the texts actually displayed to the public (such as slabs and stelae), besides the fact that literacy was not very common in ancient Mesopotamia, if the king Ashurbanipal boasted he was able to read, note that most stelae were set up in bordering areas, where Akkadian was not understood (see the conclusions of Machinist, JAOS 103, 732). The hypothesis that they could have been diffused *via* Aramaic (cf. *ibid.*, 733f) is quite unlikely — if the Assyrian monarchs wanted this, they would have made bilingual texts, as the later Persian kings did. Only the Tell-Feḫeriye stela is bilingual, and it is not a royal text, nor did it succeed in conveying this practice to the royal chancellery. Cf. also n. 118 here below and n. 15 on p. 119).

116 Despite the many efforts, a contemporary "audience" for these inscriptions has still to be found. Thus, B. Nevling Porter, in her recent book on Esarhaddon (cf. IPP, 105ff), shows that one text only (*Nin. A*) *could* have been intended for dual use — presentation to a living audience and burying as a deposit — and hence concludes that the "contemporary audience" was actually represented only by the writers of the inscriptions themselves ("the scribes who composed the texts, and their associates at court," p. 109). She also suggests that the prisms found in the Ashur temple at Ashur, probably placed there as presentations to the god, served to indoctrinate the temple personnel (p. 112). Indeed, this conclusion is rather perplexing — obviously, such temple "exhibitions" confirm that the texts were mainly intended for the gods (other examples of "texts exhibitions" are attested, cf. e.g. Ellis, FD, 113).

117 In this sense, the concept of "self-indoctrination" is often recalled: cf. e.g. Liverani, *Ideology*, 302 or Fales, RAI 25, 433f.

posterity.[118] Besides the assyriologists, this category is represented mainly by later kings who might have decided to carry on newer works on the buildings.[119] In this case, the foundation boxes were thoroughly searched for, opened, and their contents read,[120] and sometimes reburied in the newer foundations.[121] Neither later kings nor gods, however, had many ways to check the authenticity of the facts therein narrated — as it is known, in ancient Mesopotamia the gods were not considered omniscient (except Ea) or omnipresent.[122] As part of the "posterity," if we are willing to check the digits presented by the inscriptions, since the possibilities of "internal" counterchecking (other Assyrian royal inscriptions) are quite small, we can fundamentally follow three ways:

1. Checking against other documents of Assyrian provenance, possibly not so much ideologically conditioned. These could be:

a. Letters and reports sent by the provincial governors and by other officers to their kings. Generally speaking, these documents bear information that is difficult to compare with that contained in the royal inscriptions.[123] State letters are interested mainly in the administration of the empire, and those concerning war matters or "foreign" policy contain mostly information on minor operations, such as movements of enemy contingents or frontier clashes, echoing from afar the wars and deportations narrated by the royal inscriptions. The main campaigns, subject of the latter, were usually led or directly followed by the king, who therefore did not need to receive any news. The reports concerning the tribute and booty, and especially the prisoners, and their transfer to Assyria or to other new destinations, usually record limited groups of people (e.g. the prisoners in the charge of a particular officer).[124] Only rarely do they mention quantities of prisoners that could be compared to those of the royal inscriptions.[125]

---

118 Ellis, FD, 167. Renger (RlA 6, 74) maintains that the main royal inscriptions are basically commemorative texts, intended to record the deeds of the kings for posterity; cf. also his *Introd.* [26f], where (p. [39ff]) he remarks that many stelae were as well addressed to (and therefore intended for) posterity. After all, most inscriptions, either buried as deposits or publicly exposed, carried similar texts (as noted e.g. in *Introd.* [13, 23, 57 and n. 111]) thus supporting the view that the addressee was the same. Actually, there are no detailed studies of the relation between text and external features and setting of the documents (cf. the remarks of Oppenheim, *Propaganda and Communication*, 114 and 139 no. 9 and Renger, *Introd.* [37ff] on the neo-Assyrian stelae; cf. also, in general, RlA 6, 74).

119 Russell, *Sennacherib's Palace*, 223-40, lists twelve categories of actual audiences for the reliefs (mostly formed by foreigners — messengers, prisoners, etc.), but, with regard to the inscriptions, he concludes that the *intended* audience included only future kings (p. 240, cf. also 254f).

120 This is confirmed by the fact that Assyrian inscriptions often describe the construction history of a building in great detail (cf. p. 18 for the *Distanzangaben*, and cf., in general, Ellis, FD, *passim*, and Porter, IPP, p. 90 n. 207).

121 Many examples are mentioned in Ellis, FD, 96ff; cf. also p. 111: a barrel-cylinder of Ashurbanipal reburied by Nebuchadnezzar.

122 Yet Mayer, MDOG 112, 14ff argues that, since Sargon's "letter to the God" is addressed to the national deity, we can expect from this document a high degree of historical accuracy. Nonetheless, its geography has created a good deal of trouble — cf. below, p. 132 n. 76 (end).

123 Cf. the remarks of Parpola, ARINH, 123-26; Brinkman, *Prelude to Empire*, 113f; Fales, SAAB 4, 23; Zimansky, *Ecology and Empire*, 5, and Oded, *Deportations*, 8ff (evidence on the deportations in administrative documents). The letters are unevenly distributed through the reigns (for instance, there are no letters from the reign of Sennacherib), and are most of the time difficult to date, and therefore their historical context can hardly be established with certainty.

124 For the displacement, the deportees had to be divided into not excessively large groups: epistolary and administrative texts usually make reference to such small groups, while the royal inscriptions give the total of the deportees of a certain city or area. For the implementation of deportations cf. Oded, *Deportations*, 36-39 (in particular, 37f for the epistolary evidence).

Some administrative documents record lists of prisoners, subdivided into classes according to age and sex, but the quantities are usually very small: cf. e.g. ADD 783 (SAA 11, 173), mentioning *1 less than 30*

b. Some administrative documents have been classified as booty or tribute lists, and, it has been hypothesized that they could have served as sources for the royal inscriptions.[126] Most of them actually record the income of *tribute* (*madattu*),[127] but they cannot be easily linked to any large-scale military episode or to any wide area, such as those mentioned by the royal inscriptions.[128] Even in the cases in which the quantities are thoroughly recorded, they are usually quite limited.[129] The Kalah administrative tablets published by Parker, *Iraq* 23, are the most interesting in this respect — yet, sometimes their purpose is far from clear, not least because of the frequent gaps. Some texts deal with persons and may record the distribution of prisoners of war,[130] while others record the income of tribute (*madattu*), audience gifts (*nāmurtu*) and/or *iškaru*-tax[131] — anyway it is again very difficult to link them to any specific military episode or area.[132]

Similarly, the middle-Assyrian archive of Ninurta-tukul-Ashur records the audience gifts (*nāmurtu*) given to the king (?), consisting of very small amounts of commodities (mostly sheep). The only exception is a text[133] mentioning a total of *1,714* sheep, out of which *800* had been received "from the provinces" — note that this text represents a general account covering about half a year!

---

people (rev. 5), ADD 882 (SAA 11, 174), with a total of *20* people (rev. 3), and ADD 895 (SAA 11, 170), mentioning a total of *84* deportees (rev. 10). ADD 1099 (SAA 11, 167) shows somewhat higher figures, recording a total of *977* deportees from Cilicia, of which *334* are classified as "able-bodied men" (cf. ADD, vol. 3, pp. 523f).

125 E.g., ND 2634 speaks about the food supplies for *6,000* prisoners (cf. Saggs, *Iraq* 36, 200ff); ABL 304 (SAA 1, 11 or Pfeiffer, SLA, 100) mentions *1,119* able-bodied men that, together with their families, added up to *5,000* souls. Cf. also Pfeiffer, SLA, no. 40-43, 99, 101 and 103 (the last three re-edited as SAA 1, 34, 240 and 128) or SAA 5, 88.

126 See, e.g., Grayson, Or 49, n. 120 or Renger, *Introd.*, n. 120.

127 Most of passages mentioning *madattu* are quoted in Postgate, TCAE, 111ff. Texts recording *booty* are very rare and, if they exist, they are very fragmentary (cf. also Zaccagnini, *Opus* 3, 238 — for the distinction between booty and tribute see above p. 11). One such specimen might be represented by ADD 910 (SAA 11, 176), a fragmentary tablet mentioning some groups of deportees (*30* in 14, and *50* in rev. 1) and some modest quantities of precious metals. The use of "booty lists," however, may be reflected by such detailed listings as those found in Sargon's "letter to the God" (cf. below, p. 128).

128 Note that the royal inscriptions quantified the tribute only when established for the future or received on the spot (or, when a previously fixed tribute is increased, they specify, occasionally, only the increased amount). By contrast, the administrative documents usually report the regular income of tribute from the provinces *after* it has been established — therefore, they cannot have been used as sources for the royal inscriptions. For some examples of tribute lists, cf. CT 53, 722 (SAA 11, 34; very fragm.) and ADD 908 (J.D. Hawkins - J.N. Postgate, "Tribute from Tabal," SAAB 2 (1988), 31-40 = SAA 11, 30).

129 See e.g. KAJ 314 (examined by Martin, *Tribut*, 21ff), recording a *madattu* of *134* sheep comprehensively.

130 ND 2443, but the numbers are very low. ND 2485, on the contrary, records a grand total of *1,727* men, plus *277* "missing" (i.e. dead on the way to Kalah — their place of origin is not specified). ND 2497 registers groups of people according to their profession and age, mentioning even *15,097* workmen (ERÍN), but the context is unclear. Other texts either record very low numbers of people (e.g. ND 2744), or are very fragmentary (ND 2447), or, for instance, represent lists of sick people (ND 2629 and 2679).

131 ND 2490+ is an inventory of different commodities, but its purpose is not clear. A number of texts bear lists of animals, quantified and subdivided into species and categories, e.g. ND 2458, relating a "total of *730* horses of/from the merchants," or ND 2727 (cf. TCAE, 393f), a record of horses received as *iškaru* tax or *nāmurtu* payment. ND 2754 (cf. TCAE, 394f), recording the income of a yearly tribute (*madattu*) in oxen and sheep, is very fragmentary.

132 ND 2619 lists contingents of chariot drivers and cavalrymen (*pēthallu*) subdivided by place of origin (regions and towns in the south and west). This points, according to Parker (*Iraq* 23, 15), to the end of the 13th campaign of Sargon, when, among others, *150* chariots and *1,500 pēthallu* of Bīt-Yakin (southern Mesopotamia) were drafted into the Assyrian army (Lie, *Sargon*, 73:11 — cf. p. 56 no. 60) — that region, however, is not mentioned by the text.

133 Weidner, AfO 10, 41f no. 95. On this archive see, in general, Pedersén, *Archives* I, 56ff (M6).

Cf. also the fragmentary text published in Millard, *Iraq* 32, 172f, and ADD 952 (SAA 11, 80), recording small numbers of sheep subdivided according to provinces.

c. Chronicles, king lists or other chronographic documents. These are very sparing of numerals, as they are of details on the wars and the events, which are narrated in a very terse manner.[134] Chronicles report very few numbers,[135] if we exclude the time indications (years, months, days). However, one of the rare time quantifications of the royal inscriptions finds requital in a Babylonian chronicle. Esarhaddon states he massacred the armies of the pharaoh Taharqa for "*15* days":

> *ultu* ^{URU}*Iš-ḫup-ri a-di* ^{URU}*Me-em-pi* URU*(āl) šárru-ti-*[*šu*] *ma-lak* 15 u_4*-me qaq-qa-ri* u_4*-me-šam la na-par-ka-a di-ik-tú-šú ma-ʾ-diš a-duk*
>
> "from Išḫupri to Memphis, [his] capital, a journey of *15* days of ground — daily (and) without pause I slew multitudes of them."[136]

The above mentioned chronicle tells of three battles fought the 3rd, the 16th, and the 18th of Tammuz, before the conquest of Memphis, which occurred the 22nd of that month,[137] thus a period of 15 days, in conformity with Esarhaddon's text, if one leaves out the "continuity" given there to the massacres.

2. Checking against documents of external provenance.[138]

a. Here too, the references to Assyria to be found in administrative or epistolary documents, e.g. from Syria[139] or Babylonia,[140] are difficult to link to any specific event referred to by the Assyrian inscriptions. Further, also the external historiographical records are unusable to check the data, or to draw correspondences of any kind[141] — obviously, the *other* royal inscrip-

134 These texts are not always as "objective" as their terse style would make us believe, and they present exegesis problems as well. For example Zawadzki, with not completely convincing arguments, inferred that the "Gadd chronicle" (= Grayson, ABC, no. 3) gives a filtered view (in an anti-Median key) of the events, cf. *Fall of Assyria*, 118ff and especially 129f and 147f. On the tendentiousness of this chronicle cf., however, also Na'aman, ZA 81, 261. It is anyway admitted without argument that the chronicles give a "parochial" view of the events (thus Grayson, ABC, 11; cf. also the remarks of Brinkman, *Fs Moran*), and that some of them are even factious — this being the case of the "synchronistic history" (on which cf. e.g. Grayson, ABC, 50ff; *id.*, Or 49, 181 and 189; RlA 6, 88; Brinkman, PKB, 32) and possibly also of the "Esarhaddon chronicle" (cf. Frame, *Babylonia*, 7f). This could be said of the Assyrian king list as well, "intended to justify Shamshi-Adad's claim to the throne" (Grayson, Or 49, 179 and cf. n. 184); the king lists, however, are devoid of numbers other than the lengths of reigns.

135 Cf. Grayson, ABC, 123:23 (*2* soldiers) and 164:12′ (= ARI 1, 996; *40* chariots), plus the fragmentary passage of p. 174:2ff. The sole text of this category to contain a fair quantity of numbers is the already mentioned fragmentary tablet of Arik-den-ili, but it is not improbable that its ample lacunae actually hide a royal inscription (cf. above p. 22 n. 10).

136 VS 1, no. 78 Rs. 38-40 (= Borger, *Asarh.*, §65 and ARAB 2, 580). Cf. p. 18 for another "journey of 15 days" by Ashurbanipal, in the land of the Medes. The latter king gives the travel from Memphis to Thebes as (*1*) month and *10* days (Streck, *Asb.*, 160:30 and cf. 164:73 = ARAB 2, 901 and 906).

137 Grayson, ABC, 85:24ff.

138 Obviously, no attempt at mentioning all the "external" documents making reference to Assyria will be made. Note that the references to be found in 2nd millennium texts have no value in this respect, since the contemporary Assyrian inscriptions did not relate quantifications (the pertinent texts from Nuzi, Ḫatti and Ugarit are discussed in Harrak, *Hanigalbat*).

139 The records of the local Aramaic rulers are collected in Sader, *États araméens* (pp. 11ff, 77ff, 120ff, 156ff, 206ff and 246). For a review of 1st millennium Syro-Hittite inscriptions cf. J.D. Hawkins, "Assyrians and Hittites," *Iraq* 36 (1974), 67-83 (most texts are edited in P. Meriggi, *Manuale di eteo geroglifico, II: Testi* [in 2 parts], Roma 1967). In both cases the references to Assyria are scanty and doubtful, most of the texts being either too fragmentary, too obscure, or simply too short to record helpful information.

140 Cf. e.g. O.R. Gurney, "Texts from Dur-Kurigalzu," *Iraq* 11 (1949), 131-49, pl. XL-XLI, text no. 10.

141 It is hard to find any reference to events connected in some manner with Mesopotamia in the contemporary Egyptian documents (which do not speak, for instance, of the obelisks taken by Ashurbanipal from Thebes: cf. p. 123). Furthermore, they are almost devoid of quantifications — see e.g. ARE 4, 892ff (Tanis stela of Taharqa), 919ff ("dream" stela of Tantamani).

tions prefer to speak about *their* "victories" and not of the Assyrians'. This applies to the Urartian inscriptions from the historical period corresponding to the reigns of Shalmaneser III and his successors, of which the longest contain numerous quantifications, often not less conspicuous than those of the contemporary Assyrian annals.[142] As for the Babylonian inscriptions, they present the kings almost exclusively as builders, while allusions to military achievements are very rare.[143] However, some years ago in 'Ana and its vicinity (middle Euphrates) were discovered a number of inscriptions written around the middle of the 8th century by a couple of local rulers calling themselves "governors of Suḫu," Šamaš-rēš-uṣur and Ninurta-kudurri-uṣur.[144] The longer texts relate also military matters, including confrontations with Assyria,[145] in a rather genuine way. The many quantifications reported, quite restrained and mostly "exact" in form, do not find any correspondence in the contemporary Assyrian inscriptions (Adad-nerari III), which are almost completely devoid of numbers.

b. Israel is an exception, the Old Testament allowing some comparisons and counterchecks.

The conspicuous tribute given by Menahem of Israel to Pul of Assyria, *1,000* tal. of silver according to 2 Kings 15:19,[146] receives no (quantitative) mention in the inscriptions of Tiglath-pileser III.[147] Other passages of his inscriptions containing references to tribute (or booty) collected in Palestine at later times are in most cases very fragmentary and have not, in turn, any

---

142 Most of the Urartian royal inscriptions appear in König, HCI and Melikišvili, UKN. They are not the subject of the present work, therefore the problems and the facts related to their quantifications will be ignored. At the end of the volume, however, a table showing their highest quantifications is included (p. 187), to ease the comparison with the homologous Assyrian *records*. Some inscriptions — cf. especially Argisti I's and Sarduri II's annals — make reference to encounters with Assyria as well. Now, as Salvini pointed out, it is noteworthy that we do not have a single mention of any of these events in the Assyrian inscriptions, nor do we have in the Urartian documentation one mention of the events there related — cf. M. Salvini, "Assyrian and Urartean written Sources for Urartian History," *Sumer* 42 (*s.d.*), 155-59, especially p. 156a, where he remarks that "the basic reason is obviously the universal tendency to praise one's own victories, and to be silent, on the other hand, about one's defeats" (cf. also Pecorella - Salvini, *Tra lo Zagros e l'Urmia*, 40b and 33b).

143 Cf. the remarks by Grayson, Or 49, 159ff. For the contemporary inscriptions, cf. also Brinkman, PKB, 24f, especially n. 110. Thus, the recently published inscriptions of Tall Bderi (middle Ḫābur), written by a local ruler, "king of Māri," yet subject to Tiglath-pileser I of Assyria, explicitly defined as "his lord," and dated according to the Assyrian *līmu* system, are devoid of any military reference — see S.M. Maul, *Die Inschriften von Tall Bderi*, Berlin 1992 (cf. the colophons at pp. 21 and 38).

We already spoke about the chronicles (Babylonian and other), cf. however Brinkman, PKB, 30ff. For the Elamite royal inscriptions, which make reference to wars or tribute, but not in relation to Assyria, cf. F.W. König, *Die elamischen Königsinschriften* (AfO *Beiheft* 16), Graz 1965 (cf. e.g. no. 28 C I (a)).

144 A. Cavigneaux - B.K. Ismail, "Die Statthalter von Suḫu und Mari im 8. Jh. v. Chr.," BaM 21 (1990), 321-456, Tf. 36-38, including a re-edition of a long known inscription of the former.

145 Actually, the sections mentioning Assyria do not contain numbers except for a couple of *Distanzangaben* in years (cf. *ibid.*, passages indicated at p. 338, incl. no. 17:15ff). Obviously, no reference is made to older "encounters" between the Assyrians and the preceding governors of Suḫu, which led to the payment of tribute to Tukulti-Ninurta II (on the part of Ilī-ibni) and to military collision with Ashurnasirpal II (Kudurru) — many quantifications are given by the Assyrian inscriptions in these cases, cf. p. 194 and p. 199. For a brief history of the relations between 'Ana/Suḫu and Assyria, cf. B.K. Ismail - M.D. Roaf - J. Black, "'Ana in the Cuneiform Sources," *Sumer* 39 (1983), 191-194.

146 Also mentioned in Josephus Flavius, *Ant. Jud.* IX, 11:1. The passage has recently been discussed by O. Loretz - W. Mayer, "Pūlu – Tiglath-pileser III. und Menahem von Israel nach assyrischen Quellen und Kön 15,19-20," *Ugarit Forschungen* 22 (1990), 221-31, and by Becking, *Samaria*, 4f.

147 Menahem of Samaria is mentioned by Tiglath-pileser's annals and by a stela reporting a list of western tributaries, but neither text gives any numerals nor other details of the tribute, see Tadmor, Tigl. III, Ann. 13:10 and Stele III A:5 — older editions resp. in Rost, *Tigl. III*, 26:150 and Levine, *Stelae*, 18:5′ (cf. Weippert, ZDPV 89, 29 and 34).

correspondence with the Old Testament.[148] The deportees from Samaria, *27,280* according to Sargon's prism from Nimrud, are mentioned in 2 Kings 17:6, where, however, no numeral is given.[149]

Yet a passage of the same book partially confirms the assertions of Sennacherib: "and the king of Assyria required of Hezekiah king of Judah *three hundred* talents of silver and *thirty* talents of gold."[150] The pertinent Assyrian passage states that:

> "His elite troops (*urbi*) and his best soldiers, which he had brought in to strengthen Jerusalem, his royal city, with *30* talents of gold, *800* talents of silver, choice antimony ... (there follows a list of commodities of various kinds, not quantified), and his (own) daughters, his concubines, male and female singers, he had bring after me to Nineveh, my royal city. To pay tribute and to do obeisance (Hezekiah) dispatched his (personal) messenger."[151]

3. The direct checking of some data given by the royal inscriptions, such as dimensions of the buildings, height of the walls, etc.

This is a quite impractical method, since most of the Assyrian architectural works have only partially survived, so that the measurement most frequently reported by the inscriptions, the height of the walls, is impossible to check. After all, these measures were subject to (possible) exaggerations no more than the real dimensions of the buildings, which were used as a symbol of power and for ostentation. It is therefore not surprising that the extant dimensions of the palaces confirm these measures at least in some cases.[152]

---

148 Cf., in general, R. Borger - H. Tadmor, "Zwei Beiträge zur alttestamentlichen Wissenschaft aufgrund der Inschriften Tiglatpilesers III," *Zeitschrift für Alttestamentliche Wissenschaft* 94 (1982), 244-51; Gelio, RBI 32, 135ff and, most recently, Tadmor, *Tigl. III,* supplementary study F. As he observes (*ibid.*, 276 and 171 n. 16′), the amount of silver paid by Menahem is on the same order as the payments made by Hoshea of Israel, Metenna of Tyre (cf. p. 51 no. 37) and Wassurme (Ḫulli) of Tabal (*10* talents of gold, *1,000* talents of silver and *2,000* horses; *ibid.*, Summ. 7 rev. 15). Like him, they were newly installed kings, if not usurpers, who paid for Assyrian support.

The tribute of Hoshea is mentioned in a short text published last century by Rawlinson (3 R, 10, no. 2): "The land of Bīt-Ḫumria (Israel) [...] all his people [with their properties in] Assyria I brought. Peqaḫ, their king, I/they killed(?) and Hosea [as king] over them I seated. *10*? talents of go[ld ...] *1,000*? talents of silver [... from] them I received and [in Assyria I bro]ught (?)" (Tadmor, *Tigl. III*, Summ. 4:15′ff = Rost, *Tigl. III*, 80; TUAT, 374).

For the booty taken from Ḫanūnu of Gaza, *800* talents of silver, cf. Tadmor, *Tigl. III*, Summ. 8:15′ (= Wiseman, *Iraq* 13, pl. XI and p. 23; TUAT, 376). A very fragmentary passage in Summ. 9 rev. 12 (= Wiseman, *Iraq* 18, pl. XXIII and p. 126; TUAT, 377) apparently records booty taken in Palestine (Tadmor suggests in Ashkelon), [X? +] *100* talents of silver.

149 Josephus Flavius speaks of "all the inhabitants" of Samaria, that is "*10* tribes" (cf. *Ant. Jud.*, IX, 14:1). Cf. variation no. 38 for the Assyrian number.

150 2 Kings 18:14. The numerals appear also in Josephus Flavius, *Ant. Jud.* X, 1:1.

See Cogan - Tadmor, *2 Kings*, 246ff for a detailed comparison of the two accounts (biblical and Assyrian) on the campaign of Sennacherib.

151 Luckenbill, *Senn.*, 60:56ff (= ARAB 2, 284); similar passage at p. 70:31f (= 312) and 33:39ff (= 240; TUAT 390; dupl. Ling-Israel, *Fs Artzi*, 228:34ff); also in CT 26, pl. 11:68ff (= Heidel, *Sumer* 9, col. III 95ff) and Smith, AD, 306:29f. Cf. Cogan - Tadmor, *2 Kings*, 247 for the translation of the passage (esp. nn. 2f there).

As van Leeuwen points out (OTS 14, 249 and n. 3), it is very unlikely that this difference can be explained by means of a relationship of 3:8 between the Babylonian and the Hebrew talent. Cf. also p. 174 below.

152 This view is expressed in W.G. Lambert, "The Reigns of Aššurnaṣirpal II and Shalmaneser III: An Interpretation," *Iraq* 36 (1974), 103 (cf. also Grayson, Or 49, 170). This may apply to the city perimeters as well: Thureau-Dangin (cf. JA 1909, 82f and RA 22) used the present-day perimeters of Dur-Sharruken and Nineveh to calculate the length of the Assyrian cubit, comparing the measures given by some royal inscriptions (cf. below, p. 140) — cf. however the remarks of Powell in RlA 7, 474f.

For the comparison of a measurement — height of a cultroom — given by Shalmaneser I (cf. p. 153 n. 178) with the archaeological remains of the described structure, see G. van Driel, *The Cult of Aššur*, Assen 1969, 16f and P.A. Miglus, "Auf der Suche nach dem »Ekur« in Ashur," BaM 21 (1990), 312 (the results are rather uncertain since it is difficult to know exactly to which wall the measurement refers).

We could check also the distances between given places, often specified by the inscriptions (cf. p. 17). The problem here lies in the fact that they can be checked only approximately, either because most of the time one or both of the places are vaguely indicated, and because the distances could not be measured with a high degree of precision. However, the geographical elements contained in the Assyrian royal inscriptions are, generally speaking, reliable,[153] and the exaggerations in the distance given, if they occurred, seem modest.

7. POSSIBLE QUANTIFICATIONS

To ascertain exactly what could be the highest quantification given to an object and still stay within the bounds of verisimilitude must have been a difficult operation for an Assyrian, given that, preferably, the inscriptions narrate episodes related to distant (and "exotic") lands situated on the borders of the known world. Moreover, the inscriptions often give "total" quantifications referring to places or areas very generically indicated. When dealing with complex and/or vague events, it becomes difficult to evaluate the quantifications: the "totals" referring to the whole length of the reign and to the whole extension of the empire could have been as enigmatic for an Assyrian as are the "national accounts" for a "modern." He was facing quantities transcending his everyday experience.

It is much easier to try to establish the feasibility of data referring to a single event, as the capture of booty or the killing of some enemies during a battle. Even in this case the factors involved will be manifold: the ecological/demographic reality of the area (which not always can be reconstructed, not even approximately), its contingent historical situation, the exact development of the events (even if an army has been defeated, this does not mean that all its components have been either killed or captured), etc. Given this high number of variables, then, the general context has to be reconstructed rather empirically. A "minimal" approach could be to compare each quantification to the other homologous quantifications given by the Assyrian royal inscriptions. In this respect one should consider the highest figures, or even the "averages," given for each object (cf. the tables on pp. 181ff). Also data from other sources, or from other periods (e.g. from the Persian or Hellenic era), can be used for comparison, given that they are in turn critically evaluated and verified. In this way, however, one can only fix a "maximum limit" for a given quantification, and, since the subjective factor is preponderant, conflicting results may ensue.[154]

The discussion of specific quantifications is postponed to the appendix to this chapter (pp. 97ff), since some of the arguments of the following paragraphs are presupposed, while a topical and detailed examination will often be necessary.

---

153 Cf. the remarks of Brinkman, PKB, 25.

154 See e.g. Delbrück's estimates for the Persian army, p. 109 n. 265, and cf. p. 3 n. 14 on the method of *Sachkritik*.

8. DISTRIBUTION OF NUMBERS AND DIGITS

The formal characteristics of the numbers reported by certain royal inscriptions (frequency of each digit, frequency of round/exact numbers, etc.) have already been dealt with in chapter 2. Starting from the tables of that chapter (see especially pp. 42-43) a number of general statements can be made, valid, except where otherwise specified, for every inscription taken into account, ranging from Tiglath-pileser I to Sennacherib.

Concerning the distribution of the digits, it will be immediately noted the regular prevalence of "1" and "2" in almost all the inscriptions, far more frequent than any other digits. The "1," in turn, is in most of the cases more frequent than the "2": exceptions are the annals of Tiglath-pileser I and Sargon, where "2" is more frequent than "1," while the two digits appear the same number of times in the "black obelisk" (edition "F") of Shalmaneser III, in the annals of Shamshi-Adad V and in the final edition of those of Sennacherib. Only in Tiglath-pileser III's annals the most frequent digit is "5." Actually, one may note the tendency, over the course of time, to use also higher digits — thus, "6," "7," "8" and "9" are rather rare, except in the later reigns.[155]

If the use of roundings might explain the higher frequency of the digit "1," it cannot explain why "2" is almost equally frequent. As a matter of fact, it is natural to round up to *1,000* numbers such as, e.g., "880," "950," or the like, but it is far more difficult to carry out a similar operation having as a result the digit "2." It often appears in numerals like *200*, *1,200*, that are not "more round" than *300* or *1,300*. We must assume that the quantities related by the royal inscriptions were (in origin) represented by equally recurring digits (see the quantifications appearing in the administrative documents). A certain frequency of round numbers, employing the digit "1" (*100*, *1,000*, etc.), might be justified by the presence e.g. of quantifications of Assyrian military contingents, which probably were structured on a decimal basis,[156] or of tribute, which, we may presume, was imposed in round measure.[157] Yet,

155 This applies especially to Sargon's successors, whose annals, however, contain few numbers and thus have not been included in the table under discussion.

156 It is known that the Persian army was divided into units of *10, 100, 1,000* and *10,000* men (Herodotus, VII:81). Now, some facts seem to indicate that also the Assyrian army was similarly structured, for instance there existed the LÚ.GAL.10 = *rab eširte* "chief of 10," as well as the "chief of 50" and the GAL*(ráb)* 1 *lim* "chief of 1,000," cf. Henshaw, *Palaeologia* 16, 21 and Malbran-Labat, *L'armée*, 119-122 — note that the "chief of 100" as well as the "chief of 70," are not attested: cf. the review by J.N. Postgate in BiOr 41 (1984), 422.

Some letters, however, speak about (Assyrian) contingents not certainly quantified in decimal measures: cf. e.g. ND 2366 (Saggs, *Iraq* 21, 171); ND 2437 (SAA 1, 176); ND 2631 (SAA 5, 215); ABL 563 (SAA 1, 241); ABL 56+ (SAA 5, 251).

A Nimrud letter, ND 2631 (cf. Saggs, *Iraq* 28, 185ff, pl. LVI; SAA 5, 215; LCA 7), according to J.V. Kinnier-Wilson, *The Nimrud Wine Lists* (*Cuneiform Texts fron Nimrud* I), London 1972, 50ff, indicates that the basic chariot unit was made of 106 men, possibly forming the crew of 53 chariots, 50 chariots plus three command-chariots. This conclusion, however, is based on very scanty facts and causes some doubts: cf. P. Garelli, "Remarques sur l'administration de l'empire assyrien," RA 68 (1974), 137f; Parpola, JSS 21, 172 and Fales, SAAB 4, 31-34 — note also the total number of chariots as given by some inscriptions of Shalmaneser III, p. 50 nos. 29f, definitely not divisible by 53.

157 But cf. how both tribute and booty are almost always given in exact numbers by the Egyptian annals of Tuthmosis III (ARE 2, 391-405). These annals are especially interested in booty and tribute (cf. ARE

this is not necessarily the case with the tribute of submission[158] or the "audience gifts," and is excluded in the case of booty, even if it is conceivable that it could have been assessed at a glance. It is possible that also the prisoners were deported in units of decimal basis, even if this seems rather unlikely,[159] while it is obviously totally excluded that the Assyrian army restricted itself, say, to conquer *100* cities or to kill *1,000* enemies during a battle: see how, for instance, the figures of prisoners or killed given by certain old Akkadian royal inscriptions,[160] or by the Bisutun inscription of Darius the great,[161] or even by some Egyptian annalistic texts[162] are most of the time "exact". In any case, quantifications of "regular" tribute and of Assyrian military contingents are quite rare within the *corpus*.

Undeniably, a component of casualness can partly explain the exact form and the details of the royal inscriptions and of the numbers appearing there. Even so, to round the numbers was apparently a very common practice, in view of the high frequency of "round" numbers, especially in the inscriptions of the reigns between Ashurnasirpal II and Shamshi-Adad V. In most cases, thus, the digits "6," "7," "8" and "9," were replaced with a "10," and alternatively with a "20." Now, the fact that in certain inscriptions also the digits "3," "4" and "5" are quite infrequent, makes us suppose that also the numbers formed with these digits were subject to rounding, and, since it can be presumed that the roundings had a single direction (*up*), we should therefore conclude that sometimes "real" quantities of "500," "400," etc. were brought up to *1,000* or *2,000*.[163] The higher frequency of the digit "5" with

2, 394), recording even the regular income of yearly imposts: cf. e.g those of Wawat (Nubia), varying from one year to another both in composition and in quantity, yet always represented by exact numbers (ARE 2, 475, 487, 495, 503, 515, 523, 527 and 539).

158 Thus, Tiglath-pileser III's stela from Iran records the tribute (*ma-da-ta*), represented by horses only, paid by some Median kings who submitted. Most of the 14 preserved numerals are "round" (the end of the list is broken off): 8 times *100*, once each *200* and *300*. Yet, also tribute of *120*, *132?* (cf. line 30′), *33* and *32* horses is recorded (Tadmor, *Tigl. III*, Pl. XXXVII and p. 106:30′-40′). A later *summary inscription* reports booty captured in Media and made up of "*5,000* horses, people, oxen and sheep without number (*ana lā māni*)" (Rost, *Tigl. III*, Pl. XXX:11′ = Tadmor, *Tigl. III*, Pl. XLVIIIf:14′ - Summ. 3; ARAB 1, 812). The attribution of this quantification to the same event is quite likely (no other tribute from Media is recorded by the annalistic inscriptions of this king — cf. also the dates suggested in Tadmor, *Tigl. III*, p. 129), while it is self-evident that the above specified numbers necessarily summed up to an "exact" numeral: most probably, then, we are facing a "rounding" — unfortunately, it is impossible to check whether the total had been rounded *up* or *down*. The extant numerals add up to 1,617 (or 1,616), while, after the listed tributaries, three or four lines, including some other names, are partly readable. It is not easy, however, to estimate how many lines are lacking at the end of the column: judging from Tadmor's reconstruction of the stela (*op. cit.*, Fig. 6), I would say no more than 10. Assuming the list continued until the end of the column (the next column begins with a list of western tributaries), one might suppose some 15 other quantifications were included, which would possibly double the "total" we have calculated.

159 Some letters mention prisoners just captured (e.g. ABL 280 and 520:11; cf. Pfeiffer, SLA, nos. 40 and 43), or generic groups (ABL 839, 520 rev. 19, 792, 794; cf. Pfeiffer, SLA, nos. 16, 43, 46 and 45 respect.), using mostly round numbers, just as the royal inscriptions would do in the same context. Thus, the *6,000* captives of ND 2634 and the *5,000* of ABL 304 (cf. n. 125 here above) represent intentional approximations. Administrative documents recording lists of prisoners usually show rather uneven quantities, mostly given in "exact" numbers: cf. some instances in n. 124 here above.

160 This is the case with some texts of Rimush and Naram-Sin, recently re-edited in I.J. Gelb - B. Kienast, *Die altakkadischen Königsinschriften des dritten Jahrtausends v.Chr.* (Freiburger altorientalische Studien 7), Stuttgart 1990, 191ff and 226ff.

161 E.N. von Voigtlander, *The Bisitun Inscription of Darius the Great, Babylonian Version*, (Corpus Inscriptionum Iranicarum, Pt. I, 2/1), London 1978, *passim*. Cf. also, in the same series (No. 5/1), the (later) Aramaic version of this inscription, showing many variants in the digits: J.C. Greenfield - B. Porten, *The Bisitun Inscription of Darius the Great, Aramaic Version*, London 1982.

162 These texts, at the same time, explicate the system employed to count the killed: cutting off their extremities, cf. below p. 118 n. 8, and cf. also n. 157 above for the booty.

163 The lower frequency of these digits is not due to the rounding of digits of low order (tens, units), as e.g. would be "14" > *20*, or "440" > *500*, in view of the fact that just among the digits of low order appear most often the "4," "5," "6" etc. This does not appear in the above mentioned table, which is not

respect to the "4" can be explained on the basis that it is half of "10," and therefore, in the decimal system, it is "almost round."

If we take into account all the numbers between "100" and "1,000," representing all the possible "real" quantifications, it seems that the tendency was to round up to *1,000* all those higher than "700," and in some cases those lower as well, from "400" or from "300" up.[164] The high frequency of the digit "2" is due to the fact that sometimes these numbers were rounded up to *2,000,* since it is difficult to think of a rounding of the kind "3,000" > *2,000.*

These considerations obviously should be taken with reservation, since they concern a very wide period of time and a great number of inscriptions of different natures. The theoretical framework thus delineated does not exclude other possibilities, as, for instance, that most of the numbers given by the inscriptions are pure fabrication. Actually, there are some arguments in favor of this last hypothesis, as, to mention only one, the fact that in the lists of booty the commodities always show well-proportioned quantifications (or, better, quantifications in inverse ratio to the intrinsic value of the object, cf. below p. 127). Is it possible that so rarely has the booty consisted of a similar number of oxen and sheep, or that there were so few of the latter? Is it possible that their numbers had to conform to or approach the ratio 1:10 so often? We may observe that the lower digits are, in the cuneiform system, "shorter" than the others. This might let us in turn suppose that a scribe, if willing to "create" a number, would prefer one with a low digit, not simply to "save work," but rather for the sake of simplicity, in view of the fact that what really matters is the presence of the number, of any number (with a certain order of quantity) in that specific context.

9. EVALUATION OF THE NUMBER BY ITSELF

A number might seem artificial even if taken into account for its pure graphic appearance, independently from what is quantified, from the context, etc. This is the case with certain high "round" numbers reported by the pre-Sargonid inscriptions, but of some "exact" numbers as well, like the *800,600* sheep mentioned by Sennacherib (p. 58 no. 69). The factors to take into consideration might be numerous: for example the fact that the varying digits frequently surpass by one unit other digits preceding them (cf. n. 103

---

detailed enough in this respect. Putting aside the idea of making further tables or diagrams that would be very complicated and would excessively broaden the discussion, let it suffice only to underscore two facts that are easily verifiable for most of the inscriptions taken into account:

• the higher digits appear mostly in low numbers, in sentences like "for the *18th* time I crossed the Euphrates" (ARAB 1, 576), or "*19* districts of Hamath I brought within the border of Assyria" (*ibid.*, 770), and not in "high - round" numbers;

• the low order digits surpass, in most cases, those of higher order; in other words, the numbers often show digits arranged in increasing order, and this is especially true for certain inscriptions. For instance, in the "black obelisk" of Shalmaneser III (cf. Table 1 on pp. 42-43) there are 14 numbers with 2 digits: 11 times these are "increasing" (*20,500, 12, 12, 16, 18, 24, 27, 46, 23, 250*), only once are they "decreasing" (*21*, counting the crossings of the Euphrates), while in three cases they are "even": *22, 55, 22*. There are, then, two numbers with three digits, one is *470* (written 4 *me* 1,10) while the other one is *89* (1,30+9, also "increasing"), and one with four: *1,121*.

164 Cf. editions "C" and "F" of the annals of Shalmaneser III, where the digits "1" and "2" are far more frequent than all the others (Table 1).

at p. 71) is possibly an indication of an unintentional "accounting" (*1, 2, 3,* etc.). Thus a number formed by digits in an "increasing" series should be taken with a certain suspicion (cf. also n. 163 [end] to this chapter).

Obviously, this assumpion of "artificialness" is rather subjective, and can have a certain importance only in the case that other elements make us doubt the veracity of the numerals under examination. Sometimes it could be useful to check how the signs expressing the numbers have been fixed in the document. The possible presence of erasures or blank spaces can make us think of hesitations on the part of the scribe, and this leads to the suspicion that copying errors could have been made. Another factor is the possible interaction between the signs of the numbers and other signs. A passage of a short inscription of Tiglath-pileser III can serve as an example — it is rather fragmentary and looks thus:

> ...*a*]-*duk* 1 *lim* (?) UN^{MEŠ}*(nišē)* 30 *lim* ANŠE.A.AB.BA^{MEŠ}*(ibilē)* 20 *lim* GU$_4$.NÍTA^{MEŠ}*(alpē)* [...
>
> "... I] killed, *1,000* (?) people, *30,000* camels, *20,000* oxen [...."[165]

Now, the number "30" is written immediately after the (similar) sign MEŠ, with which it can be confused: 𒈨𒌍 𒌍 𒇷 (MEŠ - 30 - *lim*), and the same applies for the "20." Considering that the number of camels is really exceptional (even if the booty referred to is from Arabia), one might wonder whether the scribe had confused the signs.[166]

10. EVALUATION OF THE NUMBERS: CONCLUSIONS

In view of the fact that only a very limited number of quantifications can be checked through other sources, the factor that in most cases will be evaluated is their verisimilitude. Yet the issues that may result in this case will be rather vague and certainly not demonstrable. However, taking into consideration all the royal inscriptions of the period between Tiglath-pileser I and Sargon (and also, partly, the others), some general statement can be made.

Most of the quantifications prove themselves sufficiently realistic already from a surface examination. Numbers of prisoners or of soldiers or of animals in the order of hundreds or thousands, as well as of cities in the order of tens or few hundreds, etc., are not to be seen as necessarily exaggerated, considering that they are mentioned also in texts of other kinds, and above any suspicion (e.g. administrative documents). These kinds of quantifications probably hide only minor (and thus difficult to single out) inflations. Roughly speaking, all the round numbers should be considered the result of alterations to numbers originally lower, thus *1,000*, and often also *2,000*, could represent

---

165 Tadmor, *Tigl. III*, Summ. 4:20′ (Pl. XLIXf; cf. 3 R, 10 no. 2; Rost, *Tigl. III*, p. 80:20 / Pl. XXV:last line). Note esp. "Fr. e" (Pl. L), possibly reading 20 *lim* or 21 *lim*. Cf. Ephᶜal, *Ancient Arabs*, 33ff.

166 Similar oversights made in modern times during the copying of cuneiform texts for publication are quite common, cf. e.g. nn. 24 and 28f. here above.

an original "400," "450," etc.[167] This fact, naturally, does not come as a surprise, since exactness should not be asked of these inscriptions. Besides, one should consider that many "round" quantifications can originate in rough estimates (which eventually may also represent some exaggerations) made "on the spot" by the officers responsible for the accountings.

This does not exclude the possibility that the numbers may have been completely invented, but this issue cannot be demonstrated except, possibly, for a few cases (cf. the *28,800* Hittites, pp. 150ff below). Where "exact" numbers are used, on the contrary, we have to maintain that, as a matter of principle, these are authentic (it was obviously possible to alter these too), especially if within the same inscription "round" quantifications also appear. In such cases, while the exact numbers represent genuine data (possibly slightly "revised"), the round ones could be the result of the following:

a) data originally approximated (for instance, it is difficult to establish the exact size of an enemy army);

b) data rounded-up by the author of some inscription used as source (since the author of the inscription under discussion uses also exact numbers);

c) data invented, because the context necessitated a quantification, while no data was available.

One may note that in certain inscriptions both "round" and "exact" numbers appear, but the latter are used only for particular quantifications, such as the *Distanzangaben*, or the perimeter of a city, that cannot be subject to gross roundings (nor have they particular ideological significance).

From these remarks are excluded certain "exact" quantifications, mostly from a later period, particularly relevant from the numerical point of view ("high"). Here most often the verisimilitude of the number can be doubted. Anyway, it is a matter of a few cases that have to be dealt with individually (cf. appendix to this chapter and, more generally, chapter 5).

11. SOME CONSIDERATIONS ON THE "EXACT" NUMBERS

In some texts of the Sargonid period certain particularly high "exact" numbers appear, as *800,500* or *200,150* (cf. list at p. 188), formed essentially by a round number on the order of hundreds of thousands and a smaller number appended to it — the opposite never happens: numbers like "199,850" or similar do not exist. An explanation, suggested by Millard, *Fs Tadmor*, 215, could be that these numbers represent the sum of partial quantities, some "round" and some others "exact," and that would explain their final form. But the case of the sheep from the booty of Muṣaṣir (passing from *1,235* to *100,225*) clearly shows that this might not be true if only in some cases.[168]

---

[167] Even if the biblical sources are not to be considered more reliable than the Assyrian royal inscriptions, being, as they reached us, a much later composition, the silver given by Hezekiah to Sennacherib can represent a typical inflation, as it is in the Assyrian inscriptions (*300* > *800* — but for this difference cf. another explanation below, p. 174). Cf. also a (possible) example of rounding-up, showing a similar degree of inflation, here above, n. 58.

[168] Most likely when the numerals represent "totals" — cf. for instance, Shalmaneser III's totals for the first 20 years, mentioned below, p. 93. In any case, this does not necessarily imply that these numerals are (or were) genuine, since their truthfulness depended on the truthfulness of the source material.

Two other facts must also be considered, namely, that these numbers make up very often some exceptional quantifications, hardly credible (if not impossible), and the fact that they vary quite often from one inscription to another. As we saw (p. 73), the just mentioned passage *1,235 > 100,225* has given rise to the hypothesis that these numbers could represent the alteration of a genuine number, brought about by the multiplication by *100* of the first digit. But if this applies to the cited case (*1,235 > 100,225*), it did not necessarily apply to all the other similar "exact" numbers. It could be that, as suggested by Ungnad, the number *200,150*, referring to the deportees from Judah during Sennacherib's reign, "hides" a "2,150" (no counter-evidence exists) — but then, were the sheep taken during the first campaign of Sennacherib (cf. above, variation no. 69) "8,100," "8,600," or "6,600"? And were the oxen (no. 68) "805," "810," or "2,010"? This method obviously cannot be valid always. If one is willing to inflate the number *1,285* to obtain a much larger one, obviously, one must act at the level of thousands. It is not surprising therefore that only these have been multiplied, reusing the lower order digits for a different reason than that of preserving a "portion" of reality. Why use "real" numbers then, to alter in such a way? Consider also that not much care has been expended on the exactness of the tens in the passage under discussion (also the tens or hundreds of the other numbers relating to the same booty vary). The very fact that these kinds of numbers are easily subject to variations (as in the case of Sennacherib's 1st campaign) indicates that they were not employed to give genuine or more specific data.

Millard's hypothesis (*Fs Tadmor*, 219) that in the case of the sheep from the booty of Muṣaṣir the original quantity (*1,235* var. *1,285*) represents a sort of estimate, written in the field in haste and subsequently updated when complete data were available, is very unlikely, since it does not explain the striking similarity between the digits (*235 > 285 > 225*). It should be supposed that the original quantity had been "updated" with the addition of 98,990 head of animals, a rather singular figure indeed. How then, is it possible that the first "estimate" had been so low? There is hardly any doubt that these digits have been purposely reused, though inexactly copied. In short, whether the number *100,225* is genuine or not, feasible or not, in any case it represents nothing else but the result of some editing rule which called for the using of a "high - exact" number in that context.

We have seen that up to the Sargonid period the use of round numbers was quite common, while during the reigns of Sargon and Sennacherib, this use had been given up. Thus, it was decided not to round up the *1,235*, e.g. to "100,000," but rather to write *100,225*. The reason for this procedure lies in the number itself. Most of the roundings up, in the inscriptions of the earlier periods, were the result of quantitatively modest alterations: we spoke, above, of passages "400" > *1,000* or, at most, *2,000* — even if we cannot exclude bigger alterations. However, if someone states that he uprooted *300* enemies after a battle, this number cannot be *much* exaggerated. By contrast, the very "high" numbers may not be easily believed, especially if they are "round." The higher a round number is, the more it will appear approximated: if *100* or *1,000* can pass as minor roundings up, "100,000" would be noticed for its size as much as for its degree of approximation.

The "unusual" exactness of certain very high numbers of this period is therefore due to the objective, on the part of the writer, of making them credible. It is not the case that these "exact" numbers are typical of the inscriptions of Sargon and Sennacherib, at the time when the very high quantifications were frequent:[169] their absence from the texts of Esarhaddon and, especially, Ashurbanipal is due to the general scarcity of numbers in the inscriptions of these later kings, where examples of particularly relevant "round" numbers are almost totally missing (cf. below, ch. 5 §15).

12. SOME CONSIDERATIONS ON THE "TOTALS"

When the quantification refers to the whole reign, this fact is usually specified. A good example is given by Tiglath-pileser I, when he affirms:

> ŠU.NÍGIN*(napḫar)* 42 KUR.KUR^{MEŠ}*(mātāti) ù mal-ki-ši-na* (...) *iš-tu* SAG (var. *re-eš*) *šarru-ti-ia a-di* 5 BALA^{MEŠ}*(palē)-ia qa-ti* (*lu*) *ik-šud*
>
> "Altogether I conquered *42* lands and their rulers, from ... to ... (specifies the geographic extent of his conquests) from my accession year to my 5th regnal year."[170]

This total is taken up also by a later *summary* inscription, yet updating it to the 10th year, but the passage is fragmentary and the number of the conquered lands is not readable.[171] It does not appear in two later analogous inscriptions, where, however, it is replaced by the "total" of the Euphrates crossings:

> 28-*šu* EGIR*(arki)* KUR *Aḫ-la-me-e* KUR *Ar-ma-a-ia*^{MEŠ} ^{ÍD}*Pu-rat-ta* MU 1.KÁM 2-*šu lu e-te-bir*

---

[169] In some inscriptions of Sennacherib all the numbers reaching or surpassing a thousand are "exact," e.g. in the "Bellino cylinder" (campaigns 1 and 2, cf. ICC, 63f. = Luckenbill, *Senn.*, 55-60, 94f, 99-101) the numbers equal to or higher than "1,000" are: *208,000, 7,200, 11,703, 5,230, 80,100* and *800,500*. In the final edition of the annals (cf. p. 38 n. 82) we meet only *208,000* and *200,150*. In general, the "high - round" numbers are very rare also in the other inscriptions of this reign: cf. the *150,000* warriors mentioned here below, p. 116, which, however, appear in a few texts only (cf. *loc. cit.*, n. 308). Thus, also the *80,000* bowmen of ARAB 2, 257 (report of the 1st campaign) are not mentioned by the later "Bellino cylinder" nor by the subsequent inscriptions. On the contrary, there seems a desire to preserve the high "exact" numbers. The only other round numbers appearing in the inscriptions of Sennacherib are those that enumerate the bows and shields collected during his wars. These numbers vary from one inscription to another, being "totals," and actually in the last version they become "exact" (cf. p. 89). As for the *11,400* and *6,000* talents, the weight of the colossi placed in the palace of Nineveh, note that these numerals are written in the sexagesimal system (cf. p. 61 no. 75).

[170] AKA, 82:39-45 (variants in nn.) = RIMA 2, 25 and ARI 2, 40. Note that the number *42* is apparently contradicting a statement appearing at the beginning of the inscription (in the "titulary"): "I vied with *60* crowned heads and achieved victory over them in battle" (*it-ti* 1 *šu-ši* LUGAL.MEŠ-*ni šu-ut* TÚG.SAGŠU *al-ta-na-an-ma li-i-ta šit-nun-ta eli-šu-nu al-ta-ka-an*; AKA, 34:54ff = RIMA 2, 13:54f = ARI 2, 11). Unless this does mean that 18 of the *60* defeated kings were not (subsequently) subdued or that *60* kings ruled over *42* lands, the numeral "60" ("round," since written 1 *šu-ši*) is symbolic and is simply indicating a "high" number.

[171] AKA, 126:6 (RIMA 2, 35; ARI 2, 72).

"for *28* times in pursuit of the Aḫlamû Arameans the Euphrates I crossed, *2* times in one year."[172]

As any other multiplicative number, it is necessarily a "total." The position occupied by the new passage, judging from what remains of the inscriptions, both quite fragmentary at this point, is substantially the same: in the appendix to the narrative of the military campaigns. With the new inscriptions they wanted to respect the same structure, and, if the number of lands for some reason was not suitable anymore, they used that of the crossings. It may be that the number of lands had not increased much after the first conquests (one cannot continuously find new lands to conquer), or simply the *28* crossings seemed a much more relevant deed — in any case, one datum had to summarize the king's achievements.

Some inscriptions of Sennacherib report the number of bows and shields deriving from looted lands (*ina šallat mātāti šātina ša ašlula*) collected (*kaṣāru*) and added (*radû*, D) to the royal equipment:

	(3)	(4)	(5a)	(5b)	(6)
^{GIŠ}PAN *(qašti)*	*10,000*	*20,000*	*20,400*	*30,000*	*30,500*
^{GIŠ}*arīte*	*10,000*	*15,000*	*20,200*	*20,000*	*20,200*

References:[173] (3) cylinder "Rassam," 3 campaigns: Evetts, ZA 3, 312:59 (= Luckenbill, *Senn.*, 60); (4) "cylinder C," prism with 4 campaigns: Smith, *Senn.*, 76 (= AD, 308 col. V 10);[174] (5a) unpublished fragment of hexagonal prism with 5 campaigns;[175] (5b) prism "Heidel-King," 5 campaigns: Heidel, *Sumer* 9, 151 [printed "251"]:45 + CT 26, pl. 17 and p. 16:16 (Luckenbill, *Senn.*, 63); (6) "Bull 4," 6 campaigns: 3 R, 13 [*slab* 3]:18 (Luckenbill, *Senn.*, 76:103).[176]

In these cases as well, each passage appears at the close of the historical section, regardless of how many campaigns it included. It is thus clear from the context that the quantifications refer to the whole reign, and one therefore is not surprised if the numbers increase in the later texts. This applies, generally speaking, also to the quantifications of the beasts killed or captured by the king. It is quite rare that, in these cases, reference is made to a single hunting campaign, or even to several hunts taking place, say, throughout one year: most of the time the hunting reports, appearing in the closing section of

172 Weidner, AfO 18, 350:34 (RIMA 2, 43; ARI 2, 97). The version included in a similar inscription is slightly different, having *ša-at-ta* instead of MU I.KÁM, but a lacuna conceals the number: cf. Weidner, AfO 18, 344:29ff (RIMA 2, 37; ARI 2, §83 and cf. n. 108 for the translation "twice in one year" instead of "twice a year").

173 For the chronology of the main inscriptions of this king cf. here above, n. 52.

174 Luckenbill considered it a duplicate of BM 103000, which is the "Heidel-King," prism with 5 campaigns — cf. (5b) — and did not note the different number of bows and shields. Smith read [*15,000*?] bows and *15,000* shields: however, according to E. Frahm's collations (cf. n. 52 above) the bows are *20,000*. One unpublished fragment, BM 121011+, apparently does not have the two *lims*, thus recording "*20* bows and *15* shields," which would be an obvious mistake (the passage is, however, difficult to read). Still according E. Frahm, "cylinder D," an octagonal prism with 5 campaigns, has these same quantifications, thus differing from the (contemporary?) hexagonal prism here marked as (5a). Prism "Heidel-King," (5b), is slightly later, having the *Eponymenfeldzüge* next to the first 5 campaigns.

175 83-1-18,599 (cf. E. Frahm's forthcoming work). The prisms marked as (4) and (5b) are both octagonal.

176 According to A. Salonen, *Die Wasserfahrzeuge in Babylonien* (Studia Orientalia VIII.4), Helsinki 1939, 183, the shields are *20,200* and not *30,500* (thus also BAL^2, 77).

the inscriptions, refer to the whole reign, even if this is not explicitly stated.[177] The variation of these numbers will therefore be considered an "updating" and not an alteration.

In other cases, the nature of a quantification as a "total" is in doubt. Or, since the "totals" may encompass more or less extensive periods of time, it might not be clear what the period actually taken into consideration is.

The text of the 1st campaign of Sennacherib gives a clear example of this situation. The later version of the annals reports the following:[178]

line
43: *i-na ta-ai-ar-ti-ia* ^{LÚ}*Tu-ʾ-mu-na* (...)
49: (...) ^{LÚ}*A-ra-mu la kan-šu-ti*
50: *mit-ḫa-riš ak-šud*^{ud} 2 *me* 8 *lim* UN^{MEŠ}*(nišē)* TUR*(ṣeḫer)* GAL*(rabi)*
51: NITA*(zikar) ù* MUNUS*(sinniš)* ANŠE.KUR.RA^{MEŠ}*(sīsē)* ANŠE. KUNGA^{MEŠ}*(parē)* ANŠE^{MEŠ}*(imērē)*
52: ANŠE.GAM.MAL^{MEŠ}*(gammalē)* GU$_4$^{MEŠ}*(alpē) ù ṣe-e-ni ša la ni-bi*
53: *šal-la-tu ka-bit-tu áš-lu-la a-na qí-rib mat Aš-šur*^{KI}

"On my return (march) the Tuʾmuna ... (various other populations are named), Arameans (who were) not submissive, all of them I conquered. *208,000* people, small and great, male and female, horses, mules, asses, camels, oxen and sheep without number, a heavy booty, I carried off to Assyria."

The quantification is located in the final part of the first campaign, which, however, concludes after another 11 lines (54-64) including the narration of some other events of a military nature. Therefore, nothing would let us think that in some manner this quantification summarizes the whole campaign: it would rather seem that the *208,000* deportees represent a partial "total," referring to the Arameans only, encountered during the return march.

The version given by the cylinder containing only the first campaign[179] is a little clearer in this respect:

it-ti 2 *me* 8 *lim šal-lat* UN^{MEŠ}*(nišē) ka-bit-tum* 7 *lim* 2 *me* ANŠE.KUR.RA^{MEŠ}*(sīsē)* ANŠE.KUNGA^{MEŠ}*(parē)* 11 *lim* 73 ANŠE^{MEŠ}*(imērē)* 5 *lim* 2 *me* 30 ANŠE.GAM.MAL^{MEŠ}*(gammalē)* 80 *lim* 50 GU$_4$^{MEŠ}*(alpē)* 8 *me lim* 1 *me* U$_8$.UDU.ḪÁ.MUNUS^{MEŠ}*(ṣēni) a-tu-ra a-na qé-reb Aš-šur*^{KI}

"With *208,000* people, a heavy booty, *7,200* horses and mules, *11,073* asses, *5,230* camels, *80,050* oxen, *800,100* sheep I returned to Assyria."

Here, in the earliest annalistic inscription of Sennacherib, the passage is located at the very end of the campaign, entirely occupying one single line,

177 Cf. Olmstead, JAOS 38, 250. The quantifications referring to the royal hunts will be dealt with on pp. 143ff.

178 Lines are numbered according to the "Oriental Institute" prism (Luckenbill, *Senn.*, 165 and 25, col. I); for the "Jerusalem prism" cf. Ling-Israel, *Fs Artzi*, 222:38ff.

179 Smith, *First Camp.*, 44 line 60 (Luckenbill, *Senn.*, 55).

and it is clearly stated that the booty is that with (*itti*) which the king returned (*târu*) to Assyria.[180]

Without this text, and basing ourselves only on the later inscriptions, it would have been very difficult to understand that the *208,000* deportees actually represent a "total." Already in the "Bellino cylinder," containing two campaigns and therefore slightly later than the original version,[181] two episodes are moved after the passage with the "totals," which no longer appears at the end of the campaign, but within it (compare lines 57-59 of the first version with lines 54-64 of the "Oriental Institute" prism).

A similar example, even though referring to only part of a campaign, comes from some texts of Tiglath-pileser III. A *summary inscription* from Nimrud offers the number of the people captured within a group of cities: from Sarrabānu were taken *55,000* people, from Tarbaṣu and Yaballu *30,000*, from Dūr [Baliḫāya] *40,500*, and from Amlilatu another group (no number is specified).[182] A shorter *summary inscription*, still from Nimrud, which partially duplicates the previous one, leaves out many details and lists the above mentioned cities one after the other, as if they were conquered within a single episode. The text then gives us (the total of) the captured: *155,000*, though it does not explicitly refer to this number as a "total":

> [*mat Bit-*ᵐ*Sila*]*-a-ni mat Bit-*ᵐ*Sa-ʾ-al-li a-na paṭ gim-ri-šú-nu as-su-am-ma* ᵐᵈAK*(Nabû)-ú-šab-ši* ᵐ*Za-qi-ru* LUGAL^MEŠ^*(šarrāni)-šú-nu qa-ti ik-šud* [^URU^*Sarrabā*]*-nu* ^URU^*Tar-ba-ṣu* ^URU^*Ia-bal-lu* ^URU^BÀD.AN*(Dūr)*.ILLAT*(Baliḫ)-ai* ^URU^*Ma-li-la-tu* URU*(āl) šarru-ti-šu-nu* GAL^MEŠ^*(rabûti)* [...]-A *ù* ^GIŠ^*šu-pi-i ak-šud* 1 *me* 50 *lim* 5 *lim* UN^MEŠ^*(nišē) a-di mar-ši-ti-šú-nu* [...]*-šú-nu* MÁŠ.ANŠE*(būl)-šú-nu a-na la ma-ni áš-lu-la*

> "[Bit-Sil]āni (and) Bīt-Sa'alli I tore up to their farthest borders. Nabû-ušabši (and) Zaqiru, their kings, my hands captured [Sarrabānu] Tarbaṣu, Yaballu, Dūr-Biliḫāya, Malilatu, their great royal cities, I captured [by means of earthworks] and siege engines, *155,000* people, together with their possessions, [...] their, their livestock without number I took as spoil."[183]

One may thus doubt whether many quantifications contained in the Assyrian royal inscriptions actually represent wider events than those to which they refer. The number of deportees from a certain city, at the end of a military

---

180 *târu*, "to return," obviously indicates the conclusion of the campaign, differently from *šalālu*, "to carry off," used in the other case. The passage is followed by another two lines of text, bound to it and also characterized by a recapitulative tone: "that does not include the men, asses, camels, oxen and sheep which my troops carried off and parcelled out among themselves, nor the enemy warriors, strong and proud, who had not submitted to my yoke, cut down with the sword and hung on stakes." Then follows directly the building record.

181 Cf. ICC, 63 (Luckenbill, *Senn.*, 57:14ff), its text will be taken up by all the later annalistic inscriptions.

182 Tadmor, *Tigl. III*, Summ. 7:16-22 (Rost, *Tigl. III*, 58; ARAB 1, 789ff); the nature of the booty captured on the same occasions is not specified, nor it is quantified. The numeral referring to Dūr-Baliḫāya is *40,500* (cf. Tadmor, *loc. cit.*, n. to line 21; so also ARAB) and not 50,400 (2 R, 67:21; Rost, *Tigl. III*, Pl. XXXV).

183 Rost, *Tigl. III*, Pl. XXXIV:[12ff] (Tadmor, *Tigl. III*, Summ. 11; ARAB 1, 806). See further p. 156 on these deportees.

One similar "total" may appear in a passage of the first of the mentioned *summary* inscr., reporting *6,500* captives from several eastern lands (Namri and Media). This numeral, however, has no other parallel (Tadmor, *Tigl. III*, Summ. 7:29ff; passage reported at p. 128 below).

campaign, could, for instance, include all the deported in that campaign. Or, they may include also the deported from that city (or from that region) during the subsequent campaigns.[184] Thus, in practice, we may suppose that they limited the quantifications to certain situations, using numbers that actually comprised far wider accountings. Such an expedient would permit the use of fewer, yet "higher," numbers giving the impression that they do not represent *all* the quantities. If this hypothesis is correct, then some variations could represent the updating of "totals" not explicitly declared as such. The expedient can be more easily carried out within the so-called *summary inscriptions*, which present the events in a synthetic fashion and do not give exact temporal references. This possibility, on the contrary, seems to be excluded for the annalistic texts, where there are no sure examples of "totals" representing very long periods (the entire reign) but refer rather to a single episode or at least to a much more limited period of time. The case of the 1st campaign of Sennacherib rather represents the result of an editing operation, dictated by stylistic considerations,[185] or at most aimed at concealing a "partial-total" (one single campaign) within one specific episode.

One may also doubt that this device found use within most of the *summary inscriptions*: these texts, in some cases, represent a simple selection of the events reported by the annalistic texts, displaying exactly the same wording (and quantifications), in such a way as to give a review, an "extract" of the most significant episodes of the reign.[186] In consequence, often they do not offer any new information in comparison with the longer texts. This is possibly demonstrated also by the fact that the quantifications rarely differ, except in later epochs (especially Sargon): in fact, generally speaking, the variations between homologous texts are more frequent than those between different texts.[187] However, these inscriptions show a certain variety,[188] and in some (other) cases they display the events in a way that makes it is very difficult to link them to any specific chronological context.

One concrete example is given by variation no. 6 (p. 46): Tiglath-pileser I's annals name the kings of some lands of the region of Nairi, "in all *23* kings of the lands of Nairi."[189] The battle that followed their advancing against the Assyrian army involved, in addition to other things, the capture of "*60* kings of Nairi including those who came to their help."[190] Now, in some slightly

---

184 Millard, *Fs Tadmor*, 219f. suggested — yet rather skeptically — that the "updated" numerals of the killed at Qarqar could include also the slain from other battles of that campaign (cf. also below, p. 107 and n. 252 there). Sauren, WO 16, implicitly assumed that the deportees of the 1st and 3rd campaigns of Sennacherib, from Babylonia and Judah respectively, represent a "total" referring to the whole reign (but cf. below, p. 113 n. 294).

185 Cf. further at p. 169)

186 Though perhaps farfetched, one is reminded of the sample-book of a salesman or representative or of a pop-rock compilation album (of the type "greatest hits" or "the best of").

For the practice of the "extract" cf. pp. 129ff.

187 Cf. above p. 69 (note that the "Kurḫ monolith," responsible for variations 10-14, is not at all a *summary inscription*, as Sargon's "letter" is not).

188 The characteristics of the *summary* and/or *display* inscriptions and their relation to the annalistic texts have never been systematically studied (cf., however, Tadmor, *Tigl. III*, 117f for the *summary inscriptions* of that reign, and cf. also n. 5 at p. 22 above).

189 RIMA 2, 21:71-83. Actually, the text names the lands and not the rulers: "the king of the land X, the king of the land Y," etc.

190 1 *šu-ši* LUGAL.MEŠ-*ni* KUR.KUR *Na-i-ri a-di ša a-na né-ra-ru-te-šu-nu il-li-ku-ni*, lines 96f.

Note that the event is recalled within Tiglath-pileser's titulary on the same inscription: "with 60 (1 *šu-ši*) crowned heads I vied and I achieved victory over them in battle" (cf. above n. 170)

later *summary* inscriptions,[191] in a much more vague context, but still within the military narrative, the submission of *30* kings of the lands of Nairi is mentioned. The characteristics of these inscriptions, presenting a very synthetic *résumé* of the military achievements, do not make very certain the identification with the *23* (or the *60*) kings of Nairi from the annals and the *30* could simply represent a "total" not referring to that specific episode. However, in this case it is rather difficult to say whether the *30* may represent an "updating," accounting for the capture of another 7 kings. It seems rather clear that the passage *23* > *30* represents a rounding up.[192]

The "totals" given by Shalmaneser III's inscriptions represent a quite complicated matter. His annalistic texts report, at times, some total clearly referring to the entire reign, since they appear at the very close of the historical sections or even, if there are no building reports, of the whole inscription. Thus, editions "C" and "E" of his annals give us the total of chariots and horses kept for the Assyrian army (the digits diverge from one text to the other), cf. p. 50 nos. 29-30, while the latter also states that

> 1 *me lim* 10 *lim* 6 *me* 10 *šal-lu-tu* 80 *lim* 2 *lim* 6 *me di-ik-tu* 9 *lim* 9 *me* 20 ANŠE.KUR.RA^{MEŠ}*(sīsē)* ^{ANŠE}*ku-di-ni* 30 *lim* 5 *lim* 5 *me* 65 GU$_4$^{MEŠ}*(alpē)* 19 *lim* 6 *me* 90 ANŠE^{MEŠ}*(imērē)* 1 *me lim* 80 *lim* 4 *lim* 7 *me* 55 UDU^{MEŠ}*(immerē) ḫu-ub-tu ša* TA*(ultu) reš šárru-ti-ia a-di* 20 BALA^{MEŠ}*(palê)-ia*
>
> "*110,610* prisoners, *82,600* killed, *9,920* horses (and) mules, *35,565* oxen, *19,690* asses, *184,755* sheep (this is) the booty from the beginning of my reign to my 20th year of reign."[193]

The earliest version of his annals, containing the report of the first two campaigns (accession year and 1st actual year), relates the deportation of *22,000* troops of Ḫatti.[194] This passage also appears at the very end of the inscription, and is devoid of any geographical indication — unless it refers to the last narrated episode, which tells of the tribute of Arrame received by the Assyrian king, but this is very unlikely. Obviously enough, this *22,000* represents the total of the campaign.[195] However, it does not appear in the next version, the "monolith" inscription (edition "A"), which takes up, even if with some variants, the text of those two campaigns, while adding some others. It is a clearly annalistic text, since each campaign is introduced by the name of the eponym. In contrast, edition "B" of his annals actually contains only a *résumé* of the events of the first years of the reign, not dated in any manner, followed by the short reports of the 3rd and 4th year (not dated) and by a report of the campaigns of the 8th and 9th *palû* (dated with eponyms; cf. p. 31). Within the first of the mentioned sections, at the end of the passage

---

191 Cf. p. 46.

192 For this reason the "variation" is considered as such and is mentioned in its proper place.

193 Safar, *Sumer* 7, 13:34ff = Michel, WO 2, 40:34ff. Safar reads *35,555* for the oxen, but cf. photograph in *Sumer* 7, Pl. II = WO 2, Tf. 3. Cf. above, n. 168.

194 20 *lim* 2 *lim* ERÍN.ḪÁ *mat Ḫat-ti a-su-ḫa a-na* URU*-ia Aš-šur ub-la*, Mahmud - Black, *Sumer* 44, 148 and 142:last line (they actually correct the "2 *lim*" to "2 *me*," thus obtaining *20,200*, but the writing "X *lim* + X *lim*" is not so unusual).

195 Only this campaign concerned the land of Ḫatti: that of the accession year took place in another direction.

that lists the extension of the king's conquests, "from the sea of Nairi and the sea of Zamua ... to the great sea of Amurru, the land of Ḫatti to its farthest border," the deportation of *40,400* troops is recalled:

> 40 *lim* 4? *me* LÚ.ZAB.GAL.ḪÀ*(ṣābē) ina* KUR*(māti)-šú-nu a-su-ḫa a-na* UN^{MEŠ}*(nišē)* KUR*(māti)-ia am-*[*nu*]
>
> "*40,400* soldiers from their land(s) I uprooted (and) with the people of my land I counted"[196]

That this quantification represents a "total," and not a single episode, is thus obvious from the context. In consideration of the exceptionality of the number, it is not so unlikely that it refers to the first 9 years of the reign, or, better, to all the military actions that, during the first 9 years, concerned the mentioned lands. The quantification would then appear in that place for a simple geographical association — keep in mind that, in general, these texts tended to antedate as far as possible the "victories".[197]

In a *summary* inscription carved on a throne-base Shalmaneser III states: "I carried off *87,500* soldiers of the land of Ḫatti."[198] In this inscription the events are not dated,[199] and in any case are presented in a very approximate chronological order. That short statement appears at the end of the passage referring to the 1st campaign, and could represent the same "total" referred to above, since it recalls, more or less, the same geographic area. It is followed immediately by the news of the deportation of Aḫuni, an event that, according to the annals, took place during the 4th *palû*. Now, also the above mentioned edition "B," within a passage clearly referring to the 4th *palû* (though not dated), associates the deportation of Aḫuni with that of (another) *17,500* soldiers — a number showing some analogy with that given by the throne-base inscription (*87,500*). These *17,500* deportees of the 4th year in the later editions "C" and "E" of Shalmaneser's annals will become *22,000* (cf. p. 49 no. 24), exactly as many as the deportees reported by his oldest annalistic text. At this point, to make things less confusing, a small table summarizing Shalmaneser III's quantified deportations will be necessary (inscriptions in chronological order):

196 Michel, WO 2, 410 col. II 3 (ARAB 1, 617). The passage does not appear in Pinches or Rasmussen's copies: cf. however Unger, *Balawat*, 24f and *id.*, MDAIA 45, 90f, or the photograph in WO 2, Tf. 11 (3rd fragm.).

197 Cf. above p. 74 n. 111.

198 Hulin, *Iraq* 25, 54:25f.

199 Except for a reference to the 13th *palû* and to the 10th Euphrates crossing; cf. p. 137. This inscription, however, clearly aims at presenting a general "synthesis" including all the events of the reign and not just a series of single episodes. It contains some passages so vague and generic that are difficult to date, for instance the journey to the western sea (line 42) or the references to the western lands (lines 35f: "the splendor of my lordship I spread over the lands of Ḫatti, Meṣri, Tyre, Sidon and Ḫanigalbat"), while even when it makes reference to events which occurred at specific moments (exactly datable thanks to other documents) it often uses original words and expressions.

Inscr.	Number	Year*
Sumer 44	*22,000* troops from Ḫatti	(1st - a "total")
Ed. "B"	*40,400* soldiers	(summary of conquests)
Ed. "B"	*17,500* troops of Aḫuni	(4th)
Throne-base	*87,500* troops from Ḫatti	(1st?)
Ed. "C"	*22,000* troops of Aḫuni	4th (dated)

* The Year to which the directly preceding events belong.

What then, does the *87,500* of the throne-base represent? The interpretations that could be pointed out are: (a) it is a "total" referring to the 1st campaign (and in this case it would represent a "variation" of the preceding *22,000*); (b) it is a "total" referring to several years (and thus including also the *40,400* and the *17,500* from above); (c) it makes reference to the deportation of the 4th *palû* (and in this case too it would represent a "variation" in respect to the preceding *17,500*). In consideration of the structure and characteristics of this inscription, the second hypothesis seems to be quite likely.[200] It would be a "total" referring to the entire reign, and inserted here in order to respect the discursive and geographical continuity of what was being narrated in that place. It does not represent, however, a simple displacement of the "totals" (from the close of the text to within it, as in the case of Sennacherib), nor the result of the desire to concentrate all the "results" in a single, yet "higher," quantification — this would imply a substantial honesty in using "genuine" data, which does not seem to be the case. In fact, if in the narrative this quantification appears as a total, thinking of the analogies shown by the numerals (*22,000 - 40,400* and *17,500 - 87,500*) one may have the impression that they could be the invention of a not particularly creative mind, rather than the result of a real "sum." To say more, one would need to know something about the compositional processes of these kinds of inscriptions.[201] It seems likely to me that the scribe who composed this *summary* text, while gathering the information according to his aims and procedures (whatever they might have been), and wanting to include one quantification of deportees and one of killed enemies,[202] decided to use the former in reference to the Hittite deportees (the most relevant deportation of the reign: cf. the table at p. 196). He then put forth the highest numeral available to him — i.e. a "total" of *87,500*, the result of some kind of "sum" — or perhaps the most suitable approximation coming to his mind, thus taking "inspiration" from the mentioned *17,500*.

In any case, the "totals" appear limited to those contexts making clear reference to the entirety of the reign, or at most to some texts that clearly exhibit a *summary* fashion, as this throne-base inscription of Shalmaneser III.

---

200 Schneider, *Shalm.*, 221 suggests the deportation refers to the 2nd *palû*, yet on p. 99 she interprets the *87,500* as a "total" for the whole reign.

201 Cf. here above, n. 188. The more general subject-matter "composition of the royal inscriptions" will be taken up at the beginning of the next chapter.

202 The text has one of each such quantifications. That referring to the killed also represents a variation: *13,500* soldiers of Aramu killed, which in ed. "B" was reported to be *3,000* (3rd *palû*; cf. p. 49 no. 23). These same remarks can partly apply to this quantification too (which has been inserted into the variations list mostly for reasons of completeness).

In contrast, the using of "totals" hidden within the narrative of the campaigns seems to be excluded. In this case one would presuppose that the editors of the inscriptions were bound to use real data, though merged together and concentrated in a single place.[203] If this were so, the quantifications (or some of them) would constantly increase, which happens rather rarely. Furthermore, in this case the inscriptions would present, at least as a general rule, non homologous quantifications, that is, one quantification only for each object, situation, or geographical area. But apparently the opposite is true: the inscriptions tend to quantify certain situations rather than others. While the "annals" of Ashurnasirpal II and those of Shamshi-Adad V often quantify the enemies (killed, captured or other), those of Tukulti-Ninurta II never do. Similarly, the annals of Tiglath-pileser III often report quantifications of deported, listing digits and names of cities one after the other.[204] And, in general, it would not be difficult to find homologous quantifications within a single inscription: for instance, going back to Shalmaneser III, editions "C" and "E" of his annals quantify enemies killed or captured only in relation to the Syrian campaigns, while they report other massacres, not quantified, made in other regions. Edition "F" preserves only one of these quantifications (the killed of Qarqar), yet without "carrying forward" the other numerals suppressed, in respect of edition "E," and referring to the same area and even to the same enemies, the coalition of the *12* western kings.[205] In conclusion, thus, it seems not so likely that devices of this kind have been employed.

---

203 See further below chap. 4 §13. Cf. also the remarks at p. 107.

204 Cf. chapter 2 in general, and pp. 100ff for Tiglath-pileser III.

205 Quantifications of enemies in editions "C" and "E": 4th *palû*: *22,000* prisoners (Aḫuni son of Adini); 6th: *25,000* killed (at Qarqar); 11th: *10,000* killed (against the western coalition); 18th: *16,020* killed (Ḫazael of Damascus). To ed. "E" is appended the total of *110,610* prisoners made between the beginning of the reign and the 20th *palû* (cf. p. 93), while only on three occasions are prisoners mentioned: the *22,000* of the 4th *palû*, and some others captured during the 13th and the 18th, simply defined as *šallatu* (they seem not very relevant episodes). To these add some references to enemies killed but not quantified during the 9th, 12th and 20th *palû*, in Babylonia, Syria and Anatolia respectively (cf. Cameron, *Sumer* 6, and Safar, *Sumer* 7, *passim*). Edition "B" is ignored here since it does not report all the campaigns, while "D" is fragmentary. For edition "F" cf. Michel, WO 2, 137ff or ARAB 1, 555ff.

APPENDIX TO CHAPTER 3

SOME "ORDINARY" QUANTIFICATIONS

In this appendix the argument of "possible quantifications" will be discussed, mainly in relation to tribute, yet leaving out any all-inclusive evaluation (evidence from administrative documentation, etc.), which indeed should be necessary, but which is beyond the scope of this study. In the first part, three "realistic" situations will be isolated. This limitation stems from the fact that, as is obvious, the royal inscriptions are not the most appropriate documents to examine for unquestionably authentic information. These three situations may have an exemplary value, since they deal with specific situations and restricted contexts, such as a single campaign or part of it. In this case, a lower number of interpretational problems arises. Expanding the studying of quantities to a whole reign or more, for example adding up the data given by the inscriptions to obtain the total of deportees, the total of income from the war booty, or the population of a certain (wide) area, etc., would inevitably lead to very uncertain and approximate results.[206] It also would need to face several textual and historical problems to avoid attaching too much value to what are no more than standard phrases. Each inscription or each part of an inscription has its own "literary" or compositional history.[207]

1. The **annals of Tukulti-Ninurta II** contain, as already mentioned (p. 27), a long passage of "show of strength," actually formed by a detailed report of a campaign in the south (towards Babylon) and in the south-west (middle Euphrates). This "itinerary" is much more detailed than other such itineraries, still taking place in the same area, included in inscriptions of Adad-nerari II and Ashurnasirpal II, that are not even reporting many quantifications.[208] Some amounts of *nāmurtu* tributes are recorded, all presented in a rather stereotyped way.[209] All the listed items are quantified except

206 This is the case with the recent study of Liverani, *Asn.*, where (p. 133ff, "demography"), on the basis of the data given by the "annals," are calculated the killed, the deportees, etc. made during the reign of Ashurnasirpal II, as well as the population of certain peripheral areas of the empire. The results obtained are highly hypothetical and roughly approximated. Only a comprehensive study of the quantifications in the "annals" of Ashurnasirpal II would allow the elaboration of such a statistical approach to the complex question of demography, complicated as it is (note that the tables at p. 133 and 135 include some oversights: e.g., the "population" of Bara should be corrected to 620 = *320* + *300*). Cf. also below, n. 212 (end).

207 Annalistic inscriptions usually show sections composed at different times and in different places, since the later annals often re-used the text of older editions for the part of narrative covered by them. This obviously implies that each part of these inscriptions has, or could have, a different compositional background, thus presenting diverging textual problems.

208 Cf. Russell, *Iraq* 47, 60f and Kühne, BaM 11, 48, on how the geography of this "itinerary" is the most detailed.

209 Cf. below, p. 200 for an overview of the tributes (the lists are there slightly abridged). Eleven times the tribute is indicated with the term *nāmurtu*: X X X *na-mur-tu ša* NP and/or NL *i/at-ta-ḫar* (with X X X = list of commodities, NP = pers. name, NL = city or region; cf. ll. 76ff, 85, 86f, 88f, 90ff and 103; in ll. 98ff no verb is used). In another two cases the list is placed between *nāmurtu ša* NP / NL and *attaḫar* (ll. 93ff, 105ff), as in two further cases: ll. 109ff, where the verb is missing in a lacuna, and lines 69ff, where the verb is *am-ḫur* (passage quoted below, p. 122). This case, which is the first actual tribute, shows another deviation: the *nāmurtu* is labeled ḪI.A.MEŠ ("numerous"), even if it does not seem to be the most considerable, at least as far as the quantities are concerned. The passage in ll. 113f presents only the list without *nāmurtu* and without any verb. Only (perhaps) in l. 105 do we have a tribute not quantified (the passage is fragmentary: it could be booty as well). Note that in two cases for a single individual two tributes are recorded: Ḫamatāya of the land Laqû (ll. 86f and 98ff) and Ḫarānu still of Laqû (88f and 93f) — and probably this is also the case with Mudadda of the land Laqû (85) or of the city of Sirqu (90ff). In all three cases the first tribute is composed of oxen, sheep and supplies only, while the second is much more sizable. Thus it appears that, after having received an ordinary tribute, the king expected from his tributaries another (more consistent) payment, received in another place, and probably agreed upon during the former meeting (however, both are labeled *nāmurtu* = gift, extraordinary tribute, cf. Postgate, TCAE, 154ff).

in a few instances.[210] This allows them to be added easily — besides the various metallic or wooden commodities, here omitted for brevity,[211] the total thus includes:

commodity	not qtf.	qtf.	total	[calculated total]
silver	1	6	*3* t *57* m	[*4* t *331/2* m]
gold (incl. *liqtu*)	-	6	*70*? m	(= *1* t *10* m)
tin	-	5*	*51* + *x* t	[*63* t *40* m]
tin bars	-	1	*18*	
bronze	-	2	*130* t *3*? m	
iron (AN.BAR)	-	2*	*2* + *x* t	[*4* t]
myrrh (ŠIM.SES)	-	3	*4* t	
antimony prepar.	-	3*	*x* t *14* m	
antimony	-	1	*8* m	
purple wool	1	1	*1* t + *x*	[*2* t]
dromedaries (*udru*)	-	1	*30*	
oxen	3	8**	*430* + *x*	[*670*]
asses	-	3	*70*	
birds	-	4*	*134* + *x*	[*179*]
sheep	3	8*	*2,800* + *x*	[*4,342*]

The number of times in which each object is not quantified and the number in which it is are shown (each asterisk marks a totally or partly missing number). The entry "birds" includes ducks and geese. The last two columns report the sum of the readable digits and, in brackets, the "total" obtained applying the mean of these quantifications to the cases not quantified, and summing up everything.[212]

These sums can be considered indicative of how much a campaign of this kind could have "yielded" in terms of tribute. The context seems very realistic, considering the numerous geographical details there related,[213] the variety of the commodities mentioned by the lists[214] and the relatively modest level of the numerals (the presence

---

210 Outside the lacunae, the following are not quantified: the booty (*šallatu*) of l. 50 (only people? — cf. n. 39 on p. 27), a tribute (?) or perhaps booty at l. 105 (fragmentary), booty at ll. 122f (*šallassunu* NÍG.ŠU.MEŠ-*šunu* GU_4.MEŠ-*šunu* UDU *ṣēni*.[MEŠ-*šunu* ...]).

From the first tribute, garments, wool, oxen and sheep are not quantified (l. 72), in another case the quantity of the alabaster (NA_4.GIŠ.NU_{11}.GAL) and that of the "refined oil" (Ì.GIŠ DÙG.GA) are not specified. Never quantified are "bread, beer, grain, fodder" (which represent the supplies for the troops).

211 Comprehensively, *4* objects of ivory, *47* in *meskannu* wood, *150* bronze utensils (*u-da-e*), *60* bronze BAD (?), *70* bronze casseroles, *3* bath tubs (of which 1 is of bronze and *1* of silver), *100* iron daggers (GÍR), and *150* garments. The horses, one of the most valued goods, are mentioned only once, without being quantified (l. 105).

For the *130* (or *140*?) talents of bronze, cf. p. 185 n. 5. Note that RIMA 2, 177:106 translates "10 talents of silver" (so also ARI 2), instead of "10 minas," as in the transliteration (correct).

212 This same system has been employed by Oded to calculate the total of the deportations referred to in the Assyrian royal inscriptions, which produced a grand total on the order of four and a half million people (*Deportations*, 19ff and n. 5 there). Actually, Oded's conclusions are not warranted, since he applied his calculations to a very wide context, where different kinds of quantifications ("totals," etc.) are added up regardless of the context. Furthermore, since only a few cases are quantified (43 out of 157), and since it is possible that these represent the biggest deportations (see chapter 4 §14), the average of the given quantifications could not apply to the instances not quantified.

213 The text, as it appears, presents some *gaps* in its final part, in the concluding section of the "itinerary" (cf. E.I. Gordon, "The meaning of the Ideogram ^{d}KASKAL.KUR ...," JCS 21 (1967), 86). However, the geographical references make it seem a substantially realistic text that has not been summarized, and this is possibly demonstrated by the numerous studies on the historical geography of the area it has stimulated (e.g. those mentioned in n. 208).

214 Cf. footnote 111 here above. Note that in the last three cases of tribute no oxen or sheep are mentioned (ll. 106ff), this is quite surprising, since these animals are almost ubiquitous. However, there is no need to hypothesize that these lists were abridged, since they do not appear at the end of the inscription and not even at the end of the "itinerary" passage. Thus, possibly, this "itinerary" represents a genuine field-record, devoid of significant chancellery revising. The fact that some items of *booty* have not been

of several round numbers is probably explainable by the fact that the quantity of items of tribute were fixed in advance). Obviously, a stronger military engagement — that in the case of the "show of strength" was modest — would have yielded higher quantities of booty. But remaining in the field of tribute, that is of pre-fixed and/or "voluntary" payments, it is obvious that it was difficult to draw from a given area for long periods of time without causing its fast decay. The tribute was obviously established in a measure somehow proportioned to the "productivity" of the area, and could not vary so much in the course of time. In the case of booty, which, as a rule, implied the looting of everything that had been accumulated, the quantities could be much higher, but the possibility of repeating the act in the same place after a short time was excluded.

2. The "**Kurḫ monolith**" of **Shalmaneser III** quantifies tribute only within the passage containing the report of the campaign of the second year (cf. p. 30). The other campaigns, though containing frequent references to tribute, of which at times also the composition is specified, do not relate quantifications, if we except *2* and *7* dromedaries of Asû and Asāu (cf. p. 123) and some general indications such as *ma'du*, *ana lā mēni* and the like. In the passage under discussion 4 tributes received and 4 imposed on a total of 5 different Syrian principalities are fully quantified.[215] They have without doubt a lower "exemplifying" value if compared with those of Tukulti-Ninurta referred to above. The context is not so clear (the narrative is very synthetic), and some tributaries that one would expect to find do not appear. During a former campaign (1st full year of reign) taking place in the same area, after having received the tribute (*madattu*) of Kummuḫu and Gurgumu (col. I 37 and 40f), Shalmaneser had twice engaged with a coalition formed by Sam'al, Patinu, Bīt-Adini and Karkemish. Later on, he received also the tribute (*madattu*) of Bīt-(A)gusi (col. II 12f). During the 2nd year, at the end of a short campaign in Syria that involved the submission of "all the kings of Amurru (?)" (cf. col. II 20), the above mentioned items of tribute had been received from the kings of Patinu, Sam'al, Bīt-Agusi, Karkemish and Kummuḫu. The text presents the events as if they happened on a single occasion: the tribute lists are directly juxtaposed (only the names of the tributaries separate each of them). For each principality the tribute received on the spot[216] and the tribute imposed[217] are reported, the latter for all except for Bīt-Agusi, the former for all except Kummuḫu. This could perhaps be explained by the fact that this country was a tributary since the preceding year, and thus was not required to pay any "tribute of submission".[218] Yet one would ask why Bīt-Agusi, which also was a former tributary, pays a tribute of submission,[219] while Gurgumu is absent from the list.[220] It is possible, therefore, that the related tribute represents only an "example," an "extract" (cf.

quantified (cf. n. 210 above) is not an indication of incompleteness. Possibly, "in the field" the booty, contrary to the tribute, was probably recorded according to very rough estimates (except for the "treasure," appanage of the king), if it was at all. Rather, the texts reporting exact quantities for the booty should be viewed with a little suspicion.

215 3 R, 7f II 13b-30a. Two groups of booty cited at lines 18 and 20 (*šallassunu ašlula*), at the beginning of the campaign, probably refer only to "prisoners" (cf. above p. 8 n. 46).

216 Neither *nāmurtu* (as one would expect) nor other terms are used; there is not much doubt, however, that "extraordinary" tribute is intended: cf. Peñuela, *Sefarad* 9, 16 n. 31.

217 In two cases termed *madattu* (regular tribute); in the third case no term appears: 1 *ma-na* GUŠKIN 1 GUN KÙ.BABBAR 2 GUN SÍG.ZA.GÌN.SA$_5$ *ina eli-šú áš-kun*, "*1* mina of gold, *1* talent of silver (and) *2* talents of purple wool upon him I established." In the last case (3 R, 7 II 29f) the verb used is *maḫāru* (Gt), but it is in any case an imposed tribute: 20 *ma-na* KÙ.BABBAR 3 *me* GIŠ.ÙR GIŠ.*e-ri-ni* MU-*šam-ma am-da-ḫar*, "*20* minas of silver (and) *300* cedar logs yearly I received."

218 This is the explanation given by Peñuela, *Sefarad* 9, 24.

219 Still, Peñuela (*Sefarad* 9, 20 n. 42) suggested that the tribute imposed on Bīt-Agusi had been left out by the scribe by mistake. Note how also the list of tribute *received* from Bīt-Agusi is much shorter than the others — possibly, instead of an error, we may hypothesize a deliberate abridgement. This could apply to the list of Kummuḫu as well.

220 Cf. the remarks of Grayson in CAH^3 3/1, 260.

chapter 4 §6), simply aimed at concretely demonstrating the (positive) end of a campaign rather poor in military successes.[221]

In any case, leaving aside these reservations,[222] we can calculate the sum of the received tribute:[223]

commodity	n. qtf.	partial amounts	total
gold	3	*3* t, *10* m, *3* t	*6* t *10* m
silver	4	*100* t, *10* t, *6* t, *70* t	*186* t
bronze (ZABAR)	3	*300* t, *90* t, *30* t	*420* t
iron (AN.BAR)	3	*300* t, *30* t, *100* t	*430* t
bronze cauldrons (ÚTUL)	1		*1,000*
multicol. garments	2	*1,000, 300*	*1,300*
weapons (TUKUL)	1		*500*
princesses (w. dowry)	-		*3*
daughters of nobles	1		*100*
purple (ZA.GÌN.SA$_5$)	2	*20* t, *20* t	*40* t
oxen	4	*500, 300, 500, 500*	*1,800*
sheep	4	*5,000, 3,000, 5,000, 5,000*	*18,000*
cedar logs (GIŠ.ÙR)	1		*200*
cedar resin (ÚŠ *erini*)	1		[*X*$^?$+] *2 emāru*

The imposed tribute amounted to:[224]

gold	1		*1* m
silver	4	*1* t, *10* m, *1* t, *20* m	2 t *30* m
purple	2	2 t, 2 t	*4* t
cedar logs	3	*200, 100, 300*	*600*
cedar resin	-		*1*$^?$ *emāru*

These amounts do not really seem exceptional, especially if we think that they refer to the entire Syrian area. However, one has to keep in mind also the possible exclusion of some tribute for narrative or stylistic reasons, even if, after all, it was not in the authors' interests to lower the (total) amount of the tribute, not without having "revised" some numeral to make up for the (possible) deficiency.

3. The main subject of the **8th *palû* of Tiglath-pileser III**'s annals from Kalah is the subjugation of Syria and the annexation of some of its provinces. This section is

221 The tribute represents the *climax* of the narrative, and its detail, unusual for the inscription, is probably due to the need for enriching a campaign rather lacking in events and devoid of armed engagements (= "victories") and thus of amplifying its results (note also that it was the year of the eponymate of Shalmaneser himself — as specified by the inscription, col. II 13).

222 Olmstead thought that these digits are truthful, since "their very modesty is the best proof of their authenticity" (JAOS 41, 357). Peñuela too believed the numerals are credible (*Sefarad* 9, 12-14), even if there are some reasons to be skeptical. The main argument in support of their credibility is, according to Peñuela, the differentiation of the digits, a fact that does not seem so evident to me. The inscription shows a clear preference for the low digits ("1," "2" and "3," in order) and contains a very high percentage of round numbers (cf. also below about the battle of Qarqar). The livestock is constantly given in the ratio of 1:10 (1 ox:10 sheep), which could represent a rule for the imposed tribute, but yet a rather singular coincidence for the "extraordinary" tribute (cf. also nn. 219 and 221 here above).

223 The *90* talents of bronze at l. 24 could be read *30*, the signs being rather uncertain. For the *30* talents of iron (l. 25) cf. Craig, *Monolith Inscr.*, p. 32 nos. 87 and 89. Three times a "daughter with her (rich) dowry" is mentioned (no numeral is used).

224 Cf. Craig, *Monolith Inscr.*, p. 32 no. 89 for the 1 *me* cedar logs of line 26. For the cedar resin no digit is given: probably the scribe has forgotten the number (in view of the fact that he uses the plural), otherwise we should understand "(one) *emāru* of resin." Craig reads *1 emāru*, but transliterates it as *100* ("C") at p. 12 (*ibid.*, no. 90; cf. also Rasmussen, *Salm.*, pl. V and pp. 16f). He also interprets the *1* mina of gold of line 29 as "60" — this is very unlikely since the scribe would have written *1* talent.

concluded by the account of the deportation of *83,000* [+ X$^{?}$] Syrians to Tuš[ḫān] and of another *1,223* to Ulluba.[225] Subsequently, the annals describe a campaign to southern Babylonia led by some high-ranking officials. Then follows a passage including several quantifications of deportees originating from certain southern Babylonian tribes or cities, not mentioned in the preceding section. They are resettled in some Syrian cities or provinces, which, in this case correspond — at least partly — with those mentioned earlier. The passage is not introduced in any manner,[226] being directly appended to the narrative of the campaign in Babylonia. It is followed, in turn, by a list of western tribute-bearers, also devoid of any introductory formula, whose tribute is subsequently itemized, but not quantified.

The passage pertinent to the deportees — the subject of this discussion — can be given directly in translation in view of its simplicity. It may be divided into four sections:[227]

(1)
600 captives of ^{c}Amlate of the ^{p}Damunu
5,400 captives of ^{c}Dēr
(in ^{c}Kunalia) [...] ^{c}Ḫuzarra ^{c}Tae ^{c}Tarmanazi ^{c}Kulmadara ^{c}Ḫatatirra ^{c}Irgillu, [cities] (of ^{l}Unqi, I settled)

(2)
[...] captive highlanders[228] (of) ^{l}Bīt-Sangibūti
1,200 ^{p}Illilāya
6,208 ^{p}Nakkabāya, ^{p}Būdāya [^{c}Dunāya(?)
in ^{c}Ṣi]mirra, ^{c}Arqā, ^{c}Usnū, ^{c}Siannu of the seacoast, I settled

(3)
588/9 ^{p}Būdāya ^{c}Dunāya
[XXX (?) ^{p} ...][229]
[X+(?)] *252* ^{p}Bilāya
554 ^{p}Banītāya
380 ^{p}Nergal-andil-māti
460 ^{p}Sangillu
[... ^{p}]Illilāya
458 captive highlanders (of) ^{l}Bīt-Sangibūti
in the district ^{c}Tu'imme I settled

(4)
555 [captive] highlanders (of) ^{l}Bīt-Sangibūti
in ^{c}Til-karme I settled
With the people of ^{l}Aššur I counted them. Corvée (and) labour like that of the Assyrians [I imposed upon th]em.

^{p} = people (LÚ); ^{c} = city (URU); ^{l} = land (KUR); XXX = missing numeral

---

225 Tadmor, *Tigl. III*, Ann. 19:11f (Rost, *Tigl. III*, 22:132f; ARAB 1, 770). The first numeral has been previously read as *30,300* (cf. Tadmor, *Tigl. III*, 62 n. 11). It appears in a very fragmentary context, and it is not clear whether it refers to people or not.

226 For example with a temporal reference, such as *ina ūmēšuma* ... "in those days ...", employed within the subsequent slab of the Kalah annals (Tadmor, *Tigl. III*, Ann. 14:8).

227 Cf. Tadmor, *Tigl. III*, Pl. IX = Ann. 13:3ff (Rost, *Tigl. III*, 24:143ff; ARAB 1, 772). The beginning of each line is broken: up to line 7, some 10 signs are lacking, from line 8 on the missing signs are about 5 (*ibid.*, p. 31). Passages appearing in a parallel fragm. text (*ibid.*, Pl. II = Ann. 2:5ff + Ann. 3:1f) are placed in parentheses.

228 "Highlanders" stands for ^{l}Qutē.

229 The lacuna allows enough space for a further group of captives to be quantified.

All the deportees come from southern Mesopotamia,[230] except the "highlanders" of Bīt-Sangibūti, which is in the Zagros region.

In (1) two different groups of deportees are settled in 8 cities of Unqi, on the seacoast of northern Syria. The total is 6,000.

In (2) three groups of deportees are settled in 4 cities of the Phoenician coast. Total: 7,408 + X.

In (3) seven or eight groups of deportees are settled in the province of Tu'imme (northern Syria, in the country of Pattina). Total: 2,440/1 + X.[231]

In (4) one group of deportees is settled in the city of Til-karme (northern Syria),[232] 555.

The "highlanders" of Bīt-Sangibūti appear three times, totalling 1,013 + X. In (2) they are listed as the first group, in (3) they are appended at the end of the list, in (4) they represent the whole group of deportees.[233] Also the Illileans appear in (2), after the mention of Bīt-Sangibūtu, and in (3) before it (in this case the number is broken). The Budaeans appear twice as well, as (apparently) the last group of (2), in this case quantified together with the Nakkabaeans, and as the first group of (3).

Because of the terse style and of the abrupt inception of this passage, one may suppose it relates genuine data, drawn from some kind of official record.[234] In any case, given the context in which it occurs, it *had to appear* as an official record. It was inserted in this position, after the account of the annexation of certain Syrian provinces, to underscore how the "normalization" of Syria had been throughly carried out, by resettling deportees from the opposite part of the empire, which was simultaneously[235] subdued by the royal officers. In such a context, to relate exact figures, names and places in a somewhat technical mode, devoid of any rhetoric, was thought to be more effective.

As for the numerals, there are some arguments in favor of their authenticity. First, their modest magnitude, then, the variety of the digits involved.[236] In this sense, the fact that (1) and (2) record round digits, in contrast to (3) and (4), has to be positively viewed. These figures are, obviously, approximations: it depended on the officer responsible for each group of deportees to calculate their number, thus some of them recorded exact figures, while some others reported approximated quantities,[237] an easier approach if one had to deal with larger groups of deportees.[238]

The grand total is 16,403 (or 16,404). The original total was higher, since two or three numerals are missing, while another, *252*, possibly has to be restored as *352* —

---

230 As far as it is possible to establish the location of each city. Cf., in general, Oded, *Deportations*, 116ff or, for the towns mentioned in neo-Babylonian texts, R. Zadok, *Geographical Names According to New- and Late-Babylonian Texts* (Tübinger Atlas der Vorderer Oriens Bh., Reihe B/7, 8), Wiesbaden 1985, *passim*. For the regions of destination cf. E. Forrer, *Die Provinzenteilung des assyrischen Reiches*, Leipzig 1920, 56-59.

231 One or two quantifications missing, cf. n. 239 below.

232 On the border between Unqu and Gurgum (cf. Tadmor, *Tigl. III*, 67 n. 9).

233 Note how, in (2) and in (3), the term *šallat* is prefixed, in a different manner from the other peoples listed.

234 Cf. also below, p. 171.

235 According to the annals, the booty taken in Babylonia had been brought before the king who was in Ḫatti at that moment: cf. Tadmor, *Tigl. III*, Ann. 19:20 (cf. also lines 14 and 18).

236 The passage shows the following digits: 1 (3 times), 2 (5 or 6 t.), 3 (1 or 2 t.), 4 (4 t.), 5 (9 t.), 6 (3 t.), 8 (2/3 t.), 9 (0/1 t.). Only 7 is absent, while 9 appears as a variant. The high frequency of 5 is partly explained by the fact that most of the groups tend to approach a size of 500 people.

237 The approximations possibly concerned women and children: cf. how the letter ABL 304 (mentioned here above, n. 125) records *1,119* able-bodied men and *5,000* persons in all, including their families.

238 Note also how (1) and (2) carefully record the cities of destination, while (3) is exact in recording the city of origin (and the quantities) but rather approximative in regard to the destination. Could it be that (3) depended on information given by officer(s) responsible for the dispatching of the captives, but not concerned with their relocation?

still, only with difficulty would we get more than 20,000 in all.[239] Thus, the quantities involved are not so striking.[240]

SOME "EXTRAORDINARY" QUANTIFICATIONS

In this section, some of the most "extraordinary" quantifications will be discussed, trying to explain the data in some manner — even at the cost of being somewhat reckless.

4. Shalmaneser III — Battle of Qarqar.

The forces of the Syro-Palestinian coalition are quantified as follows by the monolith inscription:[241]

Chariots	Horses	Troops	Ruler / Country[242]
^{GIŠ}GIGIR	*pētḫallu*	ERÍN	(i=I, l=KUR, c=URU)
1,200	*1,200*	*20,000*	i Adad-idri [l *ša*] ANŠE.ŠÚ (Damascus)
700	*700*	*10,000*	i Irḫuleni l Amatta (Hamath)
2,000		*10,000*	i Aḫabbu l Sir'ala (Ahab of Israel)
		500	l Gua [Byblos?]
		1,000	l Muṣri [Egypt]
10		*10,000*	l Irqanata
		200	i Matinu-ba'li c Armada (Arwad)
		200	l Usanata
30		*10$^{?}$,000*	i Adunu-ba'li l Šiana
1,000 camels (ANŠE *gam-ma-lu*)			i Gindibu' l Arbâ (Arabia)
		[X],000	i Ba'sa son of Ruḫubi l A(m)mana (Ammon)

Adding up the digits (which the inscription does not) we get *3,940* chariots, *1,900* *pētḫallu*, *1,000* camels and at least *62,900* infantry if we compute *1,000* infantry for

---

239 The number of "highlanders" from Bīt-Sangibūti was in the thousands. The other two missing numerals were, however, on the level of some hundreds — note that, within section (3), the quantities are very consistent.

A similar passage, with the quantities of several groups of deportees recorded one after another, appears in two parallel texts, also part of the Kalah annals, see Tadmor, *Tigl. III*, Ann. 18 and 24:4′ff (Rost, *Tigl. III*, 38:232ff; ARAB 1, 779). The fragmentary state of the two slabs in this case does not permit any kind of analysis; cf., however, how in this case too most of the recorded quantities approach the value of 500 (*625, 226, 650, 400* [+ X], *656*). The only exception is a *13,520* [+ X?] appearing in Ann. 24:9′, possibly representing the total of the group.

240 It is not clear whether this passage records the deportees of that specific campaign to southern Mesopotamia, or, rather, if it relates the results of operations involving the implementation of people taken throughout several years (cf. a remark in this sense in Tadmor, *Tigl. III*, 67 n. 5) — deporting many thousands of people through such a distance obviously requires a lot of time. One may, in any case, suppose that the passage under discussion represents only a kind of "sample," as suggested in the case of Shalmaneser III's tributes (above, n. 219).

241 Copy: 3 R, 8:90ff. Chariots, horses and troops are always listed in this order, unless absent — directly following is the determinative pronoun *ša*, "(that) of," and the name of the ruler and/or of the region of provenance. On this coalition of "*12*" kings, cf. also pp. 134ff.

242 The identification of some of these lands is rather disputed (cf., in general, Gelio, RBI 32, 122f n. 7): the name of KUR Gua, according to the emendation *Gu-<bal>-ai* proposed by Tadmor (IEJ 11, 144f, cf. also 149f), hides that of Byblos (KUR is here restored according to Craig, *Monolith Inscr.*, p. 32 no. 172a). For KUR *ša* ANŠE.ŠÚ = Aram (country of Damascus), cf. Sader, *États araméens*, 260ff or Pitard, *Ancient Damascus*, 14ff. For A(m)mana see, recently, G.A. Rendsburg, "Baasha of Ammon," *Journal of the Ancient Near Eastern Society* 20 (1991), 57-61.

Ba'sa.[243] During the battle *14,000* of these warriors will be killed by the Assyrians (more according the later versions, cf. variation no. 25). Needless to say that, if these numerals are true, it had to be one of the biggest military concentrations ever seen.

The fact that the forces furnished by Ahab of Israel (= Aḫabbu of Sir'ala) are particularly relevant (especially as concerns the chariots) has raised some doubt. Na'aman[244] thought that *2,000* (chariots) is an excessive quantity for a land such as Israel (in the middle of the 8th century), and that it could be a scribal error for *200*. Elat[245] on the contrary maintained that the scribe of the "monolith" was particularly careful with the numbers, in view of the frequency of the quantifications appearing there, while, on the other hand, Israel was capable of having considerable forces, since its political and economic situation was favorable.[246] However, the "monolith" has many scribal errors (a fact noted also by Elat). Olmstead had faced the issue some years ago, showing himself skeptical of these numerals: he observed that the O.T., besides not mentioning the episode, lets us think that during the wars with Ben-Hadad Israel was lacking in battle chariots.[247] In any case, these are the reasons that may allow us to think that these numerals had been altered or fabricated:

1. These numbers, if added, give a total of almost 75,000 men-at-arms: at least 62,900 foot (perhaps thousands more), 1,900 horse, almost 8,000 charioteers (allowing a crew of two)[248] and 2,000 cameleers (*idem*), and this does not include those assigned to logistics and supplies, followers, etc. This seems excessive, considering they are all coming from Syria (the Arab and Egyptian contingents, according to the inscription, were composed only of *1,000* camels and *1,000* troops respectively).

2. It was presumably difficult for the Assyrians to know exactly the consistence of the enemy forces. It could have been possible through the use of informers, or comparing the enemy and the Assyrian armies as displayed on the battlefield (knowing the size of the latter). Or, after having victoriously concluded the battle, it was possible to count the enemies killed and captured, to which add those escaped (also estimated "at a glance"), or to obtain the number from the enemy chiefs possibly captured. This last seems not to be the case at Qarqar (which was not an Assyrian victory),[249] while the first two ways show ample margins of uncertainty (the informers, or "spies," were not always reliable) or of approximation (rough evaluation).

3. Within texts that after all do not aim at presenting "objective" information, it is quite natural to augment the content of the enemy forces (especially in cases where

---

243 Craig's collations gave no results for this number, cf. *Monolith Inscr.*, p. 32 no. 174 (the lacuna is 8 cm. long!). Adunu-ba'li's soldiers were *10,000*, according to the copies in 1 R, 8:93 and Rasmussen, *Salm.*, X, while Craig read [...] *lim*.

244 *Tel Aviv* 3, 97ff, followed by T.C. Mitchell, CAH³ 3/1, 479.

245 M. Elat, "The Monarchy and the Development of Trade in ancient Israel," in E. Lipiński (ed.), *State and Temple Economy in The Ancient Near East* (Orientalia Lovanensia Analecta 6, Leuven 1979), 542 n. 61, cf. also his earlier article in IEJ 25.

246 In Megiddo have been excavated some large stables, probably built by king Ahab. Actually, these stables could have had enough room for some 450 horses, that is, allowing three horses per vehicle, 150 teams of horses (cf. Yadin, *Warfare*, 298). It seems rather difficult to consider this fact as a confirmation of the *2,000* — cf. also T.C. Mitchell, CAH³ 3/1, 477. Note that the O.T. credits king Solomon with having *40,000* (1 Kings 4:26) var. *4,000* (2 Chron. 9:25) stalls for horses and chariots, or *1,400* chariots (1 Kings 10:26 — the three passages agree on a total crew of *12,000* horsemen).

247 JAOS 41, 366. Note also that, according to 1 Kings 20:15 king Ahab employed only *7,000* men, "all the people of Israel," in a war against the Arameans, just a few years before the battle of Qarqar (on *7,000* as a round typological number within the O.T. literature cf. Cross, HHI, 154f). In Jehoahaz' times the army of Israel had *50* horsemen, *10* chariots and *10,000* troops (2 Kings 13:7).

The battle is also ignored by the extant Egyptian sources, cf. Kitchen, *Third Int.*, p. 325 n. 456.

248 The crew of an Assyrian chariot, according to reliefs, was 2 men in the ninth century. Later, in Sargon's time, it carried 3 or even 4 men (cf. Reade, *Iraq* 34, 103; Noble, SAAB 4, 64). Cf. Yadin, *Warfare*, 298 for the other countries (crew of two).

249 Several factors make us think that Shalmaneser suffered a defeat rather than being victorious at Qarqar that year: cf. e.g. Olmstead, JAOS 41, 367; *id.*, *History*, 124ff; *Fisher Weltgeschichte* IV (Frankfurt 1967), 33; Elat, IEJ 25, 25ff; Gelio, RBI 32, 113f n. 9; Pitard, *Ancient Damascus*, 128; Grayson in CAH³ 3/1, 261. After all, further campaigns had to be undertaken against the coalition of the *"12"* Syrian kings, cf. p. 134.

the contest did not conclude with a clear Assyrian victory) according to a "justification" logic. In other instances the Assyrian royal inscriptions describe the military defeats suffered by their rulers in very vague terms, or completely ignore them. Here, however, we are not dealing with a defeat (rather it had been a "draw"), while on the other hand to ignore that battle would be to ignore completely the Syrian campaign of the 6th year, which in any case yielded some significant results.

4. The quantities given by the inscription are rather dissimilar: the contingents vary from *10* to *2,000* chariots, from *200* to *20,000* infantry. This is quite curious: how could the Assyrians know that amongst almost *4,000* enemy chariots there were *10* of Irqanate? What sense is there in representing such small quantities, when the other contingents are much more relevant? The obvious roundings to which the other numbers had been subjected far surpass these quantities. Even from an ideological point of view, if the Assyrians were interested in stressing the consistence of the enemy forces (*ad maiorem* Salmanassar*is gloriam*), there was no sense in making reference to such small contingents, and then presenting "*12*" kings allies of Irḫuleni after having shown how the forces of some of them were ridiculously small. What was the purpose of mentioning armies of *200* soldiers (Arwad and Usanate), within the context of a battle in which over *100,000* men participated, including the Assyrians?

5. The monolith inscription shows a high percentage of round numbers: 45 out of 75 (the highest amongst the inscriptions taken into consideration in Table 1 on pp. 42-43), while there are no "exact" numbers at all. Some 62 numbers are formed by a single digit, while the low digits (1, 2 and 3) are also very frequent: these two facts let us think that several roundings had been undertaken. Indeed, the very numbers relating to the battle are roughly rounded.

Substantially, four hypotheses could be formulated concerning these numerals:

A. They are authentic. This conflicts with all the 5 points mentioned above.

B. They are the result of a guess made "in the field" (how did they notice the *10* chariots of Irqanate?) or, rather, they are the result of information given by spies or informers. Still, an army of over *70,000* Syrians seems excessive (above, no. 1).

C. They are the deliberate alteration of originally different (= lower) data.

D. They have been invented. This looks rather unlikely, since contingents of homogenous consistence could have been invented more easily (above, no. 4).

Hypothesis C (alteration of data) seems most likely. The most practical way to proceed in this case is not that of consistently altering all the data, for instance multiplying all the numbers by 10. This could yield excessively high numerals: if the "real" size of the Syrian army was 30,000, it would have become 300,000. To multiply all the data by 4, or by 5 etc. would be rather complicated. But note how the quantities relating to the first three contingents are much higher than the others: the author of the inscription could well have given tenfold their original quantity, leaving unaltered the subsequent ones (since he had already reached an adequate comprehensive total), except for, perhaps, multiplying the numbers referring to the troops of Irqanate and to the Arab camels (these latter were the sole contingent of camels, and it seemed that also their number had to be sufficiently "high"). It seems thus likely that the author, once he had decided what had to be, approximately, the size of the Syro-Palestinian army (that is about 70,000), tenfolded some numbers until he got this value. This explanation, needless to stress how hazardous it is, would presuppose the will to preserve a part of the reality, altering only some quantities and no digits (the passage 200 > 2,000 does not involve the alteration of the digit "2"). If true, it would involve the following "original" numerical consistence of the Syro-Palestinian army: 430 chariots (120 + 70 + 200 + 10 + 30), 190 cavalry (120 + 70), about 9,000 infantry (8,900: 2,000 + 1,000 + 1,000 + 500 + 1,000 + 1,000 + 200 + 200 + 1,000 (?) + 1,000 (?)) and 100 camels. The resulting numbers may seem perhaps too low, yet one has

TABLE 2 CONTINGENTS FROM CONQUERED LANDS ADDED TO SARGON'S ARMY

Excluded are the quantifications of infantry, or generically defined (e.g. the *1,000* ERÍN.MEŠ and *2,400* ERÍN.MEŠ, respectively in Lie, *Sargon*, 66:447 and 450). Chronological indications according to the Khorsabad annals.

Chariots	*petḫallu*	infantry	
50 / 200			1st year, Samaria; var. no. 39
200 / 300	*600*		(2nd year) Hamath; var. no. 61
50	*200 / 500*	*3,000*	5th year, Karkemish; var. no. 42
100			9th year, Tabal; Lie, *Sargon*, 32:202; Winckler, *Sargon*, 102:32; Gadd, *Iraq* 16, 182:27
150	*1,500*		*20,000* bowmen, *1,000 / 10,000* shield-bearers - 13th year, Bīt Yakin, var. no. 60; in charge to Assyrian provincial forces.
	600	*4,000*	12th year, Lie, *Sargon*, 44:276

The last group (12th *palû*) represents enemy forces faced by the Assyrians. The *4,000* ERÍN.MEŠ, as it seems, after the battle, had been gathered together (*ašāru*) and *1* out of *3* had been "taken" (*ṣabātu*), obviously to serve under the Assyrian army (cf. Lie, *Sargon*, 46:285 and no. 3).

Another *30* chariots were taken as booty (4th year; Winckler, *Sargon*, 100:28) — add also some quantifications of *zīm pāni* (men of the guard):[250] *4,[000*?] "received" during the 7th year, (Lie, *Sargon*, 20:112), another *2,200* "received" during the 9th year (*ibid.*, 28:168) and *1,000* "brought" to the king during the 13th year (*ibid.*, 68:451). *1,000* assyrian *pētḫallu* and infantry are also mentioned for the 8th year, (*ibid.*, 26:150; TCL 3, 50:319).

to keep in mind that Sargon, two centuries later, will take into the imperial army *200* chariots of Samaria (only *50* according to a later source) and *200* (var. *300*) chariots and *600* cavalry from Hamath (see Table 2) — obviously forces worth mentioning. Other mentions of contingents on the order of some hundred horses or chariots can be found in the inscriptions of this ruler, during a reign in which certainly the inscriptions were not much more sober in representing quantities (cf. Table 2, end).[251] However, the numeral referring to the cavalry (190) seems actually too low in view of the fact that they were usually more numerous than the chariots.

It is possible, obviously, that the "alteration" applied (still within hypothesis C - alteration of data) did not consist of tenfolding some numerals, but was the result of a different operation. One alternative explanation could be that some data was available (including most likely the lower numerals) and some others were plainly invented to fill in the gaps in the information — it seems in any case probable that only some data had been manipulated. Obviously, the scribe writing this inscription

250 Fuchs, *Sargon*, translates this expression with "Tropäen" ("Aussehen des Gesichtes," cf. p. 319 n. 257).

251 On the Samaritan contingent cf. S. Dalley, "Foreign Chariotry and Cavalry in the Armies of Tiglath-Pileser III and Sargon II," *Iraq* 47 (1985), 31-48. Note also Becking, *Samaria*, 4f. For the size of the "regional" forces at Sennacherib's time cf. p. 89). An inscription of Ashurbanipal speaks of *16,000* bows (very fragmentary and questionable passage, cf. CT 35, 22:3 = Bauer, IWA, 95).

had to use exact data for some contingents, including those represented by very insignificant quantities, which he did not alter (e.g. writing 100 instead of *10* and 300 instead of *30*), while he felt free to manipulate the others.

The number of **enemies killed** during the ensuing battle (cf. above variation no. 25 and pertinent notes) is certainly more credible in the version given by the monolith (*14,000*: perhaps this as well coming from 1,400?). It is only with difficulty that the later versions (having *25,000* and *29,000*) could be interpreted as a "total" referring to all the Syrian campaigns,[252] since groups of Syrians killed are quantified at other times.[253]

5. *Size of the Assyrian army*

Some inscriptions quantify the Assyrian forces at *circa 2,000* chariots and over *5,000* cavalry units (*pētḫallu*).[254] The variations are undubitably due to oversights. Yet, the form of these numerals may give rise to some doubt: the number of chariots (*2,000 + 1 / 2*) is perhaps conventional or approximated, while that of the cavalry seems more credible, in view of its "exactness." These forces do absolutely deserve respect, yet they are not impossible for an expanding empire.[255]

The size of the army that crossed the Euphrates lead by Shalmaneser III during his 14th campaign, *120,000*, is indeed remarkable. The number actually appears within a rather vague passage, and that very campaign did not involve conspicuous results (according to the inscriptions themselves), notwithstanding the displaying of such an army:

> *ina* 14 BALA^{MEŠ}*(palê)-ia ma-a-ti* DAGAL*(rapaš)-tu a-na la ma-ni-e ad-ki it-ti* 1 *me lim* 20 *lim* ERÍN.ḪI.A*(ummānāti)* ^{ÍD}A.RAT*(Puratta) ina mi-li-ša e-bir ina* u_4*-mi-šu-ma* ^{md}IM*(Adad)-id-ri ša mat* ANŠE-*šú(Dimašqi)* ^{m}*Ir-ḫu-li-na mat A-mat-ai a-di* 12 MAN^{MEŠ}*(šarrā)-ni ša ši-di tam-di* AN.TA*(elīti) ù* KI.TA*(šaplīti)* <<ḪI>> ERÍN.ḪI*(ummānāti)-šú-nu* <<*u*>> ḪI.A^{MEŠ}*(ma'dūti) a-na la ma-ni-e id-ku-ú-ni a-na* GABA*(irti)-ia it-bu-ú-ni it-ti-šú-nu am-daḫ-ḫi-iṣ* BAD$_5$.BAD$_5$*(abikta)-šú-nu áš-ku-un*
>
> "In my 14th year I mobilized the wide and numberless lands (of the empire, and) with *120,000* troops I crossed the Euphrates at its flood. At that (very) time Adad-idri of Damascus, Irḫuleni of Hamath with *12* kings of the seacost, upper and lower, mobilized their vast (and) numberless troops (and) advanced against me. I fought with them and I defeated them."[256]

---

252 E.g. including the *10,000* killed during the collision with that very Syrian coalition taking place in the 11th *palû* (below, p. 136) — yet in ed. "C" and "E" of the annals both the *25,000* killed of the 6th *palû* and the *10,000* of the 11th are reported. Cf. also above, p. 92 n. 184.

253 Cf. the *16,000* killed during the 18th *palû* (p. 50 no. 27).

254 Cf. variations nos. 29-30 — *a-na* ID^{MEŠ} KUR-*ia ak-ṣur / ar-ku-un*: error for *aškun* or *arkus*?
These numerals are discussed also by Elat, IEJ 25, 27-29. He pointed out that *2,001* chariots would have required 8,004 horses, in view of the fact that in the days of Shalmaneser III each chariot was drawn by four horses. The grand-total would thus include at least 13,245 harnessed horses, or even 18,486 if we assume that each cavalry unit (*pētḫallu*) included two horses, since they operated in teams of two men.

255 They may be compared to the quantifications of foreign contingents added to the imperial army appearing in Sargon's times, cf. Table 2 above. These quantifications, however, refer to single contingents, and are by no means "totals."

256 Cameron, *Sumer* 6, 15:24ff = Michel, WO 1, 468 (ed. "C"); the passage appears also in ICC, 16:43f = Delitzsch, BA 6, 148:99ff (ARAB 1, 658f - ed. "D"). The subsequent editions of the annals give the account of the campaign in a much more synthetic form and do not relate the number under discussion.
For the superfluous ḪI, cf. the note in Michel, WO 1, 468: it is very unlikely that we should understand *3,600* or *360,000* (as in Pettinato, *Semiramide*, 125).

The passage concludes relating the destruction of their weapons and the flight of the defeated kings.[257] It had been an exceptional effort for the Assyrians, since Shalmaneser himself emphasizes it was the result of a general call-up of all the lands he boasts of controlling, thus the usual size of the army must have been much smaller. Nevertheless, this does not allow us to take the number as genuine. This is not the place for discussing thoroughly the problem of the size of the Assyrian army, which would excessively thicken the plot.[258] Still, it would be worthwhile to try to fix a maximum limit for it. In this respect, there is not much need to emphasize how difficult it is to find mention of an army of similar size, at least if we take into account only the direct sources (contemporary and "internal"). Ancient Mesopotamia rarely witnessed such huge hosts,[259] and never did their contemporary neighbors,[260] if one does not take into account the literary texts, such as, e.g., the "Cuthaean legend of Naram-Sin"[261] or the Greek tradition. The quantifications given by the O.T., at times

---

257 Similarly to what is quoted at p. 136 for the analogous episode of the 10th *palû*.

258 The problem is that the information we get from sources other than the royal inscriptions gives us only partial accounts of men, troops, etc. In a recent study (SAAB 4, 23ff), F.M. Fales tried to estimate the medium size of the Assyrian army, on the basis of some letters dating to Sargon. The letter SAA 5, 250 is first discussed: it constitutes the answer to the king who asked data on the grain stored in the city of Kār-Ashur and in its province — probably in view of a major campaign. The quantities given, then, have been compared with another letter (SAA 1, 257 = LCA 6) suggesting that the minimum daily ration was 1 *qa* of grain. Assuming that the ordinary ration was about 2-3 *qas*, the author concludes that the storage of the above mentioned province could supply from 23,000 to 35,000 men (including the local population) — note that Engels, *Alexander*, 125f gives 3 lb. (= about 1,3 *qas*) as the minimum daily ration for the troops, but 20-24 lb. or more for the animals, half grain and half fodder. In any case, these calculations give only approximate results and are difficult to use, as Fales himself points out (p. 34) — note that the first mentioned letter simply reports to the king the total supplies available in the area, and does not indicate whether that was enough or was exceeding the needs of the army.

259 Cf. the *254,000* men killed according to a fragmentary text of the middle-Assyrian king Arik-den-ili, mentioned above, p. 22 n. 10, and the *150,000* warriors killed by Sennacherib during the battle of Ḫalulē (cf. below, p. 116).

For Mari, see Sasson, *Mari*, 8 and pertinent n. (a *30,000* man army mentioned). Cf. also the army of *60,000* (1 *šu-ši li-mi*) in force to Išme-Dagan, son of Shamshi-Adad I of Assyria, mentioned by a letter from Tell-Šemšāra (it is actually a passage quoted from another letter sent by an enemy of the Assyrians asking for help, thus it could be a gross approximation if not an exaggeration), cf. J. Læssøe, "IM 62100: A Letter from Tell Shemshara," *Fs Landsberger*, 191 and 193:12.

For ancient Sumer cf. Gelb JNES 32, 93 and *passim*. The old Akkadian royal inscriptions of Rimush and Naram-Sin do not mention such large quantities (cf. n. 160 above).

260 The size of Near Eastern armies of the second millennium BC is discussed by Y. Yadin in B. Mazar (ed.), *World History of the Jewish People, Vol. II: Patriarchs*, Tel-Aviv 1970, 138ff. For the old Hittite kingdom add the remarks of P.H.J. Houwink ten Cate, "The History of Warfare according to Hittite Sources: The Annals of Hattusilis (Part II)," *Anatolica* 11 (1984), 47-83 (esp. 72f), and for the later periods cf. Götze, *Iraq* 25, 127 — about the Hittite-Egyptian war he states that "this was probably a supreme effort; one has calculated that the Hittites gathered about 30,000 men from all over their Empire and concentrated them in Northern Syria." In fact, the Hittite historical inscriptions do not refer to (numerically) very big armies: cf. the *10,000* soldiers mentioned in A. Götze, *Die Annalen des Muršiliš* (MVAG 38), Leipzig 1933, 122:73 (cf. also the *66,000* prisoners mentioned in 76:33).

For the Urartian army not much data is available (cf. Zimansky, *Ecology and Empire*, 55ff). A text speaks of *352,011* foot soldiers but the passage, which is not easily comprehensible, probably refers to people subject to a sort of *corvée*, and the number may represent the people eligible to serve (*ibid.*, 57).

Not much information is available on the size of the Egyptian armies: cf. R.O. Faulkner, "Egyptian Military Organization," *Journal of Egyptian Archaeology* 39 (1953), 44; A.R. Schulman, *Military Rank, Title, and Organization in the Egyptian New Kingdom*, Berlin 1964, 79. According to the inscriptions of Ramesses II, at Kadesh he faced a Hittite army of *18,000* var. *19,000* foot soldiers and over *2,500* chariots, each with a crew of *3* (Gardiner, *The Kadesh Inscriptions of Ramesses II*, Oxford 1960, p. 9:P85 and 41:R43f) — numbers probably exaggerated (cf. *ibid.*, 56), but that would imply a total of about 26,000 men. Breasted estimated that the Egyptian army had some 20,000 men, four divisions 5,000 strong (ARE 3, p. 153 n. and p. 127, followed by Faulkner, *loc. cit.*).

261 An old Babylonian text mentions three armies set out by Naram-Sin, of which the second and the third had *120,000* and *60,000* men respectively. The numeral relative to the first army is broken, but a "total" of *360,000* men killed is given (the numerals are written 2/1/6 *šu-ši li-mi*, cf. J.J. Finkelstein, "The so-called «Old Babylonian Kutha Legend»," JCS 11 (1957), p. 85 col. III 2, 4 and 6). The Hittite version of the narrative confirms that the first army had *180,000* (var. *190,000*) men, and that the other two had *120,000* and *60,000* ERÍN, but omits the "total" (cf. H.G. Güterbock, "Die historische Tradition und ihre literarische Gestaltung bei Babyloniern und Hethitern bis 1200 - Zweiter Teil: Hethiter," ZA 44 (1938), 54:4ff). The neo-Assyrian version (edition in O.R. Gurney, "The Sultantepe Tablets: IV. The Cuthean Legend of Naram-sin," AnSt 5 (1955), p. 98ff) gives, for the three armies, *120,000*, *90,000* and *60,700*

boldly hyperbolic,[262] are not certainly originating from a source contemporary with the events. Likewise, also the Greek sources, relating similar (or more) astronomical figures for the Persian armies, are, obviously indirect as well as subsequent to the narrated events — unless one wants to believe that Alexander faced a Persian host of over 100,000 at the Granicus river, over 500,000 at Issus, and over 1,000,000 at Gaugamela, according to the earliest writer of a surviving account.[263] Thus, Herodotus tells us that the army of Xerxes consisted of *170* myriads of soldiers (= *1,700,000*), plus *10* of special troops (cavalry, etc.), and, comprehensively, over 5 million people counting the followers, the people assigned to the supplies, etc. These numbers have generated a broad discussion among modern scholars, and while some believe that Herodotus' account is basically true and the numbers are derived from a misunderstanding,[264] others completely ignore these figures, thought to be simply the result of a gross exaggeration. In this last case especially, the "corrected" numbers given by modern scholars have a disarming range of fluctuation.[265]

The Greeks tended to exaggerate the numbers in accordance with the territorial extent of the Persian empire, imagined immense and boundless in the proper meaning.[266] By contrast, the quantifications given for the Greek mercenaries in Asia

---

(2 UŠ *lim*, 90 *lim* and 60 *lim* 7 *me*, cf. CT 13, 39 II [20'ff]; STT 1, 30:85ff). It also states that the *7* enemy kings had *6,000* troops each (6 *lim* A.AN), and were allied with another *17* kings, having *90,000* men in all (90 *lim*, cf. CT 13, pl. 41:[17'] and pl. 44 [II, 18]; STT 1, 30:38 and 61). The first numeral, 6 *lim* A.AN, is usually taken as *360,000* (e.g. Gurney, *loc. cit.*), on the basis that bigger armies are credited to Naram-Sin (P. Jensen in *Keilinschriftliche Bibliothek* 6/1, Berlin 1901, 552 n. to line 20), but this argument seems rather speculative to me (cf. how the *17* allied kings had an army of less than 5,000 each).

262 See, e.g., the *400,000* Israelites at war against Benjamin (Jud. 20:17) or the *300,000* (plus *30,000* of Judah) of the war against the Philistines (1 Sam. 11:8).

Still according to the O.T., Sennacherib's army suffered a great disaster during a campaign in Palestine: "and that night the angel of the Lord went forth and slew one hundred and eighty-five thousand in the camp of the Assyrians" (2 Kings 19:35; cf. also Isaiah 37:36). Josephus (*Ant. Jud.* X:I 5 quoting Berossus), reporting this numeral, credits the calamity to a pestilence — for a discussion of this episode cf. von Soden, *Fs Stier* (the numeral meaning simply "sehr viele" in this context), Gonçalves, *L'expédition de Sennachérib*, 484 n. 169 and Cogan - Tadmor, *2 Kings*, 239 and 250f.

263 Diodorus, who lived in the first century BC. The exact figures involved in this *crescendo* are: Granicus — more than *10,000* cavalry and not less than *100,000* foot (XVII, 19, 4); Issus — over *400,000* soldiers, not less than *100,000* cavalry (XVII 31, 2); Gaugamela — *800,000* infantry, no less than *200,000* cavalry (XVII 53, 3).

264 Mostly on the basis that Herodotus relates a detailed list of the various contingents of the Persian satrapies and describes the system of command. Alternative explanations for the Herodotean numbers have been found, resulting in diverging calculations: see the *180,000* of J.A.R. Munro (CAH 4, 271f) or the *150,500* of F. Maurice, "The Size of the Army of Xerxes in the Invasion of Greece 480 B.C.," JHS 50 (1930), 210-35 (followed by Burn, *Persia and the Greeks*, 326ff and CHI 2, 320). Recently, figures of about *70-80,000* found some credit (cf. the discussion and the conclusions of Hignett, *Xerxes*, 350-55).

Incidentally, note that the *170* myriads of Herodotus correspond to the *170* myriads of infantry that Ctesias (cf. König, *Ktesias*, p. 37 and 137) credits for the army of Ninus, having also *21* myriads of cavalry and (only) *1* myriad and *600* chariots for his campaign to Bactria (as reported by Diodorus Sic., II, 5 but cf. also II, 17).

265 Ignoring Herodotus, it is difficult to go beyond the pure estimates. Hans Delbrück (see above p. 3 n. 14) credited to Xerxes' army a number varying between about *65-75,000* (*Die Perserkriege und die Burgunderkriege* [1887], 210; *Geschichte*, 10ff and 88) and *25,000* (in "Geist und Masse in der Geschichte," *Preussische Jahrbücher* 147 [1912], 202) or about *15,000* men (*Numbers*, 61, and cf. 32), and this according mainly to considerations of logistics.

In the new edition of CAH, N.G.L. Hammond calculates that Xerxes had some *220,000* men under arms on land, another *22,000* for the supply service and other duties, plus *408,000* estimated for the fleet and supply-ships, a grand total of *650,000* men (cf. CAH² 4, 533f). The *220,000* result from a simple comparison with the army that Mardonius needed to stay in Europe in 480 (he engaged *100,000* Greeks at Platea, so he had *120,000* men — a number that agrees with Ctesias'). These calculations seem very disputable to me, since the size of Mardonius' army at Platea is not certain — usually he is credited with much lower numbers, see e.g. J.P. Barron, CAH² 4, 594 (= the same volume — the army probably *60,000* strong, to which add the Greek allies) or Burn, *Persia and the Greeks*, 511 (he had some *60-70,000* men, including cavalry and excluding the Greek allies, and was outnumbered by the Greeks) — Delbrück, *Geschichte*, 82 credited both armies with much smaller numbers. Ctesias' credibility is not discussed nor justified: cf. the preceding note for his numbers of Ninus' army (and König, *Ktesias*, p. 11 §25 for the *12* myriads left to Mardonius), or cf. the remarks of Burn, *op. cit.*, 11f.

266 Cf. W.W. Tarn in CAH 6, 367: "to the Macedonians a Persian army was guesswork, and both camp gossip and literary men made flattering guesses, such as seemed appropriate to the territorial extent of the Persian empire." One may wonder if this applies to modern historians as well. Not necessarily does a

or for the Hellenistic armies are quite realistic, but they never tell of an army larger than Shalmaneser's.[267] It should not be forgotten that Alexander began the conquest of Asia with *30,000-32,000* infantry and *4,500-5,100* cavalry (according to Diodorus),[268] and he had not a little logistic trouble.[269] Republican Rome never witnessed such a big army,[270] at least until the Marian times, and even in imperial Rome Shalmaneser's army would have been considered exceptional. In the 1st-2nd centuries the imperial standing army was much bigger,[271] but, obviously, the legions were distributed across the whole imperial territory, and were never employed simultaneously in a single area.[272] The Roman army grew continuously during the first centuries of the C.E.,[273] reaching a strength of over 500,000 men according to an official source of the 5th century,[274] which, however, suggests that the field army was

---

bigger empire employ bigger armies. The size of the army, on the contrary, depends on logistical and strategical factors — better equipment, training, discipline and organization could account for their superiority. M.A. Dandamaev, with regard to Xerxes' invasion of Greece, calls attention to a statement of Thucydides, according to whom the "barbarians," when undertaking long campaigns, usually had no success because the local people and their neighbors, united in their fear of the attackers, always outnumbered the invaders (*Achaemenid Empire*, 195). Similar opinions are repeatedly expressed by Delbrück: cf. e.g. the conclusion of *Numbers*, last 2 pages.

Thus, when Liverani (*Asn.*, 135) estimates the average size for an Assyrian army in the years of Ashurnasirpal II at some *20,000*, on the basis that they faced (according to the "annals") armies of *6,000* men (cf. below, p. 177 n. 85), he is bound to these kinds of preconceptions. If the inscriptions are exaggerating the numbers (possibly twice the original, see above pp. 83f), and if we allow the Assyrians to be outnumbered, say, by 20% and to come out winning in whatever manner, we can only conclude that the Assyrian army had a minimum of about *2,500* men — indeed, not a striking size.

267 The largest army in Greek history commanded by one of Greek speech was put together by Antigonos when he was king of Asia west of the Euphrates (as observes W.W. Tarn in CAH 6, 367). The army consisted of *88,000* men and *83* elephants and campaigned in Egypt against Ptolemy, suffering a defeat just owing to its enormous size and the consequent difficulties of movement and supply — cf. *ibid.*, 499, and also N.G.L. Hammond, *The Macedonian State*, Oxford 1989, p. 270 or R.A. Billows, *Antigonos the One-eyed and the Creation of the Hellenistic State*, Berkeley 1990, p. 162. On the numeric strength of the Hellenistic armies see B. Bar-Kochva, *The Seleucid Army*, Cambridge 1976, pp. 7-19 or the data tabulated in Delbrück, *Geschichte*, 203f.

268 The numbers given by ancient sources on Alexander's army at the Hellespont are conveniently tabulated in P.A. Brunt, "Alexander's Macedonian Cavalry," JHS 83 (1963), 46 (cf. also 32ff).

269 See Engels, *Alexander, passim*. His army reached, according to tradition, a maximum size of about 120,000 men during the Indian campaign (including the auxiliary services, the sailors, etc.; part of these men were Greeks, part Asiatics — Indian allies included — see *ibid.*, p. 150). Actually, in most cases he was able to display only a smaller army on the battlefield. At the battle of the Hydaspes river, in India, since he decided to make a surprise crossing by night with a part of his forces, he had only some 6,000 infantry and 5,000 cavalry (Hammond, *Alexander*, 208ff). At the Granicus river, in the battle fought against the Persian satraps, he had probably 13,000 infantry and 5,100 cavalry (Hammond, JHS 100, 83), while at Gaugamela, the last and decisive battle against Darius III, he had about 40,000 infantry and 7,000 cavalry (Marsden, *Gaugamela*, 30f and 38).

270 At Cannae, Hannibal was confronted by *48,000* Romans (according to Livy), and often it has been emphasized how the Romans relied on the *number* to defeat the Carthaginians, who in turn displayed about *35,000* infantry and *10,000* cavalry, including the allied Celts. The size suggested by Polybius for the Roman army (eight legions, possibly 80,000-90,000 men allowing an equal number of allied troops) has been recently given credit by J. Briscoe in CAH² 8, 51. But he does not explain how Hannibal with the forces at his disposal could completely encircle such a host (cf. the remarks of B.L. Hallward in the older edition of CAH, p. 52f n. 1). Hannibal crossed the Alps with about *20,000* infantry and *6,000* cavalry.

271 It included from 25 to 30 legions, each made up of 5,000 men, for a grand total of about 125,000-150,000 legionaries. To get the comprehensive strength of the army one should add the pretorian guard, the urban cohorts and the *auxilia*, which would give a nearly doubled figure — cf. G.R. Watson, *The Roman Soldier*, Bristol 1969, pp. 13ff and p. 159 n. 7 or G. Webster, *The Roman Imperial Army*, Totowa 1985³, pp. 109f.

272 Thus, Vespasian faced the Jewish wars in AD 66 with *3* legions, which, including the auxiliary cohorts, the cavalry squadrons and the militia of the client-kings, could yield some 50,000-60,000 men (see A. Momigliano in CAH 10, 859).

273 For instance, after Diocletian's reforms, it counted 60 legions, with a nominal strength of 6,000 each (W. Ensslin in CAH 12, 397).

274 The *Notitia dignitatum*, "source officielle, mais d'un certain optimisme" (J. Harmand, *La guerre antique, de Sumer à Rome*, Paris 1973, 123), informs us that the Roman army had comprehensively *262,000* effectives in the West and another *262,000* in the East: see data tabulated in L. Várady, "New evidences on some problems of the late Roman military organization," *Acta Antiqua* 9 (1961), 358ff. The document is not taken without doubts and reservations — for instance, E. Binley even suggested that it is not an official document at all, but that it had been written by an *amateur* of military matters (*The Roman Army, Papers 1929-1986*, Amsterdam 1988, 69ff).

much smaller.[275] In this sense, it is clearly necessary to distinguish between the total size of the standing army, calculated on a theoretical basis, and the size of the army as it was actually employed on the battlefield.

In the subsequent centuries such large armies were almost unknown,[276] since only rarely does some source attest an army comparable to Shalmaneser's,[277] until Napoleonic times.[278]

Thus, although I may have overlooked some evidence, having wandered into fields with which I have no familiarity, and although the comparison with other epochs and situations obviously implies a good degree of uncertainty, it is clear enough, I think, that an ancient army credited with *120,000* men should be viewed with a little doubt.[279] The number may have represented the theoretical or conventional size of the Assyrian army,[280] but was by no means the size of the army on the battlefield (and the inscription is crediting this last hypothesis).

In conclusion, Shalmaneser's army boasts of a size that is not impossible — the Assyrian male population able to carry a weapon was certainly higher[281] — but that reaches the limits of credibility. And here stays the sense of this quantification: it is not *too* high (so much so as to be unbelievable), but enough to be considered exceptional. In other words, probably we have the highest possible number of soldiers in the eyes of the authors of these inscriptions (and of their readers, obviously). A

---

275 The expeditionary forces, which also were subdivided into units stationed in different areas of the empire, were comprehensively *96,300* in the East and *123,800* in the West (cf. preceding note).

276 In the (European) Middle Ages the size of the armies was much smaller: cf., in general, Delbrück, *Numbers*, 72ff; *id., Geschichte, passim.* On the Islamic conquests, cf. e.g. F.M. Donner, *The Early Islamic Conquests*, Princeton 1981, p. 221: "perhaps the most striking fact about the armies that carried out the Islamic conquest of the Fertile Crescent was their small size." Another example from the East: according to Ch'i-ch'ing Hsiao, *The Military Establishment of the Yuan Dynasty*, Cambridge (MA) 1978, p. 11, "Cinggis Qan's army had grown to 129,000 by the time of his death."

277 According to the Islamic sources, the last Umayyad caliph fought his decisive battle against the Abbasids with an army of *120,000*, a figure which some scholars take as an exaggeration (e.g. H. Kennedy, *The Early Abbasid Caliphate*, London 1981, 47) or even as a mistake for 12,000 (P.K. Hitti, *History of the Arabs*, London 1970[10], 226 and 285). The later caliphal army added up to 160,000 cavalry and foot — these being, however, theoretical figures for the whole empire (*ibid.*, 303).

278 Napoleon undertook the expedition to Africa with 38,000 men, civilians included, and the rather unfortunate campaign to Syria with 13,000 men including sappers, artillerymen and guides (D.G. Chandler, *The Campaigns of Napoleon*, New York 1966, pp. 213 and 235). He faced the Russian campaign with a theoretical maximum of about 210,000 men for the central column, but only 133,000 actually took part in the most important battle, Borodino (see the data tabulated in *ibid.*, 1009f and 1119).

279 Modern military historians start their study from the Persian wars, without taking into consideration ancient Oriental warfare. One exception is *The Encyclopedia of Military History* by T.N. Dupuy (New York 1986[2]), a work rather uncritical with regard to the figures given by the ancient sources, and often inclined to credit the highest estimates. In spite of this, at p. 9, about the "Assyrian military organization, c. 700 B.C." (i.e. after Tiglath-pileser III's reforms), he states that "the field armies may occasionally have approached a strength of 50,000 men."

One should avoid taking seriously such very popularized publications as "the Great Conquerors" or "the Great Battles," for obvious reasons. It may be possible to read of the Assyrians credited with a standing army of about 1,000,000 or so (figures usually drawn from the Greek tradition on Ninus, cf. here above n. 264).

280 This applies to the numbers referring to the chariots and the cavalry mentioned just above.

The question whether the Assyrians had a permanent standing army or had to rely mostly on levy troops has been extensively dealt with by Manitius, ZA 24, whose conclusions are still valid (see Saggs, *Iraq* 25, 145; cf. also Postgate, TCAE, 218-29). In this study, however, as far as the problem of numbers is concerned, he had to base himself on the royal inscriptions, and these brought him to calculate the total size of the Assyrian army under the Sargonids (standing forces + potential number of levied troops) in the order of hundred of thousands (*op. cit.*, p. 129; see also Saggs, *Assyria*, 243 and 253). For a different approach, which brought lower estimates, cf. Henshaw, *Palaeologia* 16, 3ff. According to him, the Assyrian army of the Sargonid period had about 35,000 men, to which one or more similar units, provided by the local governors, joined in the field, thus adding up to a size of 60,000-70,000 or 100,000 at most.

281 E. Cavaignac ("Le code assyrien et le recruitement," RA 21 (1924), 59-64) tried to estimate the population of Assyria in this period, suggesting that the *120,000* of Shalmaneser may have represented the whole male population of the land. On this basis he calculated, multiplying that number by 4 (he observed that the "Ḫarran census" assumes that to each head of the family correspond a total of 4 persons), almost half a million for Assyria proper, and over one million inhabitants for the whole empire, including the regions of the upper Tigris, Ḫarran and Edessa.

marvelously high number, roaring in its roundness,[282] but after all *possible* in consideration of the "wide and numberless lands" that constituted the Assyrian empire.

6. *"Total" booty of the first 20 years*

This includes *110,610* prisoners, *82,600* killed, *9,920* horses (and) mules, *35,565* oxen, *19,690* asses and *184,755* sheep (cf. p. 93). These numbers are undoubtedly extremely high (especially the first two relating to people), but not impossible considering that they refer to a period of *20* years — yet the breadth of these sums makes it difficult to evaluate them, or to counterattack them in some manner. Partial "totals" referring to single campaigns, at least those that appear more realistic (cf. Tukulti-Ninurta II above or Sargon's booty from Muṣaṣir just below), if multiplied by 20, do not approach these numerals.[283] However, their exact form does not seem artificial (i.e. sum of "round" and "exact" numbers).[284] In these cases we are facing, obviously, the results of a series of partial sums relating to shorter periods (one campaign or even less), whose "original" data could, in some cases, be simple guesses or approximations (especially in the case of enemies killed) or could be deliberately exaggerated numbers.

7. *Sargon II — Booty from Muṣaṣir*

The number of sheep given by the "letter" had been obviously altered in the annals,[285] through the addition of a *me*:

 1 *lim* 2 *me* 1,20 + 5 (= *1,285*)
 1 *me lim* 2 *me* 20 + 5 (= *100,225*).

It is possible that this is due to an oversight, or even to a "misunderstanding" of the first number (as suggested by Ungnad), but this seems unlikely. Also unlikely is that the later version represents an "updating." The fact that the other numbers had varied as well (prisoners, asses and mules, oxen), yet remaining in some way connected to those of the "letter,"[286] show that this text, directly or indirectly, had been the source of the annals, and that the numbers had been deliberately altered. The fact that Armenia has always been an intensive stockraising region, as pointed out by Millard,[287] can make us conceive of the large quantity of these animals (*100,225*), but does not explain the striking similarity between the smaller digits.

Now, why have the sheep been augmented in such a way (almost a hundred times), while only limited augmentations have been applied to the other animals and commodities? Perhaps it seemed that from a region of intense pastoral activity there had

282 One need not stress how 120 (actually, two sexantines) in ancient Mesopotamia could easily be employed as a typological number: cf. e.g. above, n. 261.

283 Tukulti-Ninurta's campaign on the middle Euphrates would have yielded, if repeated 20 times
13,400 oxen, 1,400 asses, 86,840 sheep,
while a similar calculation applied to Muṣaṣir's booty (actually not representing a whole campaign) would give
10,500 oxen, 7,840 asses and mules, 24,700 sheep.
Sennacherib's "total" referring to the first campaign is, however, far more relevant, but no less suspect (cf. below).

284 Even if the digits employed show some repetitions: cf. the "1" in the prisoners' numeral, the "9" in the horses' and asses' ones, and the "5" in the oxen's and sheep's ones.

285 Cf. p. 54 nos. 49ff, p. 73 and pp. 86ff.

286 The number of mules and that of asses had been summed (*12 + 380 = 392*) and added to "300," that of the people has passed from 6 *lim* 1 *me* 10 to 6 *lim* 1 *me* 1,10 — this variation is possibly due to an oversight.

287 *Fs Tadmor*, 219. However, Zimansky (*Ecology and Empire*, 16) remarks that in the summer, the only time in the year which allowed a campaign to Urartu to take place, livestock was most likely dispersed among the upland pastures. Therefore, it is difficult to imagine that the Assyrians could find over 100,000 head of sheep in Muṣaṣir and surroundings (the inscriptions clearly state that the booty was taken in that city).

to be a huge booty of sheep. Or perhaps, if the hypothesis of "misunderstanding" is kept valid, the author has dropped an exaggeratedly high digit in the former case, having confused the signs (consider also the similarity between the signs ME and LIM and the number *1*), and then compensated for this with a slight "lowering" of the quantity of silver.

In any case, leaving aside the speculations, the numerals given by the "letter" probably are genuine and thus they indicate what could have been the real contents of a war booty.[288]

8. Sennacherib — "total" booty of the first campaign

These numbers[289] are extremely suspect, for the following reasons:

a they are "exact" numbers formed by a round number and a lower number added to it: their exactness does not seem "genuine";

b they are subject to many variations;

c the quantities are definitely very high, especially if one thinks that they refer to a single campaign[290] (compare [above] the "totals" referring to the first *20* years of the reign of Shalmaneser III). In particular, the number pertinent to the prisoners (*208,000*) is very difficult to harmonize with the archaeological estimates which point to a population of no more than 220,000 for the whole of southern Mesopotamia in this period.[291] These kinds of estimates are extremely hazardous[292] — however, it is beyond doubt that the first half of the first millennium saw a drastic decline in the level of settlement in southern Mesopotamia.[293]

These reasons, I think, are enough to conclude that the numerals under discussion are artificial.[294] It is not possible to establish if they "hide" some real quantities (for

---

288 The number of mules, *12*, is very low, and probably for this reason it has been dropped from the later version. The possibility that it represents an error for *"312"* (cf. n. 286 above) is unlikely since the "letter" reports it twice.

289 Cf. p. 90 or p. 58 nos. 67ff.

290 They appear also in a text including only the first campaign, therefore they cannot be "general" totals relating to the whole reign and referred to a single campaign (H. Sauren credited this hypothesis: cf. n. 184 above).

291 According to Adams, *Heartland*, 177f, the overall area occupied in Lower Mesopotamia was 616 hectares in the middle-Babylonian period and 1,769 in the neo-Babylonian and Achemenid periods (the neo-Assyrian period is not singled out). This would imply a maximum population of 77,000 and 221,125 inhabitants respectively, assuming an average density of 125 persons for an inhabited hectare (cf. p. 69b and 350b). Cf. also p. 333 for the Eridu and Ur areas in the post-Kassite period, 16,800-27,000 inhabitants at most.

Mallowan, *Nimrud*, 72 estimated that the population of inner Assyria in this epoch probably included between a quarter of a million and half a million people (cf. below p. 141 n. 124). Henshaw, on the basis of these estimates, viewed the numeral under discussion as hardly feasible (*Palaeologia* 16, 4).

292 See the reservations of J.A. Brinkman, "Settlement Surveys and Documentary Evidence: Regional Variation and Secular Trend in Mesopotamian Demography," JNES 43 (1984), 175f; *id.*, *Prelude to Empire*, 3f and 9f. Actually, these estimates give the *maximum* possible extent of settlement without showing whether it has been simultaneously inhabited. See also, recently, N. Postgate, "How Many Sumerians per Hectare? Probing the Anatomy of an Early City," *Cambridge Archaeological Journal* 4 (1994), 47-65.

293 During the far more prosperous Ur III period the area had about half a million inhabitants, as estimated by Adams, *op. cit.* (n. 291), 149.

294 H. Sauren hypothesized the existence of a *droit de guerre* which allowed the Assyrian king to take a fixed number of deportees from each city or land rebelling or posing resistance, and subsequently conquered (cf. WO 16, 82). This theory, which he applies to Sennacherib's deportations from Babylonia and Judah (cf. below), might be fascinating, but is not substantiated by facts. The calculations he undertakes to verify its validity are often inconsistent, as he makes the figures match by assuming each time a different number of deportees or of cities to be considered: for instance, compare *loc. cit.*, p. 83 top with n. 10 (in the latter case, it is not clear where the 15 tribes come from, since cylinders "Bellino" and "Rassam" both list 18 tribes, as many as all the other texts). In addition, he overlooks a good number of elements, e.g., he totally ignores the cylinder with one campaign (Smith, *First Camp.*), which records the

instance dividing them by 10 or 100), even if, perhaps the alteration applied to the sheep of Muṣaṣir (*1,285 > 100,225*) has worked here as well (cf. Ungnad and Olmstead, *cit.* at p. 73). In this case *80,050* oxen and *800,100* sheep would become respectively 850 and 8,100 (and the prisoners 2,080?). However, the presence of lower order digits was felt more important than their authenticity, since in the later texts the numerals become *80,100* oxen (possibly confused with that of the sheep) and *800,600* sheep. Obviously enough, the authors of these inscriptions cared much more for the size of the numerals than for the digits forming them.

As for the number of **cities of Babylonia** conquered during that campaign (cf. p. 57 nos. 65-66), it will undergo a treatment similar to that of the "total" booty, being absentmindedly recopied (in this case in "reduced" measure) in the later texts (which, of course, are not for this reason more trustworthy).[295]

9. Captives taken in Judah

Sennacherib, according to his cylinders and prisms, during his third campaign conquered *46* strong walled cities belonging to Hezekiah of Judah, together with the small cities in their neighborhood which were beyond counting (cf. above p. 15). Then follows the statement

> 2 *me lim* 1 *me* 50 UN^{MEŠ}*(nišē)* TUR*(ṣeḫer)* GAL*(rabi)* NITA*(zikar) ù* MÍ*(sinniš)* ANŠE.KUR.RA^{MEŠ}*(sīsē)* ANŠE.KUNGA^{MEŠ}*(parē)* ANŠE^{MEŠ}*(imērē)* ANŠE.GAM.MAL^{MEŠ}*(gammalē)* GU$_4$^{MEŠ}*(alpē) ù ṣi-e-ni ša la ni-bi ul-tu qer-bi-šú-un ú-še-ṣa-am-ma šal-la-tiš am-nu*
>
> "*200,150* people, young, old, male and female, horses, mules, asses, camels, oxen and sheep without number from there (= the cities) I brought out and I counted as spoil"[296]

This quantification has often been pointed out as an example of propagandistic inflation.[297] The reasons for suspecting this numeral are similar to those listed for the booty from Babylonia: its characteristic structure as a "high - exact" number and its magnitude (cf. here above *sub a* and *c*), which is really impressive if correlated to a country the size of Judah. As a comparison, one may recall the number of deportees from the more populous Samaria under Sargon — *27,280* var. *27,290* according to his inscriptions[298] — or the deportees to Babylon in Nebuchadnezzar times, given as *4,600* by Jeremiah.[299]

---

conquest of only four cities of southern Babylonia, and not five or six cities as the later texts do. Cf. also next note.

[295] Cf. also below, p. 191 n. 12. Sauren, WO 16, 97f, tried to explain how these cities had been calculated, assuming that they represent some kind of total relative to the whole reign. However, his arguments are invalidated by the fact that he ignores the cylinder with one campaign (Smith, *First Camp.*), which records four different groups of cities, each individually quantified (cf. also previous note).

[296] Luckenbill, *Senn.*, 172 and 33:24 (= ARAB 2, 240; dupl. Ling-Israel, *Fs Artzi*, 228:21); Smith, AD, 305:12; CT 26, pl. 10:47 (dupl. Heidel, *Sumer* 9, col. III 65). The "bull inscriptions," after narrating the conquest of the *46* cities, simply add *šallatiš amnu*, "I counted as spoil," (i.e. the cities), thus refraining from mentioning booty, prisoners and relative numbers (cf. Luckenbill, *Senn.*, 70:28).

[297] Cf. here above, pp. 73 and 86f. Cf. also Stohlmann, *Scripture in Context* 2, 152f or Gonçalves, *L'expédition de Sennachérib*, 115 and the bibliography there quoted.

[298] Cf. above p. 52 no. 38.

[299] 52:28-30 records *3,023* in Nebuchadnezzar's 7th year, *832* in his 18th year, and *745* in his 23rd year, in all *4,600*. The second book of Kings speaks of *10,000* captives in 24:14 or of *7,000* men and *1,000* craftsmen in the duplicate accounting of 24:16, to which should be added those mentioned in 25:11, not quantified. Cf. the remarks of Cogan - Tadmor, *2 Kings*, 312 no. 14. Josephus claims *13,832*, a numeral

Many estimates have been suggested for the population of ancient Palestine, following different methodologies, but the results are remarkably discordant.[300] Yet, according to some calculations undertaken within the last two decades, and based on the archaeological surveys carried out in the area, it seems that the population of Judah did not exceed 300,000 in this period.[301] Thus, one might venture to say[302] that a number such as *200,150* may well represent the whole population of Judah at the time of the deportation.[303] However, the (consequent) suggestion that the residents of the *46* towns were not deported but only reckoned as subjects of Assyria[304] is very unlikely and can be easily criticized on several grounds.[305]

In any case, if this quantification refers to the people actually deported, as the inscriptions would lead us to think, the only possible conclusion is that it is unfeasible.[306]

---

clearly connected to those of Jeremiah (*Ant. Jud.*, X, 6:3 and 7:1).

The deportation by Sennacherib, in spite of its pretended exorbitance, is not mentioned by the O.T.

300 They are tabulated in A. Byatt, "Josephus and Population Numbers in First Century Palestine," *Palestine Exploration Quaterly* 105 (1973), 51f.

301 Y. Shiloh, "The Population of Iron Age Palestine in the Light of a Sample Analysis of Urban Plans, Areas, and Population Density," BASOR 239 (1980), 25-35 (esp. 32f). Actually, his calculations are rather optimistic: for instance, following M. Broshi ("La population de l'ancienne Jérusalem," *Revue biblique* 82 [1975], 5-14, cf. esp. 6f), he makes use of an urban density of 50 per dunam [= 500 per ha.] which is really very high. Furthermore, he does not take into consideration the public areas, and applies the population density, which had been estimated on the basis of the residential areas, to the whole extent of the settlements (even if he is fully aware of this deficiency: cf. *loc. cit.*, p. 30a and n. 22). Note that M. Broshi considers the figure of 400-500 persons per hectare as a "maximum density in built-up areas" (M. Broshi, "The Population of Western Palestine in the Roman-Byzantine Period," BASOR 236 [1979], 1) while he, more recently, prefered a coefficient of 250 persons per ha. for the population density of early bronze age Palestine (cf. M. Broshi - R. Gophna, "The Settlements and Population of Palestine During the Early Bronze Age II-III," BASOR 253 [1984], 42). Cf. also next note.

Recently N. Na'aman remarked that in Judah the number of settlements in the 8th century was much higher than in the 7th century, and that many of these had actually been destroyed, and remained deserted for centuries. He attributes these destructions to Sennacherib's campaign and thus he concludes that thousands of people must have been deported (*Tel Aviv* 20, 112-15; cf. also *id.*, " Forced Participation in Alliances in the Course of the Assyrian Campaigns to the West," *Fs Tadmor*, 96).

302 G.E. Wright, *Isaiah*, London 1964, 77 (cf. Stohlmann, *Scripture in Context* 2, 153 n. 24). W.F. Albright, "The Administrative Divisions of Israel and Judah," *Journal of the Palestine Oriental Society* 5 (1925), 25 n. 15 suggested that Judah had 250,000 inhabitants at the time of Sennacherib's conquest — cf. also his opinion on the number given by Sennacherib's inscriptions, n. 304 below, and C.C. McCown, "The Density of Population in Ancient Palestine," *Journal of Biblical Literature* 66 (1947), 425-36, reaching similar conclusions.

303 Note that Sennacherib does not claim to have conquered its capital, Jerusalem.

304 Put forth by R.P. Dougherty, "Sennacherib and the Walled Cities of Judah," *Journal of Biblical Literature* 49 (1930), 160-71. He based his argument on the fact that the Assyrian texts simply state that the prisoners were "brought away and counted as spoil" (*ušeṣamma šallatiš amnu*), while one would expect to find *ašlul*, "I carried off," in the case of a deportation. He also pointed out that the cities were not destroyed, but simply "conquered" (KUR) — cf. also W.F. Albright, "The Fourth Joint Campaign of Excavations at Tell Beit Mirsim," BASOR 47 (1932), p. 14.

305 For instance, it does not take into consideration that also livestock had been reckoned on that occasion.

More recently Stohlmann (*Scripture in Context* 2), despite his criticism of Albright's hypothesis (p. 153f), put forth a similar explanation, i.e. that "the number of Judeans represents the population of the captured cities and towns reckoned for future deportation." Then, only some of these Judeans were actually deported. He points out that a time lag was necessary between the reckoning of the people and their deportation, which took place at the end of the campaign, and suggests that the accounting in this case (unlike the 1st campaign to Babylonia) refers to the reckoning and not to the deportation (p. 156). This hypothesis is highly speculative and is even unnecessary in light of Stohlmann's further suggestions (in turn very speculative), that the number "is no doubt an exaggeration" of an originally sizeable group of captives, "certainly in the thousands" (cf. pp. 160 and 174).

306 However, a striking feature is that the amount of tribute imposed on the king of Judah at the end of that campaign is partially confirmed by the O.T., as has been noted above, p. 80. This might let us think

10. Enemies killed during the 8th campaign.

During the battle of Ḫalulē *150,000* warriors, belonging to a coalition of Elamites and Babylonians, were killed, according to a couple of *summary* texts of Sennacherib.[307]

This quantification is definitely unrealistic, if one considers that it refers to the men killed during a single battle, which, most likely, lasted but a few hours: note how, during the battle of Qarqar, *14,000* enemies out of a total of almost 75,000 men were killed, according to Shalmaneser III's monolith inscription (cf. above pp. 103-107). Even if in this case the *150,000* killed included the whole enemy army,[308] the numeral should be considered unfeasible, as we pointed out above (appendix 5).[309]

---

that these inscriptions intended to relate genuine numerals, at least in this passage: the issue will be further dealt with below, at p. 174.

307 1 R, 43:48 (Luckenbill, *Senn.*, 89; ARAB 2, 352); VS 1, 77 Rs. 14 (Luckenbill, *Senn.*, 92; ARAB 2, 357).

The outcome of this battle, despite the Assyrian claim of victory, is quite disputed: see (recently) A.K. Grayson, "Problematical Battles in Mesopotamian History," *Fs Landsberger*, 342; J.A. Brinkman, "Sennacherib's Babylonian Problem: an Interpretation," JCS 25 (1973), 89-95; L.D. Levine, "Sennacherib's Southern Front: 704-689 B.C.," JCS 34 (1982), 49ff; Brinkman, *Prelude to Empire*, 63ff.

308 The slaughter is also described, in greater detail, by the final edition of Sennacherib's annals, which does not report numerals, but simply specifies that the enemies were "made scarce" (*ušāqir*, which Luckenbill translates "I decimated," cf. *Senn.*, 45:80f; ARAB 2, 254). A similar passage appears on a stela, which, however, is badly broken at this point (cf. Grayson, AfO 20, 92:68ff). Finally, the battle is also mentioned in the Bavian inscription (Luckenbill, *Senn.*, 82:34ff; ARAB 2, 338). If the killed represented the whole army, one may expect one of these texts (incl. those mentioned in the previous n.) to report e.g. that "not one escaped," or similarly (cf. some such instances below, p. 178 nn. 88-91).

Note that the annals compare the enormous host (*kitru rabû*) of the enemies to "an onset of countless locusts (*āribi mā'di*) in the springtime" (cf. Luckenbill, *Senn.*, 43:51 and 56; ARAB 2, 252).

309 On this quantification cf. further below, p. 175.

4. WHERE THE NUMBERS ARE USED

Every military action of a certain importance was probably followed by functionaries assigned to the computation of booty and to the collection of different sorts of information.[1] Scribes accompanying the Assyrian army on campaign are often represented on the reliefs, and most times they appear in pairs, one holding a scroll and one a tablet, in booty scenes.[2] The possibility that they were composing some kind of "war diary" has been postulated,[3] but, since no example has reached us, we must maintain that either they were of perishable material or they did not exist. However, if we admit the former hypothesis,[4] the data concerning tribute and booty would have been quite exact, for practical reasons that are easy to imagine,[5] excluding booty deriving from indiscriminate sacking by the troops, possibly subject to an approximate calculation. Certain (clay) tablets that have been indicated as "booty lists," actually a quite disputable designation, usually give exact (and rather low) figures.[6] Since the royal inscriptions almost regularly give round figures for tribute and booty, we must infer that the data thus collected have not been

---

1 Note the frequent use, in the Assyrian royal inscriptions, of the phrase *ana šallatiš amnu*, "I counted as booty."

2 They first appear in the reliefs of Tiglathpileser III, cf. for instance R.D. Barnett - M. Falkner, *The Sculptures of Aššur-naṣir-apli II, Tiglath-Pileser III, Esarhaddon, from the Central and South-West Palaces at Nimrud*, London 1962, Pl. V ("spoils of a captured city. Eunuchs taking count" — also appearing e.g. in Driver, *Semitic Writing*, Pl. 23A and C.J. Gadd, *The Stones of Assyria*, London 1936, Pl. 11). Cf. the discussion in H. Tadmor, "On the Role of Aramaic in the Assyrian Empire", *Bulletin of the Middle Eastern Culture Center in Japan* 5 (1991), 419-426 (esp. 420). One has to admit, however, the possibility that the two scribes may represent only a conventional image.

3 Cf. Reade, ARINH, 162 and Russell, *Sennacherib's Palace*, 28 and 292 no. 36. On the possible existence of such "diaries," cf., in general, Grayson, Or 49, 164-167 (rather skeptical). Note that, contrary to the common belief, war diaries were not in use in ancient Egypt: cf. D.B. Redford, "A Bronze Age Itinerary in Transjordan," *Journal of the Society for the Study of Egyptian Antiquities* 12 (1982), 58 and *id.*, *Pharaonic King-Lists, Annals and Day-Books*, Mississauga 1986, 97-126 (esp. 122ff).

Renger pointed out that "an Assyrian scribe must have been able — on the basis of his training — to write down the story of a war simply by knowing a few but essential facts about who, where, and when. Then he would draw from a repertory of stock phrases — thus it is superfluous to develop a "war diaries" theory" (*Introd.* [p. 66]). This could also imply the possibility of "inventing" the numbers.

4 It is quite conceivable that the "diaries" were written on light (and therefore perishable) materials, such as leather or papyrus, for practical reasons: cf. S. Parpola, "The Royal Archives of Nineveh," in K.R. Veenhof (ed.), *Cuneiform Archives and Libraries* (Istanbul 1986), 225f (and cf. Postgate, TCAE, 19-28 for the clay dockets used as sealings for scrolls or wooden writing boards). Henshaw, *Paleologia* 16, 13, on the other hand, suggested that the day records were written on clay, and that a separate tablet was needed for each battle's or day's statistics.

5 One may wonder, however, how the original counting of sheep could have been so exact, if they added up to some hundreds of thousands, as reported by certain inscriptions of Sargon and Sennacherib.

6 On the "booty and tribute lists" cf. above, p. 77.

used. The problem is, then, *if* the (possibly) collected data has been used for the royal inscriptions, and *how*.

Thus, either during the collection of the data on the spot or during the composition of the royal inscriptions immediately following the facts, those including the first version of the narration, the quantities to be used were submitted for a selection. In the former case, besides tribute and booty, other data might have been recorded (distance covered, geographical references, etc.), perhaps already keeping in mind their possible use in forthcoming "official" royal records. Already in this phase, then, a selection between what had to be counted and what was not of interest was carried out. In the case of a military engagement, some data that could represent the "victory" had to be collected: enemies killed or captured, booty, etc. This would have been possible even after the return from the campaign, not so easily in the case of prisoners[7] or killed,[8] but certainly in the case, e.g., of the "total" number of cities of a more or less large area.

The data thus collected were submitted to a (further) selection when writing the above mentioned "first version" of the royal inscriptions. This text might be represented by the so-called "letters to the God," of which the best known is that of Sargon. It contains the report of his 8th campaign, and is addressed to the god Ashur, to the gods of the city Ashur (not specified), to the city itself and to its population.[9] Another example comes from the reign of Esarhaddon, while similar texts by Shalmaneser IV, Shamshi-Adad V and Ashurbanipal may belong to this category as well, though their fragmentary state allows some doubt.[10] Since only a small number of these texts have reached us, it is difficult to imagine that such reports were regularly compiled at the end of each campaign.[11] Their Akkadian name, however, *lišānu rēšēti*,

7 Cf. below p. 173.

8 Some Egyptian inscriptions make it clear what system was used to count them: cutting off the extremities and bringing them back. Thus, for instance, a couple of inscriptions of Merneptah inform us that the Egyptians carried off from Libya *6,359* phalli of slain men, besides *2,370* hands, *218* living prisoners, etc. (ARE 3, 588 and 601). Similarly, Ramesses III's army killed *12,535* enemies during the first Libyan war (ARE 4, 52 [end]), a number confirmed by the matching of the totals of both hands and phalli. However, in another text the number is rounded up as *3,000* × 4 piles, three of hands and one of foreskins (mentioned together with *1,000* captives; cf. *ibid.*, 94 and n. there). In a second war, again *2,175* hands of killed people were brought back, together with *2,052* prisoners (ARE 4, 106 and cf. esp. 111). See ARE 2, 435, 532 and 854 for other accounts of "hands" (*83*, *29* and *312* resp.), other instances listed in ARE 5, 131.

This habit (counting of the foreskins) is attested also in the O.T., cf. one example in 1 Sam. 18:25-27, and may have been employed by the Assyrians as well, even if their sources make no reference to it, while mentioning repeatedly similar mutilations. Sometimes they tell of the erection of piles of heads (cf. e.g. p. 48 n. 15), which may reflect the act of accounting — a relief of Sennacherib actually shows the Assyrians building up one such pile, and one of them is apparently recording their number (cf. Driver, *Semitic Writing*, Pl. 24). However, this system would not have been necessary if they were going to use, in the inscriptions, round or (artificially) "exact" numbers for the enemies killed (but cf. below, p. 162 n. 6).

9 It was suggested that the document was meant to be read publicly during some important state ceremony, in which perhaps the losses were honored (at the end the document reports *6* losses): cf. above n. 115 p. 75.

10 E.g., Tadmor, JCS 12, 82 takes as examples of "letter to the God" only those of Sargon and Esarhaddon.

On the genre of the "letter to the God," which has a long tradition in Mesopotamia, cf. R. Borger in RlA 3 *s.v.* "Gottesbrief," pp. 575f (cf. also Grayson, Or 49, 157-59).

11 This hypothesis has been advanced by Speiser, *Idea of History*, 66. Weidner has suggested that such letters were composed only after outstanding campaigns (AfO 12, 148), and this may well be the case.

The cylinder published in Smith, *First Camp.* containing only the first campaign of Sennacherib and written, as in the case of the "letters," shortly after the narrated events, is not part of this category, but is rather a very early edition of the annals, to whose model it conforms.

"first report"[12] may indicate that they were actually intended as such. In any case, whatever might have been the relationship with the "war diaries" (if they ever existed), we must maintain that in the "first reports" we are already facing the results of a censorship policy aimed at a "correct" use of the numbers that had to conform to the aims of the inscriptions.[13]

Another category of texts that might represent some sort of "first version" is typified by the royal stelae written during or at the end of a military campaign and erected on the spot — the so-called "stelae at the world boundaries."[14] In these cases first-hand information had (necessarily) to be used. However, most such inscriptions[15] do not show particular editorial divergences, as far as the use of numbers is involved, from the later annalistic or summary texts written in Assyria.[16]

As for the "corrections" (that is the alterations) at the level of digits, in the previous chapter (§§ 1-3) we saw that in most cases they are probably due to oversights or copying errors and not to the intention of inflating the numbers. We should conclude, therefore, that the alterations, if they took place, were done during the composition of the first texts or, less probably, even earlier, during the collection of the data on the spot. After all, the reasons for the exaggeration of the numbers were as valid for the earliest inscriptions as for the latest.

A further selection took place during the creation of subsequent inscriptions. In the case of the annals, the procedure applied is quite clear at least for certain epochs: for instance, in Shalmaneser III's reign, as new events accumulated, newer editions of the annals were needed,[17] and each time they included the older narrative, yet much curtailed, together with the narrative of the latest years, which were given in more detail, since they seemed more important.[18] During the reign of Sennacherib, usually the text of the older editions was recopied just as it was,[19] simply adding the new

---

12 Borger, *Asarh.*, 107:23 (cf. CAD L, 212b). However, M. Cogan recently suggested that this expression should be translated "main/best report," ("A Plaidoyer on behalf of the Royal Scribes," *Fs Tadmor*, 127 n. 27).

13 This applies mainly to the literary aspects. Cf. above p. 76 n. 122 for an opinion in favor of the reliability of these texts.

14 Cf. Liverani, *Ideology*, 307.

15 The neo-Assyrian stelae are examined (together with the rock-reliefs) in Börker-Klähn, *Bildstelen*, 54-59 and 177-224 (catalogue); cf. also Levine, *Stelae*, 51ff.

16 An interesting example is discussed above, p. 83 n. 158.

17 A new edition was often demanded by some building enterprise, such as the foundation of a new palace, that needed cylinders, prisms and/or inscribed slabs (cf. e.g. Olmstead, *Historiography*, 22 and 45). In some cases (expecially in the reign of Shalmaneser III) the completing of some important military campaign called for a new edition. Tadmor (*Fs Finkelstein*, 210) has pointed out the tendency in the 9th century to compile a first edition of the annalistic inscriptions during the 2nd year of reign and of new editions soon after the 5th, after the 10th and after the 20th year (this is actually the case with Ashurnasirpal II).

At times, some contingent needs, such as dynastic or court problems, etc., could originate new annalistic editions: cf. Tadmor, HHI, esp. 38ff.; Olmstead, *Historiography*, 27; R. Gelio, "Caratterizzazione ideologica e politica del cilindro Rassam," *Vicino Oriente* 1 (1978), 47-63.

18 Cf. Olmstead, *Historiography*, 8 and 21.

19 It happens thus that, in the modern editions of Sennacherib's inscriptions, the section *invariata* most times is not given: the latest edition of the annals (Taylor prism or Oriental Institute prism) is transliterated and/or translated in full, while only some paragraphs of the older editions are reported, those containing passages not present elsewhere. Besides its impracticality, this system involves the danger that not all the "minor" variants are noted, and among these the numbers. Thus, to check the digits given by the "Rassam cylinder" (containing three campaigns), which in some cases differ from those of the other texts (cf. p. 57 nos. 65ff), one must resort to Bezold (in KB 2, 80ff) or to the "scholastic" edition of the campaigns of Sennacherib available in BAL^2 (pp. 64ff), since those variants are ignored both by Luckenbill (*Senn.* and ARAB 2) and by Evetts who first published (in ZA 3) this cylinder (limiting himself to the conclusion, the only part "marked by many variants and several new passages," p. 311).

events.[20] Obviously, it was always possible to "improve" the older text, and sometimes it was even filled out with "new" information.[21]

Then, simplifying, the "selection," as well as the "correction" of the data could take place in three stages: collection of the data on the spot; writing of the first versions of the texts (serving as a source for the subsequent);[22] writing of new texts, more or less shortened. Comparing the extant inscriptions, we can witness only the third of these stages. The comparison between the "letter" of Sargon and the report of the eighth campaign as given by the annals, which is much shorter, is possibly the best opportunity we have, and it shows how certain quantifications were re-employed, others not, and others were "condensed." Other possible similar comparisons can be made: for example the various editions of the annals of Shalmaneser III or Sennacherib or Ashurbanipal. But even from inscriptions having no parallels we can draw some indications, for example examining how within the same episode certain objects, certain situations are quantified rather than others, given that the Assyrian royal inscriptions could not be filled up with numbers and data as certain administrative documents are.

Some criteria used to choose the quantifications can thus be pointed out. They have a validity limited to the reign under discussion, or to the inscription under discussion — however, one may suppose they had a broader value, considering the numerous similarities shown even by texts belonging to different reigns.[23]

1. IMPORTANT OBJECTS

The disappearance of several quantifications in later inscriptions is due to the abridgments to which the lists of tribute and booty were often subjected. Thus, the booty listed in the passage on the second campaign of edition "A" of Shalmaneser III's annals[24] is not mentioned by the subsequent inscriptions: the larger space available on inscriptions concerning a limited period — the beginning of the reign — allowed the inclusion of some detailed lists.

---

20 Cf. Levine, HHI; Olmstead, *Historiography*, 43. To find space for the longer texts thus resulting, the earlier cylinders were later on replaced by prisms. During the reign of Ashurbanipal, besides the number of sides, also the dimensions (height) of the prisms could be augmented.

21 Cf. examples e.g. in Levine, HHI, 71f; Olmstead, *Historiography*, 41f; Cogan - Tadmor, *Or* 46, 77ff.

22 Not always the latest available edition was the direct source of the newer one, which could have been based on even older editions, as sometimes happened: cf. e.g. Olmstead, *Historiography*, 57; Cogan - Tadmor, *Or* 46, 77f. Furthermore, not even older inscriptions necessarily had to be the source: there are examples of passages of chronicles included *verbatim* in the royal inscriptions (cf. Tadmor in *Fs Finkelstein*, 210f), or of the use of state correspondence as a source (cf. Parpola, ARINH, 123). Sometimes passages of inscriptions of previous rulers were re-used (according to Olmstead, *Historiography*, 55; obviously the source was not acknowledged).

23 This phenomenon may cause difficulty in attributing certain incomplete inscriptions or fragments to this or that ruler, which sometimes might be very distant from each other: see the case of the "white obelisk," alternately credited to Ashurnasirpal I or to Ashurnasirpal II — the matter is summarized in E. Sollberger, "The White Obelisk," *Iraq* 36 (1974), 231-38, pl. XLI-XLVIII: he credits it to the latter, while J.E. Reade, "Aššurnaṣirpal I and the White Obelisk," *Iraq* 37 (1975), 129-50, pl. XXVIII-XXXI and W. von Soden, "Zur Datierung des «Weißen Obelisken»," ZA 64 (1975), 180-91 credit it to the former.

24 ARAB 1, 601; cf. above p. 99.

If comparing the text referring to the first 6 campaigns given by the various "editions" of Shalmaneser's annals, we get the following table, indicating the number of quantifications:[25]

	"A"	"C"	"E"	"F"
Cities	4	2	1	-
Kings	1	1	1	-
Soldiers (various terms)	16	2	2	1
Prisoners	2	-	-	-
Cavalry (*pētḫallu*)	2	-	-	-
Chariots	5	-	-	-
Camels	3	-	-	-
Commodities of various kinds (tribute)	40	-	-	-
Other (multiplicatives, numeral adv., etc.)	4	-	-	-

Also a similar comparison between Sargon's "letter" with the 8th campaign and the corresponding passage in the annals can be useful:[26]

	"letter"	annals
Cities/Strongholds	16	11
People: (total)	5	3
soldiers	1	1
prisoners	1	1
dignitaries	2	1
princes	1	-
Battle chariots	1	1
Provinces/districts	2	-
Mountains	1	-
Multiplicatives, etc.	3	-
Data related to walls, foundations, etc.	3	-
Surface measures	1	-
Animals (booty)	8	3
Commodities (booty)	47 +	4 +
"Losses" (Assyrian)	3	-

In general, the quantifications are reserved for the most important objects, those having a higher value (either from a commercial or ideological point of

25 Cf. the similar tables in ch. 2, pp. 30ff, where the references can be found.

26 For the "letter" cf. Thureau-Dangin, TCL 3 (= Mayer, MDOG 115; ARAB 2, 140ff), for the annals Lie, *Sargon*, 22:127-165 (= Fuchs, *Sargon*, 110ff and 320ff; ARAB 2, 19ff). The quantifications in the two booty lists included in the "letter" add up to 47 (those that can still be read, at least). Note that the quantifications of the "losses" placed at the end of the inscription include a number written syllabically (*iš-ten*). The booty lists and the quantifications of cities will be discussed below, pp. 129-132.

view). In the tribute or booty lists most of the time only a limited number of items are quantified, those listed first are, usually, the most important:[27]

> "I received the bountiful tribute (*nāmurtu*) of Ilī-ibni, governor of Suḫu: *3* talents of silver, *20* minas of gold, an ivory couch, *3* ivory chests, *18* tin bars, *40* furniture legs of *meskannu*-wood, a bed of *meskannu*, *6* dishes of *meskannu*, a bronze bathtub, linen garments, garments with multicolored trim, purple wool, oxen, sheep, bread (and) beer."

The non-quantification of the other objects does not necessarily imply that the pertinent data were not available, but rather that these data have been ignored (or added to those of the quantified objects), or that the objects themselves are an editorial addition, which contributes to making the lists quite stereotyped and repetitive.[28]

The final version of Sennacherib's annals states that during the 1st campaign *208,000* prisoners were captured, besides horses, mules, asses, camels, oxen and sheep "without number."[29] The older editions, by contrast, give the numbers for all the animals, with varying digits (cf. p. 58 nos. 67ff). The number of prisoners, the most important "item," is preserved throughout the various versions, without "corrections," while the digits of all the animals (except the horses) vary. A similar fact can be noted also in Sargon's *display inscriptions*, preserving only the number of prisoners (here "corrected") among all the quantifications relating to the booty of Muṣaṣir given in the "letter" and by the Khorsabad annals (cf. p. 54 no. 49).

Besides being more easily quantified, it seems that the quantities of the most important items were generally better specified. In particular, sometimes gold and silver are measured with "exact" numbers, differently than the other objects. The same also applies often to horses.[30] This use is characteristic of an early period (before the Sargonids), and can be explained by the fact that certain commodities simply were evaluated more carefully for their high intrinsic value. Thus, the tendency was to use lower, apparently more genuine, numbers for highly valued objects, and "gross" and less careful quantifications for the others. This is underscored also by the fact that in some cases the numbers relating to metals, horses and people are transmitted more carefully, showing fewer variations than the other quantifications.

---

27 Schramm, BiOr 27, pl. III and p. 151:69ff (= RIMA 2, 175; ARI 2, 471). For other examples cf., in translation: Tiglath-pileser I: ARI 2, 13, 14 (cf. also 25 and 27); Tukulti-Ninurta II: ARI 2, 472, 473 (3 ex.), 474, 475; Ashurnasirpal II: ARI 2, 584 (2 ex.); Shamshi-Adad V: ARAB 1, 720; Weidner, AfO 9, 93:12ff ("*30* of their horses (*pētḫallu*), [*one*?] chariot, their property, valuables, oxen, sheep I took them"); Adad-nirari III: ARAB 2, 740; Tiglath-pileser III: ARAB 1, 769 (end), 775 (end — cattle and sheep are quantified, while horses, mules and asses are not), 812; Tadmor, *Tigl. III*, Summ. 9 rev. 7ff; Sargon: ARAB 2, 24 (cf. below, p. 129), 70 (= 184); Sennacherib: cf. the tribute from Judah (above, p. 80) in ARAB 2, 240, 284, 312.

28 In the annals of Tukulti-Ninurta II (an example has just been quoted), several lists of animals and metals, quantified, are followed by "bread, beer, straw and fodder," placed at the end of each list and never quantified. This is a way to indicate, generically, the "supplies" for the men and the animals of the Assyrian army. On the homogeneity of these tribute lists cf. p. 97.

29 Passage quoted above, p. 90.

30 Thus, in the lists of ARAB 1, 601 (Shalmaneser III), the only numbers not using the digits "1," "2," "3" and "5" (the most frequent) are those relative to the metals (bronze and silver) received on some occasions (*90* talents, *6* and *70* talents — cf. p. 201). It is useful to compare also the lists of ARI 2, 641 (plus the "updated" version of §574 — Ashurnasirpal II) and 472ff (Tukulti-Ninurta II), where, among other things, the ducks are once given with a number "more exact" than those of the other animals (*14*, *versus 30, 50, 200*, cf. ARI 2, 472).

2. UNUSUAL OBJECTS

The number is preferably used to quantify rare and unusual objects, even better if typical of the land of origin. Speaking of the booty taken from the king Kili-Tešub of Katmuḫu, the annals of Tiglath-pileser I enumerate only the copper kettles and the bronze bathtubs: "his wives, his natural sons, his clan, *180* copper kettles, *5* bronze bathtubs, together with their gods, gold and silver, the best of their property, I carried off."[31] In a subsequent passage of the same annals, the tribute of Šadi-Tešub king of Urattinaš is thus described: "his natural sons and his family as hostages I took. *60* copper kettles, (a) bronze vat, large bronze bathtub(s), together with *120* men, oxen and sheep as tribute and tax he brought."[32]

Of the booty taken at the city of Murattaš only the copper kettles are specified, together with a quantity of copper ore: "their gods, their possessions, their property, *60* (var. *120*) copper kettles, *30* talents of copper bars, the outstanding property of their palace, their booty I brought out."[33]

The monolith inscription, an early annalistic edition of Shalmaneser III, in two cases specifies the number of the camels received as tribute: "the tribute of Asû, horses, oxen, sheep, wine, *2* camels whose humps are double I received";[34] "horses broken to the yoke, oxen, sheep, wine, *7* camels whose humps are double I received from him."[35] Elsewhere in the inscriptions of this king, the camels are not quantified (as well as all the other objects mentioned), while their characteristic of having two humps is specified[36] (thus being distinguished from the dromedaries).

The exotic animals of alien provenance, as well as the objects characteristic of faraway lands, are so significant because they represent "extracts" of the

---

31 AKA, 41:28-32 (RIMA 2, 15; ARI 2, 13). The quantifications are 3 *šu-ši ruq-qi* URUDU^{MEŠ} and 5 *nàr-ma-ak* ZABAR.

32 AKA, 43:47-53 (RIMA 2, 15f; ARI 2, 14). Quantifications: 1 *šu-ši ruq-qi* URUDU^{MEŠ} and 2 *šu-ši a-mi-lu-te*. "Tribute and tax" stand for GUN *ù ma-da-[ta]*. The number of vat(s) and bathtubs(s) is not clear, the vat being in the singular and the bathtub(s) having the plural signs only in some copies (the plural of the adjective could refer to the two objects): *nam-ḫar* ZABAR *ù nàr-ma-ak* ZABAR(.MEŠ) GAL.MEŠ-*te*. However, one each should be intended, since the text later specifies that *1* vat and *1* bathtub were donated to the god Adad: cf. below p. 125.

33 AKA, 58:102ff (RIMA 2, 19; ARI 2, 25). Quantifications: 1 / 2 *šu-ši ruq-qi* URUDU^{MEŠ} and 30 GUN URUDU^{MEŠ} *ša-bar-ta*.

34 3 R, 7:28 (KB 1, 156; ARAB 1, 598). The quantification is 2 *ud-ra-a-te ša* 2 *gu-un-gu li-pi*, "tribute" is *ma-da-tu*. The older version given by the tablet published in Mahmud - Black, *Sumer* 44 omits the number 2, but specifies the camels had *2* humps (p. 145 and 140:41).

35 3 R, 8:61f (KB 1, 166ff; ARAB 1, 607). Quantif.: 7 *ud-ra-te šá* 2 *gu-un-gu li-pi-ši-na*. Note that the tribute bearers had similar names: Asû of Guzana (year of accession of Shalmaneser) and Asāu of Gilzānu (3rd year). However, Guzana is a well known western town (modern Tell Halaf, cf. Russell, *Iraq* 47, 66), while Gilzānu probably lies in the east (cf. Russell, AnSt 34, 198ff).

36 Besides the two quoted passages, cf. also Michel, WO 2, 140:A and C (two cases — ANŠE.A.AB.BA.MEŠ *šá šu-na-ai ṣe-ri-ši-na*).

Then cf. Shamshi-Adad V in 1 R, 30 and 33 II 56 (= KB 1, 180; cf. also Reade - Walker, AfO 28, 115:7′).

Some passages in Ashurbanipal's inscriptions are eloquent on the rarity of camels in Assyria and on their character as "foreign" animals: cf. ARAB 2, 827, 869 and 948 (mentioned also below, p. 179).

In some instances the camels are not quantified (nor are the other animals): cf. RIMA 2, 200:97 (*udrāte* - Ashurnasirpal II; cf. also RIMA 2, 104:26f for the mention of *udrāte* acquired by Ashur-bel-kala). From the time of Tiglath-pileser III on, the camels (or dromedaries) are not deemed worthy of particular attention, being quantified only if also the other animals are (cf. ARAB 1, 766 end, 772 end, 778, 795, 817f for Tiglath-pileser III).

periphery brought to the imperial center, and towards them sometimes a genuine attitude of curiosity or interest is shown.[37]

In a list of goods taken away from Niʾ (Thebes?) by the army of Ashurbanipal, the obelisks are specified and quantified:

> KÙ.BABBAR*(kaspu)* GUŠKIN*(ḫurāṣu) ni-siq-ti* NA$_4$^{MEŠ}*(abnē)* NÍG.ŠU*(bušû)* É.GAL*(ekalli)-šú ma-la ba-šú-u lu-bul-ti bir-me* GADA^{MEŠ}*(kitê)* ANŠE.KUR.RA^{MEŠ}*(sīsē)* GAL^{MEŠ}*(rabûti)* UN^{MEŠ}*(nišē) zik-ru u sin-niš* 2 ^{GIŠ}*tim-me* MAḪ^{MEŠ}*(ṣīrūte) pi-tiq za-ḫa-le-e eb-bi ša* 2 *lim* 5 *me* GUN*(bilāti)* KI.LÁ*(šuqulta)-šú-nu man-za-az* KÁ*(bāb)* É.KUR*(ekurri) ul-tu man-za-al-ti-šú-nu as-suḫ-ma al-qa-a a-na mat* AN.ŠÁR*(Aššur)*^{KI} *šal-la-tú ka-bit-tú ina la me-ni áš-lu-la ul-tú ki-rib* ^{URU}*Ni-ʾ*
>
> "Silver, gold, precious stones, the goods of his palace (of the king), as much as (they) were, multicolored linen garments, large horses, the people, men and women, *2* tall obelisks made of shining electrum, whose weight was *2,500* talents each, which stood by the gate of the temple, I removed from their positions and carried them off to Assyria. A heavy (and) immeasurable booty I carried away from Niʾ."[38]

These are the only numbers appearing in the report of the Egyptian campaigns present on prisms "B" and "F," and perhaps also on the fragmentary prism "C," while prism "A" has two other numbers.[39]

3. OBJECTS IN RELATION TO THE GODS OF ASSYRIA

Even if not intrinsically important, these objects assume a particular relevance, being thus easily quantified by the inscriptions.[40]

Of the booty taken in Sugu, an otherwise unknown site, the annals of Tiglath-pileser I specify the number of gods: "*25* of their gods, their booty, their possessions (and) their property I brought out." The gods were subsequently donated to the temples of the Assyrian gods: "at that time I dedicated (*áš-ru-uk*) the *25* gods of those lands, my own booty that I had taken, to adorn (*a-na ú-tu-ʾ-ut*) the temple(s) of the goddess Ninlil, beloved chief spouse of the god Ashur, my lord, of the god(s) Anu, Adad, Ishtar of Assyria, the

---

[37] For instance towards the beasts, which were gathered in herds and displayed to the people of Assyria (cf. ARI 2, 248, 597 and 681 and p. 145 below), or of which stone replicas were made (quantified in AKA, 146:16ff = RIMA 2, 105 and ARI 2, 250).

[38] 5 R, 2:39-44, cf. dupl. in 3 R, 18:75ff (Streck, *Asb.*, 16; ARAB 2, 778); other passages in Piepkorn, *Asb.*, 40:30-40; Aynard, *Prisme*, 32:51-55; Bauer, IWA, 15 III 38-46; Nassouhi, AfO 2, 102 II 7-14; cf. also BAL^2, 92. On the two obelisks cf. the study in Aynard, *Prisme*, 23ff.

[39] Prism "B" I 50-II 40 (34 lines - Piepkorn, *Asb.*, 28ff); "F" I 35-55 (21 lines, has only the 2nd campaign - Aynard, *Prisme*, 28ff); "A" I 52-II 48 (131 lines - Streck, *Asb.*, 2ff); for prism "C" cf. Bauer, IWA, 15 col. III and Streck, *Asb.*, 140ff col. II.

Other quantifications referring to the Egyptian campaigns, not present on the prisms, appear in K 3127 + K 4435 (cf. p. 175 n. 69 below).

[40] Not every object to be donated to the gods is quantified: cf. some exceptions in Ashur-dan II (RIMA 2, 134:58f), Adad-nerari II (*ibid.*, 144:17), Tukulti-Ninurta II (*ibid.*, 172:27f — here however, it is specified that two thirds of the booty had been donated to the temples and one third to the palace).

temples of my city Ashur and of the goddesses of my land."[41] They were important objects, but in spite of that, elsewhere, they are not quantified.[42] In another text the whole campaign, which occupied 33 lines in the annals, is thus synthesized: "I conquered the entire land Lullumu. I gave (*a-qí-iš*) *25* of their gods [to ...] the gods of my city Ashur and the goddesses of my land. I gave (*áš-ru-uk*) their property to the god Adad, my lord."[43] The higher relevance given to these gods, intended to be presented to the gods of Assyria, is obvious.

One further example is taken from the annals of Tiglath-pileser I. As mentioned above (p. 123), in his campaign against the land of Katmuḫu he first took prisoner the king Kili-Tešub carrying off booty consisting of "his wives, his natural sons, his clan, *180* copper kettles, *5* bronze bathtubs, together with their gods, gold and silver, the best of their property." Subsequently, he received from the king Šadi-Tešub "*60* copper kettles, (a) bronze vat, large bronze bathtub(s), together with *120* men, oxen and sheep." These kettles, as it is soon specified, were then donated to the Assyrian gods: "at that time I donated (*a-qiš*) to the god Ashur *1* bronze vat (and) *1* bronze bathtub from the booty and tribute of the land of Katmuḫu. I gave (*áš-ruk*) to the god Adad, who loves me, *60* copper kettles together with their gods."[44] It is obvious that the *60* kettles are those of Šadi-Tešub, received together with the vat and the bathtub, while the gods come from the earlier episode. In any case, in an inscription such as Tiglath-pileser I's annals that only in rare cases quantifies booty or tribute,[45] it is significant that it carefully records the quantity of kettles and not that of gold, silver, oxen, etc.

Now, if Tiglath-pileser I takes us back to the middle-Assyrian period, we have some later examples as well, from the period of Tukulti-Ninurta II — "*60* ducks (and) wild bulls to Ashur, my lord [I sacrificed]"[46] — and of Tiglath-pileser III: "*240* sheep as a gift to Ashur, my lord, I dedicated."[47] Even if these quantifications are not explicitly defined as booty, they are nonetheless inserted within the military narrative.

Thus, although the inscriptions of Ashurbanipal, in general, are rather sparing with numbers, some of his texts employ "exact" numbers in relation to temple works.[48]

---

41 AKA, 61:23f and 32ff (RIMA 2, 20; ARI 2, 27f).

42 In the time of Tiglath-pileser I foreign gods are said to be captured, without being quantified, in ARI 2, 13 (cf. 15), 23, 25 (cf. 26), 38.

43 Weidner, AfO 18, 360:23f and cf. AKA, 117:5f (RIMA 2, 34; ARI 2, 68).

44 AKA, 41:58ff (= RIMA 2, 16; ARI 2, 15). The two numerals are 2 *šu-ši* and 1 *šu-ši*. "Booty and tribute" translates *ki-šit-ti ù ma-da-at-ti*.

45 Cf. p. 200 for an overview of the quantifications of booty/tribute in the inscriptions of Tiglath-pileser I (cf. p. 23 n. 16 for the annals).

46 Schramm, BiOr 27, 149:29 (RIMA 2, 172:29; ARI 2, 467). The quantification is 1 GAL.MUŠEN^{MEŠ} *ri-me* UDU.SISKUR^{MEŠ}.

47 Tadmor, *Tigl. III*, Ann. 9:9 (= Rost, *Tigl. III*, 4:16; ARAB 1, 764). The quantification is 2 *me* 40 UDU.NÍTA^{MEŠ}, "gift" is *kad-ri-e*.

For other examples of quantifications of objects intended for the Assyrian temples cf., in Sargon's reign, the booty from Karkemish, p. 52 nos. 40f, and cf. just after, nos. 63f. Quantifications of precious metals employed in building projects are very common, cf. e.g., still in Sargon's times, Thompson, *Iraq* 7, 86:6 (silver) or Gadd, *Iraq* 16, 175:24ff (gold and silver?). For the weight of the silver vase (*20* minas) placed in front of the statue of the god Ashur cf. Saggs, *Iraq* 37, 16:41.

48 Cf. Streck, *Asb.*, 170-173 and 296-299 (ARAB 2, 914f and 1012 resp.); the former text, relatively long, contains very few quantifications besides those relating to the work at the temple of Sīn at Harran. Also prism "C" specifies the weight of some objects intended for the temple of Ashur: cf. Bauer, IWA, 13f (ARAB 2, 883).

4. IMPORTANT EVENTS

Events considered important occupy, within the inscriptions, a proportionally higher place and, consequently, their detailed exposition allows the inclusion of quantifications. Thus, for example, the sole quantifications present in inscriptions of Adad-nerari III are those concerning the tribute paid by Mariʾ of Aram-Damascus, probably the most significant event of his reign (cf. p. 51 nos. 33ff). Also the annals of Sargon show a higher frequency of numbers in the passage containing the 8th campaign, one of the most important (cf. table on p. 38).

The numbers are concentrated in the crucial points of the inscriptions. The comparison of the annalistic inscriptions of Shalmaneser III is again useful: in the final edition (the "black obelisk"), the only quantification of enemy warriors or anyway people (rulers excluded)[49] is that referring to the battle of Qarqar against the western coalition (6th campaign). This notwithstanding some earlier editions contain other analogous quantifications[50] and notwithstanding that in the black obelisk the later campaigns have a much more ample space than the earlier ones. This inscription has only two cardinal numbers up to the 18th *palû*, if excluding the numbers of the *palê* themselves.[51] The previous editions, by contrast, present several numbers in the portions referring to the earliest campaigns, especially in the case of the monolith inscription (edition "A"). It is clear that this battle had a particular relevance: amongst the Syrian campaigns, the 6th is the one that is given more space on the black obelisk. If it is true that its outcome was uncertain,[52] we would conclude that the ("high") number of enemies killed served to affirm in a tangible manner how unquestionable the Assyrian "victory" had been, thus hiding the real issue of the battle[53] — yet testifying simultaneously to the truthfulness of the assertions through the capacity of producing data ("proofs").

---

49 The rulers belong to a different sphere: cf. e.g. the remarks of F.M. Fales, "L'«ideologo» Adad-šuma-uṣur," *Accad. Naz. dei Lincei: Rendiconti*, Serie VIII vol. 29 (1974), 475.

50 Cf., in translation, ARAB 1, 599 (2 numbers), 600 (2), 605, 607 (for edition "A"); 617, 619, 621 (ed. "B"); Cameron, *Sumer* 6, 21 and 13 (2 num.; ed. "C"); ARAB 1, 658, 663 (= 672; ed. "D"); Safar, *Sumer* 7, 16, 17, 18 and 19 (ed. "E"); for ed. "F," given by the black obelisk, cf. ARAB 1, 553-593. Some of these numbers are very relevant, cf. the *82,600* enemies of Safar, *Sumer* 7, 13:35 (= WO 2, 40) or the *44,000* of Michel, WO 2, 410 col. II 3 (= ARAB 1, 617) — actually, they are both "totals" referring to the whole reign — or the *22,000* ERÍN.ḪÁ.MEŠ captured according to Cameron, *Sumer* 6, 12 II 6 (= WO 1, 462; cf. Safar, *Sumer* 7, 7 II 4 = WO 2, 30).

51 Those referring to the Euphrates crossings are multiplicative numbers. The other cardinal number is represented by the "*11* strongholds" of ARAB 1, 562 (5th *palû*), originating from a reconstruction by Layard (cf. ICC, 89:52, based on 14:12, and Rasmussen, *Salm.*, Pl. XXIV and p. 64; cf. also Amiaud - Scheil, *Salm.*, 34 and n. 5). Michel read the passage *e-li* URU.MEŠ-*ni dan-nu-ti* KUR-*ud*, taking the sign for "11" as part of *li* (cf. WO 2, 148). Cf., however, the photograph in Börker-Klähn, *Bildstelen*, Tf. 152B1, first line (and cf. Tf. 152c, upper register, line 16): the LI sign has a rather unusual form, yet the *11* seems clear. Note also that all the older editions of Shalmaneser's annals mention the *11* strongholds within the passage relative to the 5th *palû*, which in some cases is almost identical to that of the black obelisk (cf. e.g. Michel, WO 2, 30:10; WO 1, 462:17).

52 As it is commonly thought, cf. above p. 104 n. 249.

53 To achieve this aim, perhaps, the *14,000* killed according to the first version has been augmented to the *25,000* of the others (cf. p. 49 no. 25).

5. LISTS AND POSITION OF NUMBERS

Generally speaking, the more detailed the narration, the easier it will be to find long lists of tribute or booty, and to know their composition and, possibly, the quantities. In the annals of Tukulti-Ninurta II, as repeatedly noted, the campaign including the "itinerary" contains several quantifications of tribute, while the other campaigns, narrated in a much briefer fashion, have no such quantifications.[54] In the tribute lists of this inscription, as well as of others, often only a first group of commodities is quantified. The general tendency, however, is that either all the items or none are quantified.[55]

Another clear tendency is to first list the more valuable goods, leaving the animals until last.[56] Thus, when all, or most of the listed goods are quantified, it happens that the last ones are given in higher measure.[57] This phenomenon is analogous to that of the "smaller cities," always more numerous than the "main" ones, when both are quantified (cf. p. 14 n. 88). In these cases, we may observe, the higher quantity makes up for the lower "quality."

Sometimes two contiguous items are quantified with a ratio of 1:10, and this is the most frequent ratio between small animals and big ones.[58] This is partly explicable by the high frequency of "round" numbers in these contexts,[59] but the (possible) quantifications of other goods do not show any relation, in the digits, neither between them nor with those of the animals.[60]

---

54 Cf. p. 27 as well as p. 97.

55 Note how in the "annals" of Ashurnasirpal II in certain years only quantified tribute is reported, and in other years only tribute not quantified (cf. p. 29). Obviously, the compositional processes used each year are different, at least as far the use of numbers is concerned.

56 A detailed study of the composition of the booty/tribute lists would go beyond the limits of this work — still, cf. pp. 199ff, where the quantified objects are listed in their natural order (cf. also the remarks on Tukulti-Ninurta II's and Shalmaneser III's lists at pp. 97ff). In all the lists, however, including those not quantified, the (natural) tendency to group objects according to classes is clear — metals with metals, animals with animals, etc. Among these categories, a fixed order is usually respected within a certain time span (e.g. a certain campaign, a number of subsequent campaigns, or even a whole reign). Thus, Ashurnasirpal II's inscriptions regularly give first the (precious) metals and last the animals (see example below). Within each category, a hierarchical order is often respected, thus the lists of animals always show the sheep after the oxen, and these after the horses, etc.

57 Prisoners/people are an exception, being usually given first and often in higher quantities.

The growing succession usually concerns not more than two or three consecutive numbers. Interesting in this regard is a passage from the "Kurḫ monolith" of Ashurnasirpal II (quoted hereafter, at p. 128), where the numbers, in some cases diverging from those of other inscriptions (cf. above, p. 47 nos. 12-14), constitute three consecutive series: one is *40 >> 460*, the others are *2 >> 2,000*, and *1,000 >> 5,000*:

40 (/) - 460 - 2 - 2 - 100 - 200 (100) - 300 - 1,000 (100) - 2,000 (3,000) - / - / - / - 1,000 - / - / - / - 2,000 - 5,000

(a hyphen separates each quantification, while a bar indicates an object not quantified; in parentheses the digits of the *later* version, when different). It can be noted how two out of the three increasing series, as appearing in the Kurḫ monolith, disappear from the Nimrud monolith and the later "annals" thanks to the variation of some of the numerals. A too lengthy, and therefore too obvious, growing series of numbers was unwanted.

58 In a passage of Shalmaneser III (cf. ARAB 1, 601), four consecutive times the numbers of oxen and sheep have this ratio (cf. p. 201), the other numbers having no relation. Other examples are given by Ashurnasirpal II's inscriptions (*1,000* oxen and *10,000* sheep - ARI 2, 584 and 602), Sennacherib (*1* ox and *10* lambs, cf. below n. 79; *80,050* oxen and *800,100* sheep in ARAB 2, 267 and 274). See also Shalmaneser III in Safar, *Sumer* 7, 13:37f (*19,690* asses and *184,755* sheep).

Incidentally, note how the ratio 1:10 between large and small animals recurs also in some administrative texts, e.g. in ABL 241:5 (*2:20* - cf. Postgate, TCAE, 266 or Martin, *Tribut*, 34); ABL 724:5 (*2:20* - Postgate, TCAE, 290); ADD 1057:4′ (= SAA 7, 134) (*1:10* - Postgate TCAE, 333); MacGinnis, SAAB 1, 3:24′f. But this is not, obviously, a rule — thus e.g. ADD 1139 (= SAA 7, 45) and ND 2754 (Postgate, TCAE, 343 and 394) are exceptions.

59 Certain lists, especially the longest ones, use a limited selection of repeating round numbers. The most obvious example is given by the "banquet stela" (it is not tribute actually), for which see below pp. 141f. On the frequency of round numbers with tribute, cf., in general, chapter 2.

60 ARAB 1, 735 (Adad-nerari II) mentions *100?* talents of gold and *1,000* talents of silver (cf. p. 51 nos. 33f); ARAB 2, 2, 46 and 64 (Sargon) speak of *150* chariots and *1,500* horses (cf. p. 106).

The use of "round" numbers in the lists of booty or tribute is very frequent until the Sargonid period. The following example from Ashurnasirpal II is typical:

> "I received *(40)* harnessed chariots, equipment for troops (and) horses, *460* harnessed trained horses, 2 talents of silver, *2* talents of gold, *100* talents of tin (AN.NA), *100* (var. *200*) talents of bronze (ZABAR), *300* talents of iron (AN.BAR), *100* (var. *1,000*) bronze casseroles, *3,000* (var. *2,000*) bronze receptacles, bronze bowls, bronze receptacles, *1,000* multicolored linen garments, dishes, chests, couches of ivory decorated with gold, the treasure of his palace, *2,000* oxen, *5,000* sheep, his sister with her rich dowry, the daughters of his nobles with their rich dowries."[61]

The inscriptions of Sargon and his successors prefer to quantify only the deportees, and not the booty,[62] which is often indicated in the same context as "without measure":

> "*6,500* people with (their) possessions, their horses, their mules, their camels (ANŠE*ud-ra-a-te-šú-nu*), their oxen, their sheep without number (*a-na la ma-ni*) I carried off" (Tiglath-pileser III).[63]

The number of commodities listed might be very high: one such example is given by Sargon's "letter to the God." The final part of the inscription contains two lists, one relating the goods carried off from the palace of Muṣaṣir and one, longer, relating the treasure of the god Ḫaldia and of the goddess Bagbartu (lines 352-67 and 369-405).[64] All the objects are carefully specified and are quantified, if we leave out some simplifications, as that of line 364: "*120* bronze objects, large and small, of the local workmanship, whose names are not easy to write." Most of the precious and unusual objects are quantified with ever changing numbers and digits.[65] The presentation, in the tablet, is very careful: each object (or group of objects) occupies a line, with the number placed at the beginning. Another text contains these same

---

61 AKA, 341:120ff and 237:38ff (RIMA 2, 211 and 252:113ff; ARI 2, 574 and 641). Only the second version has the *40* chariots — cf. n. 57 here above.

62 According to Elat, AfO *Bh.* 19, 249f n. 16, this is due to stylistic reasons. Cf. in this respect the remarks at p. 37, n. 77, for the annals of Sargon and at p. 39, n. 83, for those of Sennacherib.

63 Tadmor, *Tigl. III*, Summ. 7:33 and n. there (Rost, *Tigl. III*, Pl. XXXVI:[5] and p. 64:33; ARAB 1, 795 — both erroneously giving *65,000*). For other instances cf., in translation:
Tiglath-pileser I: ARI 2, 12 (end);
Ashurnasirpal II: ARI 2, 549, 554, 555, 573, 577, 640;
Shalmaneser III: ARAB 1, 599, 600, 605, 607, 611 (end), 619, 621, 654; Hulin, *Iraq* 25, 53:41;
Shamshi-Adad V: ARAB 1, 719, 721, 724, 725; cf. SAA 3 41:11;
Tiglath-pileser III: ARAB 1, 777, 789, 790 (2 cases), 795, 806;
Sargon: ARAB 2, 7 (and cf. 55), 31, 33, 60;
Sennacherib: ARAB 2, 234, 240, 352, 357.
In other instances the booty is only generically indicated ("I carried off *X* people with their booty" or similarly):
Tiglath-pileser I: ARI 2, 12, 18;
Tiglath-pileser III: ARAB 1, 769, 771;
Sargon: ARAB 2, 4, 5, 13, 15, 59.

64 Cf. Thureau-Dangin, TCL 3, 53-63 (= Mayer, MAOG 115, 104ff and ARAB 2, 172ff). The two booty lists are examined in detail by W. Mayer, "Die Finanzierung einer Kampagne (TCL 3, 346-410)," *Ugarit Forschungen* 11 (1979), 571-95.

65 Besides *1* and *2*, no other numeral appears very often: *6* appears 3 times, *12*, *5* and *34*, each 2 times. Several numbers are "exact" (a couple of them appear at p. 189, nos. 18f). In this respect, both lists seem to furnish trustworthy data — and this applies to the numbers of prisoners and animals as well, cf. p. 54 nos. 49ff.

lists, but it is very fragmentary.[66] The version given by the annals, instead, is much shorter: it lists and quantifies only the most relevant goods (cf. next paragraph). It is obvious that the nature of the original text has made it possible to include exceptionally detailed lists — the report of a single campaign, of which the "treasures" represent the tangible result, occupying thus a broad space at the end of the document. Overall, the section pertaining to the booty occupies 67 lines (348-414) out of the 430 of the whole inscription. Its collocation underscores, on a macroscopic scale, the typical position of the numbers within the annalistic inscriptions — at the end of each episode (or of each campaign).

6. NUMBERS AS "EXTRACTS"

Let us continue the comparison between the letter to the God and the other inscriptions of Sargon. In the former text, the complex section concerning the booty of Muṣaṣir can be subdivided into five sections:

1) the royal family: "his wife, his sons, his daughters, his people, the seed of the house of his father I carried off" (*šalālu*, line 348);

2) directly follows the reference to people and animals: "together with *6,110* people, *12* mules, *380* asses, *525* oxen, *1,235* sheep I counted (*manû*) and I had brought within the walls of my camp" (*erēbu* Š, 349);

3) after a reference to the entering into Urzana's palace, the treasure is listed: "[*34* talents *18*] minas of gold, *167* talents *2½* minas of silver, white bronze ... (a long list of quantified objects follows) I carried off" (*šalālu*, 350-67);

4) directly follows that "my officers and soldiers to the temple I sent, and Ḫaldia, his god, and Bagbartu, his goddess, together with the great wealth of his temple, all there was,[67] [*X* +] *4* talents *3* minas of gold, *162* talents *20* minas less *6/36* of silver, *3,600* talents of raw bronze, *6* golden shields ... (an even longer list of quantified objects follows) I carried off" (*šalālu*, 368-405);

5) directly follows the remark: "this does not include the objects of gold, silver, tin, bronze, iron, ivory, ebony, boxwood and all kinds of wood that from the city, the palace and the temples (the people of) Ashur and Marduk in immeasurable quantities (*ana lā māni*) carried off" (*šalālu*, 406f).

The annals present the booty from the city and the palace as follows: "Urzana's wife, his sons, his daughters, *6,170* people, *692* mules (and) asses, *92*[*0* oxen], *100,225* sheep I led out (*waṣû* Š). *34* talents *18* minas of gold, *160* talents *2½* minas [of silver], white copper, tin ..." (there follows a list, very fragmentary, but much shorter than the one in the letter and devoid of quantifications, as far as it can be read). The temple booty list closely follows (also very fragmentary, again much shorter than in the letter). In other words, nos. 1-2 above have been fused, forming a single list, while about 2 lines and 3-4 lines at most are reserved for the palace and temple booty respectively[68] —

66 Winckler, *Sargon*, Tf. 45 (A, B and C) (= Thureau-Dangin, TCL 3, 76ff and ARAB 2, 213).
67 NÍG.GA É.KUR-*šú ma-ʾ-at-ti mal ba-šu-ú*.
68 The whole passage occupies less than 8 lines: cf. Lie, *Sargon*, 26:154-161 (Fuchs, *Sargon*, 114ff and 321f).

there possibly follows a short declaration about the sacking by the troops.[69] Thus, the shortening of the former lists was carried out simply by extracting some objects and some quantifications, and leaving out the others. In its brevity, however, the annals still do not fail to mention each single group of booty and the most relevant objects looted in each case.

The display inscriptions reduce the whole to "his wife, his sons, his daughters, the possessions, valuables and treasures of his palace, all there was (*mala bašû*), together with *20,170* people with their properties, Ḫaldia (and) Bagbartum, his gods, with their numerous (*ma'atti*) valuables, I counted as booty (*manû*)."[70] In this version, very short, as can be seen, all the booty is fused together, while the only number retained is that of the people (slightly revised indeed, cf. above p. 122). It was originally the highest number — but the quantities of metals were possibly as relevant, and despite that, their numbers had not been reused. Anyway, each single group of booty is again referred to: the family, the palace treasure, the people with the animals[71] and the temple treasures, leaving out only the soldiers' loot.

Another example can be taken from the cities conquered during the same campaign. In the annals, many details have been suppressed — thus all the numbers referring to the thickness of the walls, to distances, etc. disappear.[72] But the references to the cities (and their numbers) are in some way amalgamated and thus reported by the annals, notwithstanding their brevity (cf. Table 3 on p. 131). The four[73] paragraphs included in lines 177-268 of the letter, each mentioning a *number* of cities, correspond to the following passage in the annals (lines 139-41):

> "[the city] of Aniaštania, at the border of the land Bīt-Sangibuti, the cities of Tarui (and) Tarmakisa, which are in the flatlands of Dalea, the city of Ulḫu, which lies at the foot of Mt. Kišpal (and) *21* fortified cities together with *140* cities [of the] neighborhood, which lie on Mt. Arzabia I conquered (and) burnt them with fire."

If we look at Table 3, we note that some groups (and numbers) of neighboring cities have simply been left out: the *17* of line 185, the *30* of 189 and the *57* of 231. The subsequent section of the annals (lines 144ff) again leaves out *87* cities ("letter," line 293). It appears thus that some quantifications are used as "samples," "extracts," or "examples." This behavior makes us think that the simple presence of the number was actually needed, and that it was not necessary to relate the exact amounts, or the exact groups, etc. It was not important if the "*140* neighboring cities" conquered together with Aniaštania, Tarui, Tarmakisa and Ulḫu, according to the annals, were actually the *exact* number of smaller cities conquered in that part of campaign (according to the

69 "the remainder of their possessions to [my soldiers I distributed (?)]," line 161, according to Renger's restoration (*Sargon*, 109).

70 Winckler, *Sargon*, 110:75-76 (Fuchs, Sargon, 215 and 347; ARAB 2, 59). The passage occupies less than two lines.

71 "their properties," here *maršītišunu*, can be translated also as "their herds," cf. CAD M/1, 296a.

72 Cf. table on p. 121 here above.

73 As divided by horizontal lines in the text. In the table, each subdivision corresponds to a different paragraph of the "letter" (note that only the cities encountered on Sargon's route are reported, not all those mentioned). As for the annals, each subdivision corresponds to a sentence.

TABLE 3 CITIES QUANTIFIED — 8TH CAMPAIGN OF SARGON

LETTER TO ASHUR (TCL 3)	ANNALS (Lie, *Sargon*, 22ff)
Siniḫini (35), Latašê (37) Sirdakka (52), Appatar, Kitpat(a) (64), Zirdiakka (71) Pānziš (76)	(apparently not mentioned, cf. ll. 127-30)
Parda (84), ***12*** c. *dannūti* (named) + ***84*** c. *ša seḫrišunu* (89)	***3*** c. *dannūti* + ***24*** c. *ša limētišunu*, Parda (131f)
many c. (URU.MEŠ-*šu ma'dūti*) of thc provincc Uišdiš (164)	(the province of Uišdiš, 136)
Ušqaia, ***115*** c. *ša limētiša* (177-82)	Ušqaia + ***115*** c. (URU-*ša*) (137f)
Aniaštania + ***17*** c. *ša limētišu* (184f)	
Tarui, Tarmakisa, ***30*** c. *limētišunu* (189ff) Ulḫu (216), Sarduriḫurda + ***57*** c. *ša limētiša* (231) ***21*** c. *dannū*[*ti*] (named, 239) ***146*** c. *ša limētišunu* (268)	Aniaštania, Tarui, Tarmakisa, Ulḫu, ***21*** c. *dannūti* + ***140*** c. [*ša*] *limētišunu* (139ff)
7 c. *dannūti* (named) + ***30*** c. *ša limētišunu* (272)	***7*** c. *dannūti* + ***30*** c. *ša limētišunu* (142)
Arbu (+) Riar, ***7*** c. *ša limētišunu* (277f)	Arbu, Riar, [...] (143 ff)
30 (named 29) c. *dannūti* (286) Argištiuna, Qallania (287) + ***87*** c. *ša limētišunu* (293)	***30*** c. [*da*]*nn*[*ū*]*ti*, Argištiuna, Qallania ***5*** *birāte* (forts) *ša limet māt Uaiais* + ***40*** c. *ša māt Uaiais nagê* (144ff)
Uaiais (299) ***5*** forts (É.BÀD.MEŠ-*ni dannūti*, named) + ***40*** c. *ša limētišunu* (305)	
Muṣaṣir (350)	Muṣaṣir (153)
("Total") ***430*** c. of ***7*** provinces (422)	

c. = city, URU(.MEŠ) + = *adi* (+) = *u* (in parentheses the line of the text)
both *ša limētiša/u(nu)* and *ša seḫrišunu* are translatable by "of the neighborhood," *dannūti* being "strong" (cf. Zimansky, *Ecology and Empire*, 40ff for a discussion of the various categories of sites mentioned by the "letter").

letter, there were another 104).[74] What was important was that together with a certain group of main cities, a certain (= any) number of smaller cities had been also conquered, and the *140* were the best choice. Adding up the smaller cities was not taken into consideration, even if a total had been calculated in the letter, to which it is appended that "*430* cities in all the *7* provinces of Ursâ the Urartian I conquered"[75] — actually, this rather deficient sum has been roughly calculated, since the cities visited by the Assyrians according to the letter are as many as 708 for the whole campaign, and anyway more than 430 for Urartu alone.[76] In any case, this "ready-made" sum has not been used, but, instead, the conquest of the various cities has been separately listed, reusing some of the "partial" data in accordance with the narrative necessities.[77]

In some cases, however, the "totals" given by older texts were reused in the newer inscriptions, thus ignoring all the partial quantifications.

If we compare the first edition of the annals of Sennacherib, including a single campaign (Smith, *First Camp.*), with the later edition, having two campaigns (cylinder "Bellino"; ICC, 63f), we would note that out of 24 quantifications only 12 have been reused, that is:

a) the total of the cities conquered in Chaldaea: the older edition listed four separate groups of main cities giving the name and the number, followed by the number of the neighboring cities. All this was then summarized as "a total of *88* fortified cities of Chaldaea, with *820* forts of their borders" — and thus we have in all 10 numbers. Only these "totals" are reported by the new edition, as well as by all the subsequent ones;[78]

b) the totals of people, horses and mules, asses, camels, oxen and sheep carried off during the whole campaign — in all 6 numbers, present in both versions (in the later editions only the number of people is given);

c) the tribute established for Ḫirimmu.[79]

---

74 That is *17* + *30* + *57*, without counting Sarduriḫurda and the *6* cities sacrificed in the passage of *146* > *140*.

75 *4 me* 30 URU.MEŠ-*ni ša* 7 *na-ge-e ša* ᵐ*Ur-sa-a* ᴷᵁᴿ*Ur-ar-ṭa-a gi-mir-tu ak-šud* (l. 422).

76 Those of the annals, instead, putting "*7* cities" in the lacuna at line 144, are *433* (the annals do not relate the total). It is rather difficult to combine the "total" given by the letter with the "partial" sums — cf. Table 2. Ursâ the Urartian, if we exclude some mentions *en passant* (e.g. at l. 56), makes his appearance on the scene of the inscription at line 92, in relation to Uišdiš, "a district (*nagû*) of the land Manna." Even if an Urartian army had been encountered at line 142, actually Urartu is first entered in line 162 (*a-na* KUR *Ur-ar-ṭi áš-ku-na pa-ni-ia*, "I turned my face to Urartu"). If we add up the cities from any of these points, however, we will get a total of 603 (not including "the many cities of Uišdiš" and Muṣaṣir, which is situated outside Urartu). Counting only the cities from Ulḫu on we get 437, yet a slightly different total. Possibly some of these cities have been left out from the accounting: if we do not count the cities named but *not* quantified, we get exactly 430. It appears, thus, that the scribe added only the *numbers* of cities. This sum, however, includes apparently 6 provinces (= 6 paragraphs of the text, still leaving out Muṣaṣir), and is inconsistent with the annals, which associate the cities of Aniaštania, Tarui and Tarmakisa with Ulḫu — cf. quoted passage. Note that Ušqaia is located, by the letter, "in front of the border of Urartu," *rēš miṣri ša māt Urarṭi* (l. 167 — cf. also 177). To try to establish which "Urartian lands" were actually crossed by the Assyrian army on the basis of our knowledge of Armenian geography would be quite useless in this respect — many studies have been dedicated to the reconstruction of Sargon's 8th campaign, yet the results are often conflicting, and not only in matters of detail: cf. lately the article by P. Zimansky, "Urartian Geography and Sargon's Eighth Campaign," JNES 49 (1990), 1-21.

77 The *display inscriptions*, speaking about Ursâ the Urartian, state that "*55* (var. *56*?) fortified cities, strongholds of his *8* districts, together with his *11* inaccessible fortresses I conquered and I burnt" (cf. above, p. 53 n. 40). These inscriptions, however, do not have a clear chronological framework, and thus the just quoted passage might not refer (only) to the events of the 8th year.

78 Cf. above, p. 57 nos. 65f for the variants in the digits. Cf. p. 191 n. 12 for the actual number of cities mentioned.

79 *1* ox, *10* sheep, *10* homers of wine and *20* homers of dates, appearing in all the later editions, cf. Luckenbill, *Senn.*, 26:61, 55:59, 57:19, 67:9; Durand, *Documents cunéiformes*, 23 (no. 322: cf. 2′); Heidel, *Sumer* 9, 124:77.

The numbers that have not been reused are those relating to the commandants of divisions and to the Elamite soldiers, a date (no. of day), a number of days (for which the enemy was sought) and the above mentioned 8 numbers representing the partial accountings of the cities.

This practice, in which the "total" quantities are reused, is perhaps more natural and simple than the previous one, that of the "extraction" It is illustrated also by some passages of Esarhaddon's inscriptions: a text lists, by name, "*12* kings of the seacoast" (Syrians) and then "*10* kings of Cyprus (Yadnana), a total of *22* kings of the land of Ḫatti."[80] Another text states only that "I summoned the *22* kings of the land of Ḫatti, of the seacoast and of the midst of the sea, all of them, and I gave them their orders,"[81] not reporting the names nor the "partial" numbers . Thus, in an inscription of the same reign two intervals of time ("Distanzangaben"), of *126* and *434* years, are totalled as *580*, a numeral that is taken up by another text.[82]

7. NUMBERS AS "MOTIFS"

A quantification can have a purely narrative function, without any (or with little) value in relation to the "counting." The most clear example is that of the "*7* kings of the land of Yā', province of Yadnana [Cyprus], who live at *7* days of travel in the midst of the sea" at the time of Sargon, according to a fair number of his inscriptions.[83] In this case the digit "7" does not have an exact numerical value,[84] but has rather a "symbolic" significance.[85] It is an integral part of the narrative becoming eventually identified with the narrated episode. A similar example is the "motif" of the king of Dilmun who lives at *30 bēru* of travel in the midst of the sea "like a fish," reported in the same inscriptions of Sargon.[86] The *Stier Inschrift*,[87] in the section containing the description of the military deeds of the king — even though it makes reference

---

80 Borger, *Asarh.*, §27 episode 21 (*Nin. A* and *F*; ARAB 2, 690; cf. TUAT, 397), cf. also §67 last line and cf. here below p. 191 nos. 43-44. For the identification of Yadnana with Cyprus cf. D. D. Luckenbill, "Jadanan and Javan (Danaans and Ionians)," ZA 28 (1914), 94.

81 Borger, *Asarh.*, §27 (*Nin. B*, p. 61 = ARAB 2, 697), cf. also §67: last line but one. Here the passage is inserted in the building report concerning the restoration of the Nineveh palace.

82 Borger, *Asarh.*, §2 episode 3 (ARAB 2, 706); §3 (ARAB 2, 702). The number varies from *580* to *586*: cf. above p. 63 no. 84.

83 Cf., in transcription: Lie, *Sargon*, 68:457f (fragmentary); Winckler, *Sargon*, 126:145, 180:28f; Weissbach, ZDMG 72, 178:17ff; Lyon, *Sargon*, 42:28; BAL², 61:42ff (= ARAB 2, 44, 70, 186, 80, 92 and 99 respectively); Gadd, *Iraq* 16, 191:26. Note that one duplicate of the *Stier Inschrift* has the variant "*4* days of travel," according to Botta, 28:30 (cf. Fuchs, *Sargon*, 64).

Each time these kings (or anyway the land of Yā'; cf. Parpola, *Toponyms*, 182) are mentioned, their number and the "distance" are not left out (only in Gadd, *Iraq* 16, 192:43 the land of Yā' is cited a second time within the same episode, obviously without specifying again the "distance").

84 The number of days of sea travel is manifestly false, see Elayi - Cavigneaux, "Sargon II et les ionies," *Oriens Antiquus* 18 (1979), 65. They remark also that the *7* kings will become *10* within the space of less than 50 years, at the time of Esarhaddon (cf. the *10* kings of Cyprus mentioned above, n. 80).

85 The symbolic use of the number is discussed below, §8.

86 In transcription: Lie, *Sargon*, 66:443; Winckler, *Sargon*, 126:145f, 180:23; Weissbach, ZDMG 72, 180:20; Lyon, *Sargon*, 42:35; BAL², 61:55 (= ARAB 2, 41, 70, 185, 81, 92 and 99 respect.); Gadd, *Iraq* 16, 191:20f. In this case too, one duplicate of the *Stier Inschrift* has the variant *20* (Botta, 52:32; cf. Fuchs, *Sargon*, 66). The motif appears in all the inscriptions in which that of the "*7*" kings appears, and not elsewhere.

87 Lyon, *Sargon*, 13-19 and 40-47 (= Winckler, *Sargon*, Tf. 41f).

to events that in the other inscriptions are accompanied by several quantifications — contains in all only three digits: the *7* kings of Cyprus, the *7* days and the *30 bēru* of travel.

Equally significant is the case of the *12* Syrian kings of Shalmaneser III's time who were allied with Adad-idri (= Ben-Hadad) of Damascus and Irḫuleni of Hamath and encountered the Assyrian king four times[88] during the 6th, the 10th, the 11th and the 14th campaign. The annalistic inscriptions[89] of this reign mention the *12* kings on the following occasions:

Monolith (ed. "A," up to the 6th camp.; cf. p. 30 n. 52)

(6th year)	Irḫulena	(and)	*12* kings	(incl. Adad-idri)

Inscr. Cameron (ed. "C"; cf. p. 32 n. 57)

6th *palû*	Adad-idri, Irḫulena	*adi*	*12* kings	*ša māt Ḫatti*
10th *palû*	Adad-idri, Irḫulena	*adi*	*12* kings	*ša šiddi tāmdi*
11th *palû*	Adad-idri, Irḫulena	*adi*	*12* kings	*ša šiddi tāmdi*
14th *palû*	Adad-idri, Irḫulena	*adi*	*12* kings	*ša šiddi tāmdi*

Inscr. on colossi (ed. "D"; cf. p. 32 n. 60)

6th *palû*	Adad-idri, Irḫuleni	*adi*	*12* kings	*ša šiddi tāmdi*
10th *palû*	Adad-idri, Irḫuleni	*adi*	*12* kings	*ša šiddi tāmdi*
11th *palû*	Adad-idri, Irḫuleni	*adi*	*12* kings	*ša šiddi tāmdi*
14th *palû*	Adad-idri, Irḫuleni	*adi*	*12* kings	*ša šiddi tāmdi*

Inscr. Safar (ed. "E"; cf. p. 33 n. 61)

6th *palû*	Adad-idri, Irḫuleni	*adi*	*12* kings	*ša šiddi tāmdi*
11th *palû*	Adad-idri, Irḫulena	*adi*	*12* kings	*ša šiddi tāmdi*
14th *palû*	Adad-idri, Irḫuleni	*adi*	*12* kings	*ša šiddi tāmdi*

Black obelisk (ed. "F"; cf. p. 33 n. 63)

6th *palû*	Adad-idri, Irḫulena	*adi*	(the) kings	*ša māt Ḫatti*
11th *palû*	Adad-idri	(and)	*12* kings	*ša māt Ḫatti*
14th *palû*			*12* kings	

As shown, it is not clear whether the number *"12"* includes also Adad-idri and Irḫuleni or not. On the basis of editions "C," "D" and "E" of the annals,

88 Two texts carved at the Tigris' source refer to the "4th time" in which Shalmaneser III faced these kings: 4-*šú it-ti-šú-nu am-daḫ-ḫe-eṣ*, "for the fourth time with them I fought," cf. Lehmann-Haupt, *Materialen*, Tf. 3:22 and Tf. 4:16 (= p. 34 nos. 20 and 22; Michel, WO 3, 152 and 154; ARAB 1, 686 and 691). For the reading 4-*šú* (multiplicative number) instead of *šá-šú* (pronoun) cf. EAK 2, 85.

89 Cf., in translation: ARAB 1, 568, 571, 611, 647, 652, 654, 659; Cameron, *Sumer* 6, 21ff; Safar, *Sumer* 7, 16ff. The *"12"* kings are mentioned in undated contexts as well: KAH 1, 30 (Michel, WO 1, 59:15; ARAB 1, 681); Hulin, *Iraq* 25, 53:29f.

One of the texts mentioned in the previous note speaks of Adad-idri, Irḫuleni and *15* cities of the coast (*it-ti* 15 URU.MEŠ *šá ši-di* [*tam-di*] - Lehmann-Haupt, *Materialen*, Tf. 3:21). However, another text from the Tigris' sources, also mentioned in the previous note, has the version with the *12* kings (*ibid.*, Tf. 4:15).

A fragmentary passage of an inscription credited to Tiglath-pileser III speaks of 13? MAN.MEŠ *šá* KUR *Ḫat-ti* "*13*? kings of the land of Ḫatti" (Tadmor, *Tigl. III*, Misc. I, 2:8 = Nassouhi, MAOG 3, 15); the context, however, is unclear, and the attribution is not certain being based only on some toponyms.

Also some inscriptions of Esarhaddon and Ashurbanipal mention *12* Syrian kings (cf. nn. 80 above [p. 133] and 161 below [p. 150]).

it seems that 14 kings, Adad-idri, Irḫuleni together with (*adi*) *12* kings of the seacoast (*ša šiddi tāmdi*), had formed an alliance to fight against Shalmaneser III and were regularly defeated *4* times (according to "C" and "D" — editions "E" and "F" ignore the encounter of the 10th year). Edition "F" is the one that most deviates from the others, ignoring the *12* allies of Adad-idri and Irḫuleni in the 6th year, Irḫuleni in the 11th ("Adad-idri (and) *12* kings of Ḫatti," *ša māt Ḫatti*) and speaking only of *12* kings defeated at the 14th year, without specifying where or who they were (the only geographical reference given for this campaign is the crossing of the Euphrates). Here it would seem that Adad-idri and Irḫuleni are included in the *"12."* Edition "A" also makes us think that the two were included in the *"12,"* at least in the case of the 6th campaign. This text, given by the "monolith inscription," is the only one to mention all (?) the members of the coalition, but it creates some confusion. The events are narrated in a very synthetic fashion: first it mentions the conquest of some "royal cities" of Irḫuleni and the destruction of Qarqar, "his royal city," and immediately after, without an introduction of any kind, it lists the forces deployed by each of the allies, including those of Adad-idri (up to that moment not named) and those of Irḫuleni himself — actually, it lists 11 arrays only (including Adad-idri and Irḫuleni).[90] It then concludes with the statement that "these *12* kings he brought to his support" (thus, Irḫuleni brought also himself to his own support!).

That Shalmaneser III faced so often coalitions of *12* Syrian kings might have been true (the coalition may have survived many years), yet it appears evident that the *"12"* in this case is a literary expedient used to present a "typical" repeating situation, easily recognizable: the episode of the 10th *palû* and that of the 11th in the Cameron inscription[91] correspond almost perfectly, only the outcome of the battle shows some differences:

> Preparatory phase: *ina* u_4*-me-šú-ma* $^{m.d}$IM*-id-ri ša mat* URU*-ša-*ANŠE*-šú(Aram)* ^{m}*Ir-ḫu-le-na* KUR *A-mat-a-a a-di* 12 LUGAL.MEŠ*(šarra)-ni šá ši-di tam-di a-na* ID.MEŠ*(emūqē) a-ḫa-meš it-ták-lu-ma a-na e-piš* MURUB$_4$*(qabli) u* MÈ*(tāḫāzi) a-na* GABA*(irti)-ia it-bu-ú-ni* "In those days Adad-idri of Aram-Damascus and Irḫuleni of Hamath together with *12* kings of the seacoast, being confident in each one's strength, advanced against me to make war and do battle" (two identical passages)

---

90 3 R, 8:87-95 (= Peiser in KB 1, 172; ARAB 1, 611). The arrays are listed above at p. 103. The presence of only 11 armies obviously has been already noted (cf. e.g. KB 1, 173 note, or the discussion in Michel, WO 1, 70 n. 13). In Weidner's opinion (reported by Michel, *loc. cit.*, cf. also Na'aman, *Tel Aviv* 3, 98 n. 19), the last two names (Ba'sa son of Ruḫubi of Ammon) actually represent two rulers — Ba'aša of Bet-reḫob and a king of Ammon (on the possible location of these lands cf. *ibid.*, n. 20). The possibility that the 12th name was that of the king of Judah, omitted because he was considered to be a vassal of Israel, is discussed in Gelio, RBI 32, 122f n. 7. However, already Dhorme (still quoted by Michel) hinted that the "12" can be viewed simply as a conventional number. Tadmor compared it to that of the tribes of Israel, whose real number oscillated between 11 and 13, but was given as "12" in respect to an "amphictyonic formula of twelve" (*Unity and Diversity*, n. 29).

91 Cameron, *Sumer* 6, 14:60-64 and 15:3-6. In general, all the passages concerning the Syrian coalition show many analogies (the passage of the 14th *palû* is reported almost in its entirety at p. 107) — cf. also the remarks of Sader, *États araméens*, 215.

Outcome of the battle (underscored what was added in the text of the 11th *palû*): *it-ti-šú-nu am-daḫ-ḫi-iṣ* BAD$_5$.BAD$_5$*(abikta)-šú-nu áš-ku-un* 10 *lim* ERÍN.MEŠ*(ṣabē) ti-du-ki-šu-nu ina* GIŠ.TUKUL.MEŠ*(kakkē) ú-šam-qit* GIŠ. GIGIR.MEŠ*(narkabāte)-šú-nu pet-ḫal-la-šú-nu ú-nu-ut* MÈ*(tāḫāzi)-šu(šú)-nu e-kim-šú-nu* (the 10th *palû* adds *a-na šu-zu-ub* ZI.MEŠ*(napšǎ ti)-šú-nu e-li-ú*) "I fought with them, I defeated them (I felled with the sword *10,000* of their warriors; 11th *palû* only) (and) I deprived them of the(ir) chariots, their cavalry (and) their battle implements (to save their lives they fled; 10th *palû* only)."

There are two kings boasting of their many crossings of the Euphrates. Tiglath-pileser I crossed it *28* times, as we know.[92] Shalmaneser III, over two centuries later, crossed it numerous times: the 22nd is the last to be "counted" by his inscriptions, even if he crossed it on at least two other occasions. In this case it is difficult to hypothesize the existence of a connection between the two records, both for the temporal distance and because Shalmaneser did not surpass his illustrious predecessor (or, at least, did not boast of having done so).

According to his main inscriptions,[93] Shalmaneser III crossed the Euphrates[94] on the following occasions:

[92] Cf. above p. 89. To be mentioned also is the crossing of the river *Bu-ú-ia*, by Sargon, *a-di* 26 A.AN, "*26* times (I crossed the river)" (Thureau-Dangin, TCL 3, 4:17).

[93] In the table, "ed." indicates the various annalistic editions (cf. pp. 30ff — the Nimrud statue, Læssøe, *Iraq* 21, 147ff, confirms the contents of the black obelisk, ed. "F" — cf. EAK 2, 79 on the latter) - st. K. = Kurba'il statue (Kinnier-Wilson, *Iraq* 24, 93ff; contains a report of the 18th-20th *palê* similar to those of ed. "E").

Some texts with annalistic or anyway military contents do not mention the crossings: KAH 1, 30 (Michel, WO 1, 57ff = ARAB 1, 680ff; fragm. statue, short vers. of ed. "F"); edition "B" (Michel, WO 2, 410ff = ARAB 1, 616ff); Michel, WO 1, 259ff (= ARAB 1, 706; gold tablet); or mention only the sources of the river: Layard, ICC, 76f (Delitzsch, BA 6, 152f = ARAB 1, 673ff).

[94] The passage referring to the crossing is always placed at the beginning of the text of each *palû* — actually the wording is practically always the same: X-*šú Puratta (ina mē/īliša) e(te)bir* (sometimes also the *šu* is omitted) = "for the Xth time the Euphrates (at its flood) I crossed." This passage directly follows the number of the *palû*, and is preceded in ed. C, 4th (also ed. F), 11th and 12th *palû* by (*ina* DD. MM.) *issu* URU *Ninua attumuš* (+ *mata adki*, ed. F, 14th). Exceptions are ed. A and editions C, D, E and F for the 2nd, 6th and 14th *palû*, (also the 28th for ed. F) in which cases the crossing is reported within the narrative of the campaign.

palû	ed. "A"	ed. "C"	ed. "D"	ed. "E"	st. K.	ed. "F"
1.	C	C+	...	C		C+
2.	C+	C+	...	C+		C+
3.		(C)	...	C		C
4.		C+	...	C+		C+
6.	2-*te-šú*	C+	[C+?]	C+		C+
10.		8-*šú*	8-*šú*	8-*šú*		8-*šú*
11.		9-*šú* +	9-*šú* +	9-*šú*		9-*šú*
12.		10-*šú*	10-*šú*	10-*šú*		10
14.		C+	C+	C+		C
15.		s.	s.	(s.)		(s.)
17.				C		C
18.			16-*šú*	16-*šú*	16-*šú*	16-*šú*
19.				17-*šú*	20	18-*šú*
20.				20-*šú*		20-*šú*
21.					20-*šú*	21
22.						22-*šú*
23.						C
25.						C+
28.						C+

C = the Euphrates is crossed + = *ina mēlīša*, "at its flood" (C) the crossing is implicit[95] ... lacuna, in brackets the passage is partly broken s. = relates the travel to the sources of the Euphrates (this may imply the crossing) (s.) makes reference to the sources of the Tigris and of the Euphrates.

Notes: 6th *palû*: ed. "A" refers to a double crossing within a single year, and not to the second (total) crossing.[96]

13th *palû*: a "summary inscription" carved on a throne base from Nimrud reports the 10th crossing: *ina* 13 BALA.MEŠ-*ia* 10-*šú* ÍD A.RAD *e-bir*.[97]

15th *palû*: similarly, a very short text reports that "in my 15th *palû* I crossed the Euphrates for the 12th time."[98]

19th *palû*: for the discrepancy in the counting cf. above, p. 50 no. 28 and nn.26f

During the 28th *palû* the crossing was made by the *turtān* Dayyān-Ashur.

---

95 Aḫuni of Bīt-Adini is pursued beyond the Euphrates. The crossing of the river, ÍD A.RAD *e-te-bir*, refers obviously to Aḫuni for the following sentence, *issu* URU *Ni-nu-a at-tu-muš* ("from Nineveh I departed," col. I 59) shows that the Assyrian king was still in his land (*contra* Michel, WO 1, 461; cf. also WO 2, 29 n. 6). In the subsequent versions, ed. "E" and "F," it is not clear whether the crossing refers to Aḫuni or to the Assyrian king.

96 *ina* GIŠMÁMEŠ KUŠ DUḪ.ŠI-*e šá* 2-*te-šú* ÍD.A.RAD *ina me-li-šá e-bir*, "with boats (or rafts) of inflated skin for the 2nd time the Euphrates at its flood I crossed" (3 R, 8:82 = KB 1, 170; ARAB 1, 610; *šanûtēšu* is a numeral adverb to which is appended a pronominal suffix, cf. GAG, §71b). Since the text in it entirety refers to two other previous crossings, both carried out with boats of inflated skin (*eleppāt* KUŠ *dušê*; col. I 36 and col. II 16), we may suspect an oversight or, more easily, we have to understand "for 2 times," or "for the 2nd time [during the narrated action, or anyway during the 6th *palû*]."

97 Hulin, *Iraq* 25, 53:34f. Even if the expression is formally identical to the usual one (referring to a single crossing), Hulin suggested translating "in my 13 years of reign I crossed the Euphrates 10 times," on the basis of the fact that no crossing for the 13th *palû* is elsewhere attested (cf. *ibid.*, 61f and cf. also next note). However, one may wonder why the inscription did not use a wording similar to that employed by Tiglath-pileser I to summarize the lands he conquered: *ištu rēš šarrūtiya adi* X *palēya* ... (cf. above p. 88 — cf. also Tiglath-pileser III in Tadmor, *Tigl. III*, Summ. 1:4, Summ. 2:3, Summ. 7:5 and Summ. 11:5). Cf. also EAK 2, 82 (which is in doubt about which interpretation to give to the passage).

98 Læssøe, *Iraq* 21, 38:4f. In this case a translation "(for) 12 times," as proposed by Hulin also for this passage (cf. preceding note), is not necessary for the crossing is referred to in other texts, if the visit to the sources implies it.

It is obvious that each time Shalmaneser went to Syria he crossed the river twice.[99] But the texts give both even and uneven digits, and it is unlikely that they refer sometimes to the outward travel and sometimes to the return,[100] and in any case, counting "2 crossings" for every year for which they are attested, we would far surpass the indicated digits. I doubt whether a detailed study of the Syrian campaigns of this reign could help to establish how many times the river had actually been crossed, or possibly it would bring us to the rather evident conclusion that the counting is inconsistent.[101] This fact is apparently due to the desire of making match the crossings with the *palê*. Note that ed. "F" refers to several other rivers crossed: at the 16th *palû* the Azaba (ignored by ed. "E"), at the 24th the lower Zab, at the 27th the Arzania and at the 30th the Zab. Now, if we add, for the 26th *palû*, the 7th crossing (*nabalkutu*) of the Amanus mountains, we would see that ed. "F" reports (almost) one crossing for every year.[102] It is quite clear, therefore, that the counting of the crossings is employed to underscore, to mark off the succession of the military campaigns.[103]

The chronology of this reign and the *palû* system, however, will be discussed in the conclusions (ch. 5 §7); here it is important to point out how the "motif" of the crossing of the Euphrates implies the (preferential) usage of quantifications. The numbers referring to the crossings are preserved by the later inscriptions of Shalmaneser III even though the text relating to the older campaigns has undergone substantial curtailing:

Numerical quantifications included in the text of the *palê* 10-20

	total	crossings	others	
ed. "D"	16	4	12	(*palê* 10-18)
ed. "E"	21	6	15	
ed. "F"	12	6	6	

These instances pertain to a later epoch (Shalmaneser III and Sargon), but we may find numbers employed as literary motifs in other periods as well:

---

99 Perhaps a text contains the explicit reference to a double crossing within a single *palû* (the 2nd): cf. Læssøe, *Iraq* 21, 150:11f and 13, but the passage at line 11 is fragmentary and the name of the river has been lost.

100 In the first case (going), the years 10-11 and 20-21 and in the second (return) the years 11-12 and 21-22 would anyway show that only a single crossing has been counted.

101 In order to give a rational explanation to the numbering one should hypothesize that up to the 12th year only the outward (or only the return) crossing has been counted, thus giving a total of *10* crossings, in 8 different years (in the 2nd and 6th year the river had possibly been crossed twice, cf. nn. 96 and 99 here above). Later on, two crossings (outward + return travel?) have been counted for the 14th, 17th and 18th *palê*, thus giving a total of *16*. Then, ignoring ed. "E" of the annals and the Kurba'il statue, which give inconsistent numerations, a double crossing has also been reckoned for the 19th and 20th *palê*, and a single one for the subsequent two *palê*.

102 Only the 29th and the 31st *palû* are excepted — about the latter note that the obelisk recalls that "for the second time I set (*karāru*) in front of Ashur and Adad" (= the second term of eponymate). The inscriptions of Shamshi-Adad V are also inclined to recording the crossing of some river: cf. e.g. Weidner, AfO 9, 92:20 and 100:13.

103 Tadmor suggested that the system of dating with *palû*, first used, as far we know, with ed. "C" of Shalmaneser III's annals, could be connected with the counting of the crossings of the Euphrates (JCS 12, 29 n. 60). Note how, in the table, the systematic counting of the crossings was introduced with ed. "C" (ed. "A" and "B" dated the campaigns through the eponym). Previously, the *palû* was used within the Assyrian royal inscriptions only to indicate the first year of reign (according to Tadmor the period between the accession to the throne and the 2nd year, which is the eponymate of the king) or making reference to several years (in fact using it for the "totals," as in *iš-tu rēš šarrū-ti-ja a-di* 5 BALA.MEŠ*(palē)-ia*) — cf. *ibid.*, 29b-30a and nn. 62f.

see the "*40* kings of the land of Nairi" used to summarize the achievements of Tukulti-Ninurta I in several of his inscriptions,[104] often representing the only extant numeral.[105] They became the "label" identifying the deeds of the ruler, being even part of his titulary.[106] Similarly, when Tiglath-pileser I boasts of having vied with *"60"* kings and of having defeated them,[107] we are facing a number that has the pure and simple function of indicating a "high" quantity. In these cases a "round" number (actually it is written 1 *šu-ši*, "one sexantine") is the most suited for an assertion that is nothing but a boast.

In other cases the number acquits the task of representing "low" quantities, yet not necessarily exact. This is the case when Shalmaneser I declares that he confronted the Qutu, of whom "like the stars in the sky, no one knows their number," with one third of his chariotry,[108] or when, to the "letters to the God" of Sargon and Esarhaddon is appended that the Assyrian losses were "*1* charioteer, *2* cavalrymen and *3* outriders."[109]

In all these cases the number is not used to amplify the deeds related: any reader of the inscriptions, Assyrian or not, would have difficulty taking literally the *"60"* kings of Tiglath-pileser I or the numerals of the losses during the 8th campaign of Sargon. Rather, the number here constitutes a literary device, and even if it pretends to represent "high" or "low" quantities, these are so unrealistic as to be immediately perceived as "symbolic."

---

104 RIMA 1, 243:5′, 244:38, 247:18, 266:24, 268:6′, 272:46, 276:32, 279:15 (resp. A.0.78.4, 5, 6, 18, 20, 23, 24, 26 and ARI 1, 710, 715, 721, 760, 803, 773, 783, 795); Deller - Fadhil - Ahmad, BaM 25, 460:27 and 464:35. The numeral is always written 40-*a* (*arbâ*).

105 In the inscriptions A.0.78.6 and 18 of RIMA 1 (ARI 1, 721 and 760), and also in A.0.78.4, 20 and 26 (fragmentary; ARI 1, 710, 803 and 795), as far as they can be read.

106 "... strong king, capable in battle, the one who took over the rule of all the lands of Nairi and subdued the *40* kings, their commanders, at his feet ...," RIMA 1, 247:16ff and 266:22ff (= Weidner, *Tn.*, 14 and 23; ARI 1, 721 and 760), mentioned in the previous note.

It is often affirmed that, in the Semitic ambit, the number *40* is used to symbolize a round figure, an indefinite one (cf. e.g. Farbridge, *Studies*, 144-56). Segal, JSS 10, 10f, however, is not of this opinion, and observes that, within the O.T., the number *40* is mostly used in reference to time-spans (some instances from the Mesopotamian ambit are cited in Yamada, ZA 84, 23 n. 38).

Cf. also W.H. Roscher, *Die Zahl 40 im Glauben, Brauch und Schrifttum der Semiten* (Abhandlungen der Sächsischen Ges. der Wiss. 27, n° 4), Leipzig 1909. Mesopotamian evidence is discussed at pp. 95-98, calling attention to the equation "*40* = *kiššatu*, d.i. Gesamtheit, Universum, etc."

107 Cf. ARI 2, 11 and above, nn. 170 (p. 88) and 190 (p. 92) to chapter 3. The passage appears within the "titulary."

108 KUR *Qu-ti-i šá ki-ma* MUL AN-*e mì-nu-ta la a i-du-ú* (RIMA 1, 184:88ff [A.0.77.1]; cf. IAK, p. 118 col. III 8ff; ARI 1, 532). J.M. Munn-Rankin (in CAH³ 2/2, 282) thinks that Shalmaneser decided, on that occasion, to hasten with one third of his chariotry since a general call-up would have caused a dangerous delay. Anyhow, it is clear that, within the narrative context, the figure "⅓" (*šu-lu-ul-ta*) is used as a literary *pendant* to "without number."

109 Sargon: Thureau-Dangin, TCL 3, 66:426 (ARAB 2, 177); Esarhaddon: Borger, *Asarh.*, 107:25 (ARAB 2, 610). Cf. also Langdon, *Babylonian Liturgies*, Paris 1913, No. 169 Rs. 1f (= Ungnad, OLZ 21, 73) for the very fragmentary "letter" of Shalmaneser IV. "Outrider" here translates *kallapu* (Parpola, JSS 21, 172).

In view of the occurrence of the same "losses" in both inscriptions, the hypothesis that those given by Sargon's "letter" apply to the raid on Muṣaṣir only (hinted at by Rigg, JAOS 62, 131 n. 10), or that they could represent the losses of the city of Ashur alone are both very unlikely.

One example of a similar increasing sequence (*1, 2, 3*) appears in a fragmentary passage of Ashurbanipal: [...] GÁ 5 *ina* GÁ 6 *ina* GÁ 7 *lu*-[...] "(in) the fifth house, in the sixth house, in the seventh house may I [emerge]" (Lambert, AfO 18, 383 and Tf. XXIII col. II 13 = SAA 3, 25 — actually this badly broken text is not a royal inscription but a literary composition). Cf. also Ashurbanipal's fragment K 6085, mentioning the killing of "a third," "a fourth" and "a fifth" lion (numbers written syllabically, cf. Bauer, *Asb.*, Tf. 43 and p. 88:1′ff; cf. n. 157 [p. 148] here below).

8. NUMBERS AS SYMBOLS

The *7* kings of Yā' or the *"7"* generations (*dāru*) that have passed from the fall (or "the flourishing") of Akkad according to an inscription of Shamshi-Adad I[110] could exemplify the "symbolic" usage of the number.

There are other examples. When Esarhaddon was undertaking the reconstruction of Babylon, he recalls that Marduk, after having set *70* years as the period of its desolation, changed his mind, turned upside down the numeral thus obtaining *11* years.[111] Another number game is hidden in the perimeter of Dur-Sharruken as given by some inscriptions of Sargon:

> ŠÁR ŠÁR ŠÁR ŠÁR GÍŠ+U GÍŠ+U GÍŠ+U 1 UŠ 3 *qa-né* 2 KÙŠ *ni-bit* MU*(šumi)-ia mi-ši-iḫ-ti* BÀD*(dūrī)-šu aš-kun*, "*16,280* cubits, the number (or "the calling") of my name (as) the measurement of its wall I fixed."[112]

On the relationship between this number and the name of Sargon — "probably the earliest preserved example of gematria"[113] — the solution has yet to be found.[114] The explanations given by Peiser[115] and by Hommel[116] at the beginning of this century are totally unconvincing . Some time later, Unger pointed out that the 13 digits actually composing the numeral (*4* + *3* + *1* + *3* + *2*) correspond to the 13 wedges making up the name ŠAR$_4$-GI-NA (= Sargon), and argued that these signs simply had been rearranged.[117] This explanation is rather simplistic, and has given rise to many doubts — for

---

110 Thompson, AAA 19, pl. LXXXI and p. 105 col. I 18; YOS 9, pl. XXIII col. I 5′ (RIMA 1, 53; EAK 1, 9f; ARI 1, 140). For the discussion of the term *dāru*, "generation," and of this *Distanzangabe* cf. J.J. Finkelstein, "The Genealogy of the Hammurapi Dynasty," JCS 20 (1966), 109ff. However, Grayson remarks (note in RIMA and ARI) that this reference is not to be taken as a precise chronological statement.

111 "*70* years as the period of its desolation he wrote (on the table of destinies). (But to) the merciful Marduk his anger lasted but a moment, (and) he turned upside down (the table of destinies) and ordered its (the city's) restoration in the *11*th year," Borger, *Asarh.*, §11 episode 10 and *id.*, BiOr 21, 144b (ARAB 2, 643 and 650); cf. also Luckenbill, AJSL 41, 166-168 and the brief notes of H. Hirsch, AfO 21 (1966), 34; J. Nougayrol, RA 40 (1945-46), 64f and A. Shaffer, RA 75 (1981), 188.

For "70 years" as a standard length of punishment in the ancient Near East cf. P.A. Ackroyd, "Two Old Testament Historical Problems of the Early Persian Period," JNES 17 (1958), 23-27 (to which add the remarks of R. Borger, JNES 18 [1959], 74) or H. Tadmor in CAH 6 (new edition 1994), 267ff where the O.T. evidence is collected.

112 4 × *3,600* (*šār*) + 3 × *600* (*nēr*) + 1 × *60* + *3 qanê* + *2* cubits, giving a total of *16,280* cubits (a *qanû*, "reed," is equivalent to 6 cubits, cf. RlA 7, 463b and 476b; AHw, 898b); text in Lyon, *Sargon*, 10 and 38:65 (there are some variants in the writing but not in the numeral: cf. the recent edition by Fuchs, *Sargon*, 42). The number is given also in Lyon, *Sargon,* 22 and 50:47 (fragmentary passage) and Weissbach, ZDMG 72, 182:40 (text: Botta, 160b = Winckler, *Sargon*, no. 61:last line; also rather fragmentary) to which add n. 120 below. Note that the fragment "found" in Jerusalem (see e.g. R. de Vaux, "Deux Fragments de Khorsabad," *Journal of the Palestine Oriental Society* 16 [1936], 129), as well as some others (Böhl, MVLS 3, 6 etc.), according to Fuchs, *Sargon*, 60, actually correspond to the Bull inscription copied in Botta, pl. 44ff, which has been cut into several pieces to satisfy the demand of the antiquities market.

The passages appear in translation in ARAB 2, 85, 108 and 121.

113 Pearce, *Cryptography*, 116.

114 A solution is also wanting for another similar game related to some prisms and to the "black stone" of Esarhaddon, which have an elaborate and mysterious decoration (see Ellis, FD, fig. 34f), "corresponding to the writing of my name" (= of the king; cf. Borger, *Asarh.*, §11 episode 40:11). The explanation suggested by Luckenbill, AJSL 41, is rather improbable (cf. the remarks of Ellis, FD, 121f and n. 79).

115 In "Studien zur orientalisches Altertumskunde," III, MVAG 1900, 50f. Starting from the number, he reconstructs the (real) name of Sargon as Ashur-šar-ukîn — in addition to this, the values given to the signs themselves are very speculative.

116 F. Hommel, "Die Zahl meines Namens in Sargons Zylinderinschrift," OLZ 10 (1907), 225-28. On this "rebus" solution cf. the comments of Pearce, *Cryptography*, 117. A similar explanation was put forth by F. Delitzsch, "Soss, Ner, Sar," *Zeitschrift für Ägyptische Sprache und Alterthumskunde* 16 (1878), 63 (cf. the remarks of Fouts, JNES 53, 207 and n. 16 there).

117 FF 7; cf. also RlA 2, 250b.

instance, as K. Jaritz remarked,[118] the name of Sargon is written in different ways (actually, LUGAL-GI-NA appears to be the most common),[119] while the numeral is not always written with these 13 digits.[120] As a matter of fact, Unger's explanation could work with any numeral having 13 as the sum of its digits and any name made out of 13 cuneiform signs (without taking into account if they are simple strokes or *Winkelhaken*).

Anyway, as Pearce remarks,[121] this passage is important since "it contains the only explicit statement in the cuneiform corpus that a number in the text is to be understood on two levels: in this case, both as the measurement of a city wall and as an orthography of a royal name."

Another interesting example is given by the "banquet stela" of Ashurnasirpal II, an almost unique text.[122] The main issues are the work at the new royal palace of Kalah, with a short report on the hunting exploits of the king. The inauguration feast is described, with the listing of all that was consumed on that occasion. As in the case of Sargon's "letter to the God," here too the commodities are listed in order, according to their nature, and are all quantified. But in this case the quantities are not distributed casually: the first 28 items are quantified in the measure of *200* (1), *300* (3), *500* (4), *1,000* (8), *10,000* (11), *14,000* (1) — the distribution of the single quantities is *not* regular, but the 11 items quantified *10,000* are all consecutive). There follow 27 items quantified *100* and then 19 quantified *10* ANŠE (*imēru*, "ass-load"), being mostly seeds or fruit. Overall, 74 different items are quantified, while to the dedication of the palace were invited *47,074* persons from all parts of Assyria, together with *5,000* foreign dignitaries, *16,000* persons of Kalah,[123] *1,500* palace dependents *zāriqu*. The total number of participants at the festivities, which lasted *10* days, is then added up to *69,574*.[124]

The number 74 must mean something, since it is rather obvious that the listing of 74 items has been an intentional choice (even if the number is not specified by the inscription). The Assyrian guests, *47,074*, represent the only exact number among those of the participants, and here too the fact seems not accidental (note that *47* is *74* reversed). How the dedication of a royal palace could be a particularly meaningful event need not be stressed. The first thing that comes to mind is that the number *74* could be related to the person of Ashurnasirpal II (as king), but it is rather difficult to understand how.[125] Or,

---

118 *Adeva Mitt.* 8, 12f.

119 Cf. Tallqvist, APN, 217f.

120 In Lyon, *Sargon*, 17 and 45:79f the number is written ŠÁR ŠÁR ŠÁR ŠÁR GÍŠ+U GÍŠ+U GÍŠ+U 1 UŠ $1\frac{1}{2}$ NÍG 2 KÙŠ (ninda — Akk. *nindakku* (?) — equal to 12 cubits, cf. ABZ, 205; CAD Λ/1, 245a).

121 *Cryptography*, 117.

122 Wiseman, *Iraq* 14 (RIMA 2, 288ff; ARI 2, pp. 172ff).

123 D. Oates, *Studies in the Ancient History of Northern Iraq*, London 1968, 43ff estimated that the population of neo-Assyrian Kalah that might have been supported on the agricultural land available in the area numbered at most 12,000 inhabitants. Ashurnasirpal's irrigation works, however, allowed an increase of the potential size of the community to somewhat over 20,000 (*ibid.*, 48). Cf. also next note.

124 Mallowan, *Nimrud*, 70ff took into consideration this data to discuss the population of Kalah, yet correcting it to 86,000, after deduction of the foreign envoys and addition of an estimated 23,000 for the children, which are not mentioned by the text. Comparing this numeral with the area of the city, 884 acres, he obtained a population density somewhat lower than 100 persons per acre [= *ca.* 250 per ha.], which he thought feasible, even if, in this way, the public space had not been taken into consideration. Cf., however, preceding note.

Cf. H. Frankfort, *Kingship and the Gods*, Chicago (Phoenix edition) 1978, p. 396 n. 23 for the (estimated) population of Ashur in the Assyrian period: 24,000 (calculated on the basis of an average of 160 people per settled acre).

125 The Assyrian king list, a document that must have had a certain notoriety, at least within the court *entourage*, reports him at the 101st place (cf. A. K. Grayson, "Königlisten und Chroniken," in RlA 6, 114

it could be in relation to his name, similar to Sargon's number game mentioned above.[126] In consideration of the alternation of *74* and *47*, however, it is much easier if we have simply to take into account the two digits "4" and "7," employed for their symbolic references. Renouncing any attempt to give a complete list of these references,[127] a hard task especially for the *7*, it suffices to stress how they may overlap. The *4* not a few times recalls, even within the *corpus* under discussion, the concept of universality (see the "4 regions of the world"),[128] while the *7* often represents completeness, universality and totality.[129] Moreover, it is the mystical number par excellence, and frequently assumes symbolic significance within the rituals[130] and in sacred representations.[131]

---

§69). The list discriminates a first group of kings, indicated as *17* kings "tent dwellers" (*ibid.*, 103 §2) and a subsequent "*10* kings who were forefathers" (cf. Yamada, ZA 84, 15). The first group is made up of legendary kings, while the latter includes the previous leaders of Shamshi-Adad I's family, who never reigned in Ashur (cf. *ibid.*, 13ff; Landsberger, JCS 8, 33f; Hallo, JNES 15, 221; *id.*, ErIs 14, 4*ff). The royal list then distinguishes a third group of rulers, "altogether *6* kings [attested on bricks], whose eponyms are [broken]," which represent the actual Assyrian royal line, and then lists the subsequent names without ever adding them. Now, if excluding the first two groups, and counting from Sulili (the 27th of the list), we may venture to say that Ashurnasirpal II was (aware of having been) *preceded* on the throne by 74 kings. Kikkja, the 28th ruler of the list, is known for having built the (first) wall of the city of Ashur (in ancient Mesopotamia, a deed like this was equivalent to the affirmation of one's independence: cf. Lackenbacher, PSR, 57f). As S. Yamada (Jerusalem) reminds me, the practice of counting backwards the names of the kings is well attested: cf. e.g. the tablet published in W.G. Lambert, "Tukulti-Ninurta I and the Assyrian King List," *Iraq* 38 (1976), 85-94.

These conjectures, however, are rather weak: note how Esarhaddon and Šamaš-šuma-ukīn linked their descent to "Bēl-bāni son of Adasi," 48th and 47th ruler respectively (Borger, *Asarh.*, §65 Rs. 16f = ARAB 2, 576 and *passim*).

126 Ashurnasirpal's name is written, in most cases (cf. Tallqvist, APN, 43f), AŠ-ŠUR-PAP-A, thus being actually composed of 11 strokes (that is, 7 + 4). Therefore, Unger's above mentioned explanation of Sargon's number game could apply to this "74" as well.

127 For the Semitic ambit cf. particularly Hehn, LSS 2/5 (especially 4-44), supplemented by J. Hehn, "Zur Bedeutung der Siebenzahl," in K. Budde [ed.], *Vom alten Testament, Karl Marti zum 70. Geburtstage gewidmet* (Beihefte zur Zeit. für die alttestam. Wissenschaft 41), Giessen 1925, 128-36. Farbridge in his discussion of Semitic numerical symbolism takes into consideration only *3*, *4*, *7*, *10* and *40* (cf. *Studies*, 114-39 for *4* and *7*). For ancient Mesopotamia see G. Furlani, *La religione babilonese e assira*, II, Bologna 1929, 365ff (mainly following Hehn), and for Ugarit cf. the study by A.S. Kapelrud, "The Number Seven in Ugaritic Texts," *Vetus Testamentum* 18 (1968), 494-99.

For a recent general framework cf. A. Schimmel, *The Mystery of Numbers*, Oxford 1993, 86ff and 127ff.

128 Cf. again Hehn, LSS 2/5, 76f: the "4" carries mainly the meaning of universality ("allseitigen," "universellen") and "in diesem Sinne wechselt sie ja auch mit der Siebenzahl."

According to Winckler, AOF 2, 361, the numbers *4* and *7* represent "die grundzahlen des mondsystems," since the ancient Near Eastern peoples derived the institution of the *7* day week from an approximate division of the lunar month into *4* equal periods of *ca.* *7* days — this view is opposed by H. and J. Lewy, "The Origin of the Week and the Oldest West Asiatic Calendar," *Hebrew Union College Annual* 17 (1942-43), 1-152c. However, they felt that the "division of the horizon into four parts apparently superseded another system in which seven winds defined seven main directions dividing the compass into seven sectors" (pp. 8ff, where some cases in which references to *7* or *4* winds appearing in similar contexts are mentioned).

129 For *7* = *kiššatu*, "totality," cf. Pearce, *Cryptography*, 106 (some references may be found in A. Deimel, *Šumerisches Lexikon* 4, Roma 1933, p. 1114 *sub* 598 c, 4). The number *7* is sometimes associated with royalty within the O.T.: cf. Segal, JSS 10, 16 (as *70* is, *ibid.*, 19).

130 The number 7 appears frequently in the royal ritual of purification *bīt rimki*. For instance, the *āšipu* priest for the purification of the royal palace prepared *7* sacrifices, using, among other things (not quantified), *7* sacrificial tables, *7* censers, *7* bottles filled with wine, *7* with beer, pours out *7* heaps of flour, etc. (cf. H. Zimmern, *Beiträge zur Kenntnis der Babylonischen Religion*, Leipzig 1901, vol. 2, no. 26 = vol. 1, 122ff). The "Šamaš cycle" of the ritual was divided into *7* sections, each representing a different "house," through which the king was expected to pass reciting particular incantations (cf. J. Læssøe, *Studies on the Assyrian Ritual and Series bît rimki*, Copenhagen 1955, 28, 80f, 99). This composition is a specifically Assyrian phenomenon originating, as far as we know, in late Sargonid Assyria, but having influences from older rites (possibly Hittite — see *ibid.*, 92f and 100f). Also in several other rituals and exorcisms, as can be expected, the number *7* plays an important role, see e.g. CAD L, 39b; E. Ebeling, "Beschwörungen gegen den Feind und den Bösen blick aus dem Zweistromlande," *Archiv Orientální* 17 (1949), 204 and 210 (Rs. 7) for some examples from neo-Assyrian texts.

131 E.g., according to Hehn, LSS 2/5, 17 the sacred tree often is represented with *7* branches (or 2 × 7 = *14*), each having *7* leaves, while Ninurta's mace has 7 heads (cf. CAD S, 204a).

9. THE HUNTING EXPLOITS

There is a circumstance in which the use of quantifications is very frequent: the hunting exploits of the kings. Across the middle-Assyrian and the Sargonid periods, several inscriptions included one or more paragraphs describing these deeds, regularly completed with the "measure" of the kingly skillfulness: the quantity of beasts of each kind killed or captured.[132] It is thus possible to compare those data — actually it seems probable that the rulers themselves made this comparison.

In general, we must distinguish between single hunting trips and "general" totals relative to the whole reign.[133] In the first case the deeds are narrated within the annalistic sections or anyway within the passages relating to military activities (mainly they appear as a sporting relaxation of which the king avails himself during a pause in the military operations), while in the second case they are related at the end of the inscriptions, or at least in a position or in a manner to release them from any specific temporal context. Especially in the latter case, it is obviously important to have killed *and* captured many animals, as well as animals of *many* different species.

(1) **Tiglath-pileser I** is the first king whose hunting reports have reached us. They appear at the end of his annals, within the sections reserved for the "summaries" of the king's deeds, preceded by the "total" (*42*) of the lands conquered during his first *5* years of reign (cf. p. 88) and followed by the listing of the building enterprises and by other résumés of his military deeds ("I added land and people to Assyria," etc.; cf. pp. 153 and 162 below). The situation is actually a little more complex than what appears at first glance. The hunting reports are made up of three distinct paragraphs, separated by horizontal lines. The first and the third are preceded by the wordings *ina si-qir* *ᵈNin-urta*, "by the command of the god Ninurta ...(I killed this and this)," a repetition that might indicate that two originally distinct passages have been juxtaposed, the first of which (including the first two paragraphs, lines 58-75) perhaps relate to a single hunting trip, since the numbers are rather low and reference to precise places is made. The killing (*qatû* Š) of *4* extraordinarily strong wild bulls (*pu-ḫal* AM^{MEŠ} *dan-nu-te*) is narrated first, and this took place in the desert of the region of Mittani and near the city of Araziqu. Then, according to the second paragraph, *10* strong bull elephants (AM.SI^{MEŠ} *pu-ḫa-li dan-nu-ti*) had been killed (*dâku*), and *4* live elephants (AM.SI^{MEŠ} *bal-ṭu-te*) had been captured (*ṣabātu*) in the regions of Ḫarran and of the river Ḫabur.[134]

132 The quantification of the beasts killed during the king's hunting trips was almost a "must": the only exception is referred to below, n. 148 (Tukulti-Ninurta II). Elsewhere, some animals usually subject to hunting are not quantified only in AKA, 203:36ff (ARI 2, 598; speaks of bulls, elephants and various other beasts captured by the king and gathered together in herds to be shown to the population of Kalah), or in very short epigraphs, as those engraved on the bronze bands of Imgur-Enlil (cf. RIMA 2, 350, nos. 92-95).
Also an inscription of Sargon mentions lions (and wolves; not quantified), but the context is not a king's hunt: cf. Gadd, *Iraq* 16, 192:71f.
On the king-lion conflict cf., in general, E. Cassin, *Le semblable et le différent*, Paris 1987, 167-213.

133 This distinction is also pointed out by Galter, *Paradies*, 243.

134 AKA, 85f:62, 70 and 72 (= RIMA 2, 25f; ARI 2, 43f). For *puḫalu* ("ram," here translated with "wild bull") probably an aurochs, a beast extinct today, is meant — cf. Wiseman in CAH[3] 2/2, 463, and cf. also Salonen, *Jagd*, 247 ("Zuchtwidder/stier/hengst," aber auch = *rīmu*).

In the last passage, in contrast, the numbers referring to the lions cannot be anything but a "total": "by the command of the god Ninurta, who loves me, I killed (*a-duk*) on foot, with my courage and with a vigorous assault, *120* lions (2 *šu-ši* UR.MAḪ^{MEŠ}). In addition I felled *800* lions from my light (?) chariot (8 *me* UR.MAḪ^{MEŠ} *i-na* ^{GIŠ}GIGIR-*ia i-na pat-tu-te ú-šem-qít*)."[135] Its generality, the absence of any kind of chronological or topographical reference and the bombastic digits could make us think of a passage added to complete the preceding one (i.e., not to forget the lions).

A later inscription recalls the killing, during a campaign on the Mediterranean coast, of a *nāḫiru* fish, "that they call sea horse."[136]

(2) The "broken obelisk," a document probably to be credited to **Ashur-bel-kala**,[137] includes the most detailed passage among those relating the hunting deeds of the Assyrian kings. The analogies with the passage of Tiglath-pileser I, from which the terminology is taken, are numerous enough to make us suspect that Ashur-bel-kala appropriated his father's deeds. The first episode narrates the killing, in the "great sea" near Arwad, of a *nāḫiru* (an unusual animal, and thus deserving to appear in this position). The passage concerning the animals follows immediately:

> ... superb wild bulls and cows he killed near the city of Araziqu which is before the land Ḫatti and at the foot of mount Lebanon, ... live calves of wild bulls he captured (and) formed herds of them, ... elephants he felled with his bow, ... elephants alive he captured (and) to his city Ashur he brought, *120* lions (2 *šu-ši* UR.MAḪ^{MEŠ}) with his courage and with a vigorous assault from his light (?) chariot (and) on foot with the spear he killed, ... lions with a mace (?, ^{GIŠ}*nàr-ʾ-am-te*) he felled.

Most of the numbers are missing, and a blank space (here represented by three dots) has been left for them, probably intended for a subsequent filling in at a later time (the inscription is unfinished).[138] Possibly new "updated" information was awaited, which would be explained by the nature of "totals" of these quantifications being placed at the end of the military narrative. But, in the first group of bulls mentioned, an exact geographical reference is given (the same city as in Tiglath-pileser I), which would made us think of a specific occasion. In any case, one may ask why only the number of lions (*120*, as many as Tiglath-pileser I) has been written.

---

[135] *ibid.*, lines 76-81. The paragraph is concluded by "I have brought down every kind of wild beast and winged bird of the heavens whenever I have shot an arrow" (82-84).

[136] Weidner, AfO 18, 344:24 (cf. KAH 2, 68:25; = RIMA 2, 37; ARI 2, 81 end); cf. also RIMA 2, 49:4′f. The *nāḫiru* (*na-ḫi-ra*) is probably a narwhal ("Schwertwal," Weidner, AfO 18, 355f; cf. A. Salonen, *Die Fischerei im alten Mesopotamien*, Helsinki 1970, 214). Subsequently a basalt replica of this animal was made, together with that of a yak (?, *burḫiš*), cf. AfO 18, 352:67ff (= RIMA 2, 44; ARI 2, 103; cf. also the fragmentary passages of RIMA 2, 46:12′, 49:4′ and 57:11′) — for *burḫiš* cf. Gadd, *Iraq* 10, 19ff. According to the broken obelisk Ashur-bel-kala also had made some replicas of certain beasts (with preference for the most unusual): *2 nāḫiru*, *4* yaks and *4* lions in basalt, *2* genii (^{d}ALAD.^{d}LAMMA) in alabaster and *2* yaks in limestone (AKA, 146:16ff = RIMA 2, 105; ARI 2, 250).

[137] The passage including the hunting exploits appears at the beginning of col. IV (cf. AKA, 138ff; RIMA 2, 103f; ARI 2, 248). A singular characteristic is the use of the third person for the narration, even though, in other places, the inscription regularly uses the first person.

[138] Cf. ARI 2, n. 233, AKA, 138 n. 1 and RIMA 2, 99. Through the integration of the passages concerning the bulls and the elephants into the context of "total," the entire narrative appears more compact with respect to the obvious source of "inspiration," the two above mentioned passages of Tiglath-pileser I. And it seems more complete too, since it includes also the episode of the *nāḫiru* and the reference to the calves captured to be collected in herds.

The inscription continues relating how the king collected, in certain mountainous areas of Assyria, some herds of gazelles and deer (here also two numbers are missing), and then reports the killing of some species of animals not referred to above (nor by any other Assyrian royal inscription): panthers, tigers (?), bears, deer, etc., all preceded by a blank space for the number, except for *2* wild boars (ŠAḪ.GIŠ.GI = *šaḫḫapu*).[139] Then it is narrated how the king sent merchants to acquire more or less exotic animals (among them dromedaries), intended to be collected in herds and displayed to the Assyrian people.[140]

Further on, are mentioned a large female ape , a crocodile and a "river-man,"[141] "beasts of the great sea" (*ú-ma-a-mi šá* A.AB.BA GAL-*te*) sent by the king of Egypt (^{KUR}*Mu-uṣ-re-e*), that were displayed to the Assyrian people as well. There follows the assertion:

> *si-te-et ú-ma-a-me ma-ʾ-di ù* MUŠEN^{MEŠ}*(iṣṣūrāte)* AN*(šame)-e mut-tap-ri šá bu-ʾ-ur* EDIN*(ṣēri) ep-še-et qa-ti-šu* MU^{MEŠ}*(šumē)-šu-nu it-ti ú-ma-me* [*an-ne-e*] *la šaṭ-ru mi-nu-su-nu it-ti mi-nu-te an-ni-te* [*la šaṭ*]*-ru*
>
> "The remainder of the numerous animals and of the winged birds of the sky, wild game, his personal achievement (or "that he acquired"), their names are not written with these animals, their numbers are not written with these numbers."[142]

It seems they worried about stating clearly that the digits (that should have been written) did not include all the animals taken by the king, but only the largest ones.[143]

In a fragmentary annalistic text of Ashur-bel-kala mention is made of "*300* (5 *šu-ši*) lions" and of "*6* strong and virile bulls (6 *pu-ḫal* G[U_4.AM^{MEŠ}])." The inscription preserves only these numbers.[144]

(3) The fragmentary annals of **Ashur-dan II** preserve some numbers only in the passage concerning the hunting deeds of the king, placed after the military narrative and reporting the "total" measures of his skill that let him

---

139 The reading *"2"* seems not so certain: the number has been written in the middle of a large space originally left blank, and possibly should be read as *2* (× 60) = *120* (cf. AKA, 141 n. 3). In all, 14 numbers are missing from the inscription, all referring to animals.
This passage gives no temporal nor local references, and the missing numbers, therefore, would have been "totals." It appears as a "supplement" to what was already reported, but not as a repetition. Indeed, the general structure of the narrative is clear enough: at the beginning the killing and the capture of the same animals hunted by Tiglath-pileser I is reported, and thus we have, so to say, the "canonical" deeds, while the subsequent sections make reference to other, rather unusual, animals (typical of the mountainous areas), partly collected in herds, partly killed, partly acquired.

140 In some other cases the king boasted of having formed herds of beasts, by (cf. in RIMA 2) Tiglath-pileser I (26:4ff), Ashur-bel-kala (98:2′-5′, fragm.), Adad-nerari II (154:126f) and Ashurnasirpal II (226:31ff and 292:97ff). The earlier of Tiglath-pileser's statements were separated from the hunting reports mentioned above by the "summaries" of his building achievements.

141 *pa-gu-ta* GAL*(rabi)-ta*, *nam-su-ḫa* and LÚ$^{(?)}$ ÍD = *amēl nāri* — the first being probably a baboon, the third possibly a *muraena* (cf. Gadd, *Iraq* 10, 21-25 and pl. VI).

142 AKA, 142:31ff (RIMA 2, 104; ARI 2, 248). An analogous passage appears in other inscriptions of Ashur-bel-kala, all very fragmentary, cf. RIMA 2, 93:33′ff (cf. n. 144 below), RIMA 2, 94:9′ (= Weidner, AfO 6, 87) and RIMA 2, 107:1′f (= Millard, *Iraq* 32, 168).

143 Cf. AKA, 143 n. 2.

144 Weidner, AfO 6, p. 91 and 86:30f (RIMA 2, 93:29f; EAK 1, 136:30f; ARI 2, 225). The fragmentary context lets us know that they should be "totals." Also in RIMA 2, 94:7′ (very fragmentary) some beasts are killed (lions?).

kill even *56* elephants.[145] Also the annals of **Adad-nerari II**, relatively long and in a good state of preservation, have only a few numbers, and among these the digits of the hunted animals.[146] The homogeneity of most of the "data" allows us to compare them directly:[147]

	Ashur-dan II	Adad-nerari II
Bulls killed (*dâku*)	*[1]600*	*240*
captured (*ṣabātu*)	*2*	*9*
Lions killed (*dâku*)	*120*	*360*
Elephants killed (*dâku*)	*56*	*6*
captured (*ṣabātu*)	-	*4*
trapped (*ina kappi* + *ṣabātu*)	-	*5*

(4) **Tukulti-Ninurta II** during a hunting campaign killed *8* bulls,[148] while the passage containing the "totals" refers only to the killing of some lions (?).[149]

(5) **Ashurnasirpal II** far surpassed the considerable exploits of Adad-nerari II. We have various "sporting" reports of his: the one appearing in the "annals" probably refers to a single hunting campaign, taking place "on the other bank of the Euphrates," because the numbers are low and the passage appears within the narrative of the military campaigns.[150] The killing of *5* lions mentioned in the "Kurḫ monolith" certainly represents a single campaign.[151] A paragraph present on some monumental lions and bulls, on the

[145] Weidner, AfO 3, 155 and 160:24ff (RIMA 2, 135:68ff; ARI 2, 369). About the elephants killed by Tiglath-pileser I (*10*, cf. above), D.J. Wiseman (in CAH³ 2/2, 463) notes that the feats of Tuthmosis III were much greater (he killed *120* elephants, cf. ARE 2, 588), and this because the number of the animals was diminishing owing to the predatory activities of the Akhlamu. Thus, either this is not true, or Ashur-dan II, after a century and a half, has exaggerated a little too much with his *56* elephants. At anyrate, the deeds of Ashurnasirpal II that are the most significant according to his inscriptions (cf. below) did not involve the killing or the capture of a very high number of elephants. Still in comparison with the pharaohs, note that Amenophis III killed *102* lions during the first ten years of his reign (ARE 2, 865).

[146] KAH 2, 84:123ff (RIMA 2, 154; Seidmann, MAOG 9/3, 34; ARI 2, 436). "Totals."

[147] The "homogeneity" goes beyond the pure numerical aspect: the passages under discussion use the same expressions and occupy the same position within the inscriptions (after the "summary" of the military achievements and before the *récit de construction*), and, obviously, refer to the same beasts in the same order (except that Adad-nerari II adds the two groups of captured elephants). The model to which they clearly refer is represented by the "second passage" of the annals of Tiglath-pileser I (that relating to the lions, amplified but deprived of what was quoted in n. 135 above).

The animals are indicated by: lions = UR.MAḪ.MEŠ; elephants = AM.SI.MEŠ; bulls killed = GU_4.AM.MEŠ; the *2* bulls captured by Ashur-dan II = ^{NÍTA}*pu-ḫal* GU_4.AM.MEŠ, the *9* bulls captured by Adad-nerari II = ^{GU_4}*pu-ḫal* GU_4.AM.MEŠ. The number *[1],600* is written [...] *lim* 6 *me* (cf. Rs. 26). All the other numbers higher than *60* are written employing the sexagesimal system (*120* = 2 *šu-ši*, etc).

[148] Schramm, BiOr 27, pl. II and p. 150:46 (RIMA 2, 173; ARI 2, 469). Also just before (line 43, fragmentary) some killed beasts are mentioned (MÁŠ = "ram"? — cf. Salonen, *Jagd*, 213ff), while at lines 81ff the killing and the capture of some ostriches (*lurmu*) and deer (*ajjalu*), not quantified, are referred to. All these actions are representing single hunting campaigns that took place during the military expeditions.

[149] Cf. line 134 (Schramm, BiOr 27, pl. VI Rs. 52). The number seems to be *60*, while the term indicating the animal is completely missing, and only the context lets us know that lions are involved: the passage shows the usual pattern, cf. n. 147 above, except it is very short (altogether it occupies only two lines). It appears in another inscription as well (cf. RIMA 2, 168:5′f, very fragm.).

[150] AKA, 360:48ff = Le Gac, *Asn.*, 101 (and n. 13 for the variant; RIMA 2, 215f; ARI 2, 581).

[151] AKA, 226:33 = Le Gac, *Asn.*, 139 (RIMA 2, 258; ARI 2, 634).

other hand, seems to refer to the whole of the reign.[152] It gives only 3 numbers, easily comparable with the homologous ones of Adad-nerari II:[153]

	Adad-nerari II (ARI 2, 436)		Ashurnasirpal II (ARI 2, 600)	
Bulls	*240*	(4 *šu-ši*)	*257*	(2 *me* 57)
Lions	*360*	(6 *šu-ši*)	*370*	(3 *me* 1,10)
Elephants	*6*		*30*	

The numbers given by the "banquet stela" probably represent "totals" as well. If that is so, they are also the most up-to-date, furnishing invariably the highest numerals:[154]

reference to ARI 2 :	581	634	600	681
Lions killed (*dâku*)	-	-	*370*	*450*
killed (*maqātu* Š)	-	*5*	-	-
Cubs taken away (*nešû*)	-	-	*50*	-
Bulls killed (*dâku*)	*40 / 50*	-	-	*390*
killed (*maqātu* Š)	-	-	*257*	-
Ostriches killed (*nakāsu*)	-	-	-	*200*
killed (*dâku*)	*20*	-	-	-
Elephants killed (*ina šubti dâku*)	-	-	*30*	-
captured (*ina šubti nadû*)	-	-	-	*30*
Bulls captured (*ṣabātu*)	*8*	-	-	*50*
Ostriches captured (*ṣabātu*)	*20*	-	-	*140*
Lions captured (*ṣabātu*)	-	-	*15*	*20*
Elephants received (*maḫāru*)	-	-	-	*5*

(6) The numbers given by the "banquet stela" are closely approached by his successor **Shalmaneser III**, who, according to the "totals" related in the appendix to a tablet from Ashur (annals edition "C"), killed (*dâku*) *373* bulls and *399* lions from the chariot and captured *29* elephants with a trap (*ina šubti*

[152] AKA, 205:70ff; cf. also ICC, 44:23f = Le Gac, *Asn.*, 177; Postgate, GPA, no. 267:41f (fuller edition in RIMA 2, 226:40ff; ARI 2, 600). The passage is not placed at the end of the inscription (in some copies it is followed by some further annalistic material, cf. RIMA 2, 223), but it strictly recalls the style previously used to relate the hunting "totals," cf. n. 147 above, as well as that of the passage on the "banquet stela" (n. 154 below).

[153] The order of listing, however, is not the same; all the animals are killed, *dâku*, except the *257* bulls, *maqātu* Š.

[154] Wiseman, *Iraq* 14, 43 and 34:86ff (RIMA 2, 291; ARI 2, 681; cf. Mallowan, *Nimrud*, 69). The order of presentation follows this text, which reports first the killed animals, then those captured and finally mentions *5* elephants received as tribute and then used during a war campaign. Under ARI 2, 600 are listed also the *15* captured lions and the *50* cubs taken away that actually appear in a preceding passage of the same inscription (AKA, 202:24ff; RIMA 2, 226:33f; ARI 2, 598).

The animals are indicated by: bulls = GU$_4$.AM.MEŠ; lions = UR.MAḪ.MEŠ; lion cubs = *mu-ra-ni* UR.MAḪ; elephants = AM.SI.MEŠ; ostriches = GÁ.NU$_{11}$.MUŠEN.MEŠ (in the "banquet stela" = GÁ.NA.MUŠEN.MEŠ).

For *ina šubti dâku* = "to kill from an ambush pit" and *ina šubti nadû* = "to drive into an ambush pit" cf. A.K. Grayson, "New Evidence on an Assyrian Hunting Practice," in J.W. Wevers - D.B. Redford (eds.), *Essays on the Ancient Semitic World* (Toronto Semitic Texts and Studies 1), Toronto 1970, 3-5 (esp. 4).

nadû).[155] A subsequent edition of the annals reports the killing of *63* bulls (while *4* are captured alive), and then of another *10* bulls and *2* calves, during two campaigns which took place near the city of Zuqarru, on "the other bank of the Euphrates," in the 17th and 19th *palû* respectively (see here below). **Shamshi-Adad V**, instead, killed *3* lions during his 4th campaign,[156] while, much later, **Ashurbanipal** once killed *18* lions within *40* minutes (10 UŠ) from the onset of the day.[157]

To return to the case of Shalmaneser III and of his two hunting campaigns on the banks of the Euphrates given by edition "E" of his annals: it is easy to note that these two passages practically replace the hunting report with the "totals" given in appendix to the tablet with the older edition "C," thus suppressed to avoid repetitions. The two "new" passages are used to cover up the lack of military activity for the 17th and the 19th *palû*,[158] whose narrations contain, besides the hunting enterprises, only vague and stereotyped references to tribute and to the cutting of cedar timbers.

The text of the 17th *palû* includes six sentences:

[1] *ina* 17 BALA^{MEŠ}*(palē)-ia* ÍD A.RAD*(Puratta) e-bir*
[2] *ma-da-tú šá* MAN^{MEŠ}*(šarra)-ni ša mat Ḫat-te am-ḫur*
[3] *a-na šade-e* ^{KUR}*Ḫa-ma-ni e-li* GIŠ ÙR^{MEŠ}*(gušūrē)* ^{GIŠ}*e-ri-ni a-ki-si*
[4] *a-na* URU*(āli)-ia Aš-šur ub-la*
[5] *ina ta-ia-ar-ti-ia ša* TA*(ištu)* ^{KUR}*Ḫa-ma-ni* 1 *šu-ši* 3 GU$_4$.AM^{MEŠ}*(rīmāni) dan-nu-te šu-ut qar-ni gít-ma-lu-te ina* URU *Zu-qar-ri ša* GÌR^2*(šēpē) am-ma-a-te ša* ÍD A.RAD*(Puratti) a-duk*
[6]4 TI.LA^{MEŠ}*(balṭūte) ina qa-te aṣ-bat*

"[1] in my 17th *palû* I crossed the Euphrates, [2] the tribute of the kings of the land of Ḫatti I received, [3] at the mount Amanus timbers of cedar I cut down, [4] to my city Ashur I brought (them), [5] during my return from mount Amanus *63* mighty bulls, perfectly horned, near the city of Zuqarri that is on the other bank of the Euphrates I killed, [6] *4* alive I caught in my hand."

---

155 Cameron, *Sumer* 6, 18 and 25:42ff (= Michel, WO 1, 472); cf. also Michel, WO 1, 9 Rs. 9ff (= ARAB 1, 631). For *ina šubti nadû* cf. previous note.
These digits represent the absolute *record*, with the exclusion of those of the "banquet stela" of Ashurnasirpal II, which are just slightly higher: cf. Table 4 at p. 149.

156 1 R, 31 and 34 IV 3 (Abel in KB 1, 184:3; ARAB 1, 723). Of the successors nobody reported (quantified) hunting accounts. It is known that Sargon had hunted through his reliefs (cf. e.g. R.D. Barnett - A. Lorenzini, *Assyrian Sculpture in the British Museum*, Toronto 1975, pl. 62f).

157 10 UŠ *u$_4$-mu ina a-la-ki ša* 18 UR.MAḪ^{MEŠ} *na-ad-ru-ti uz-za-šú-nu ú-*[...] (K 6085, Bauer, *Asb.*, Tf. 43 and p. 88:5′, cf. also n. 109 here above). For the UŠ as a time measure, equivalent to 1/30 of a *bēru* ("double hour," in turn 1/12 of a day) cf. RlA 7, 467. Another quantification of 10 UŠ appears in Borger, *Asarh.*, 88:19, this time to be interpreted as a length measure (10/30 of a *bēru* = "double hour" of march, cf. above p. 17 n. 103 for the *bēru*).
The hunting motif appears also in a fragment probably connected to prism "E" (82-5-22, 2:7′ff, cf. *ibid.*, Tf. 59 / p. 30), not reporting the quantity of lions, but specifying that "each throat of them I pierced with *1* arrow": *ina* 1$^{\text{ta-àm}}$ ^{GIŠ}*šil-ta-ḫi nap-šá-te-šú-nu ap-*⌜*ru-u*] (I thank E. Weissert for drawing my attention to this text: he is now completing a dissertation on the image of the king in Ashurbanipal's royal inscriptions, which will include a new edition of the texts connected with the motif of the hunt).
For the short epigraphs accompanying Ashurbanipal's hunting reliefs cf. Gerardi, JCS 40, 25ff (ARAB 2, 1020ff).

158 The previous annalistic edition ("D," cf. p. 32), ignores the 16th and 17th year: it simply adds the text of the 18th year to that of edition "C," but omitting the hunting report which appeared at its end.

TABLE 4 HIGHEST QUANTIFICATIONS OF ANIMALS – HUNTING EXPLOITS

	BULLS		LIONS		ELEPHANTS		OSTRICHES	
	kill.	capt.	kill.	capt.	kill.	capt.	kill.	capt.
Tiglath-pileser I	*4*		*120 + 800*		*10*	*4*		
broken obelisk	...		*120 + ...*	...	...	...	...	
Ashur-bel-kala	*6*		*300*					
Ashur-dan II	*[1]600*	*2*	*120*		*56*			
Adad-nerari II	*240*	*9*	*360*		*6*	*4 + 5*		
Tukulti-Ninurta II	*8*		*60*?					
Ashurnasirpal II	*390*	*50*	*450*	*20*	*30*	*30*	*200*	*140*
Shalmaneser III	*373*		*399*			*29*		
Shamshi-Adad V			*3*					
Ashurbanipal			*18*					

kill. = killed (*dâku*, *maqāu* Š, *nakāsu*); capt. = captured (*ṣabātu*). The dots indicate a blank space left for the number.

A similar table appears in Galter, *Paradies*, 243.

The text of the 19th *palû* shows the following differences (including those at the level of writing):[159]

[1] 19 BALA^{MEŠ}; after A.RAD inserts 17-*šú*, "for the *17th* time";
[3] *šade-e* missing; after *e-ri-ni* inserts GIŠ ŠIM.LI*(burāši)*, "(and) of cypress";
[5] *10* instead of 1 *šu-ši* 3; inserts 2 GU$_4$.AMAR^{MEŠ}*(būrē)*, "(and) *2* calves," before *ina* URU *Zu-qar-ri;*
[6] suppressed [to compensate for the above insertion?!].

All too evident is the relation between the two passages, mentioning achievements "remembered" only after some years. But, what is more relevant, is the elimination of the aforementioned concluding passage with the *quasi-record* totals, sacrificed to give place to the narration of specific campaigns, even if (quantitatively) less relevant. The objective of completing the narrative of each *palû* obviously had priority — an objective that was achieved both by editions "E" and "F" (the latest) of Shalmaneser's annals.[160]

159 The two passages appear in Safar, *Sumer* 7, 10:37ff and 12:15ff (= Michel, WO 2, 38 and 40). Note that the city is *zu-qar-ri* in both cases (contrary to the translation of Safar, *Sumer* 7).

160 The argument will be taken up again in ch. 5 §7. Edition "F" (cf. p. 33 n. 63) reports, for the two campaigns under discussion, only the crossing of the Euphrates and the cutting of the cedar near the Amanus mountains. Both the references to the hunting exploits and to the tribute of the king of Ḫatti disappear.

10. QUANTIFICATIONS AND "OVERTAKINGS"

For certain quantifications singularly similar precedents exist. The list of *22* Syrian kings, named and quantified by prism "C" of Ashurbanipal, corresponds in the cities, in the names of kings (except for two of them) and in the order of listing to those of some above mentioned inscriptions of his predecessor Esarhaddon.[161] But in this case, in a list of names, the number *22* does not have a special relevance. The two numbers are identical because the lists are identical, and it could not be otherwise. Where one may expect to find "high" numbers, instead, the later ruler can easily turn to his advantage the comparison with a predecessor. In these cases its is often difficult to establish with certainty if and when certain homologous quantifications of contiguous rulers represent some intentional "overtakings." The *28,800* Hittites deported by Tukulti-Ninurta I are a well known example. His direct predecessor Shalmaneser I boasts of having blinded and deported *14,400* soldiers of Ḫanigalbat:

> *gu-un-ni-šu-nu ú-pél-liq* 4 ŠÁR *bal-ṭu-ti-su-nu ú-né-pil aš-lu-ul* "I butchered their troops, *14,400* of them (who remained) alive I blinded (and) carried off."[162]

This number is an evident rounding, and probably represents a gross exaggeration.[163] The historicity of the event, however, is not in doubt.[164] Now, some later texts of Tukulti-Ninurta I speak of *28,800* Hittites "from beyond the Euphrates" deported to Assyria, a number that is twice Shalmaneser's.[165] The action would have taken place during the year of accession to the throne, but nonetheless it is not mentioned by any earlier text.[166]

---

161 Streck, *Asb.*, 138ff:24-46 (= ARAB 2, 876); Bauer, IWA, 14 II 60-67. The list is missing from the previous edition of Ashurbanipal's annals, prism "B," as well as from the following one (cf. BAL², 93). According to Olmstead, *Historiography*, 55, the list has simply been copied from the father's (the passage of Esarhaddon is mentioned at nn. 80f).

162 KAH 1, 13 II 33f (RIMA 1, A.0.77.1, p. 184:73-75; IAK, 118:34; ARI 1, 530). Cf. Harrak, *Hanigalbat*, 135f.

163 "the number given is beyond any doubt unrealistic" (Harrak, *Hanigalbat*, 171). Groups of deportees from Ḫanigalbat are mentioned by some contemporary administrative texts, yet always in much lower quantities: the highest numbers are *70* and *160* prisoners (Saporetti, RANL 8/25, 452 — cf. also the fragmentary tablet published in J.J. Finkelstein, "Cuneiform Texts from Tell Billa," JCS 7 (1953), p. 132 and 162 no. 49, mentioning some levy troops sent *to* Ḫanigalbat).

164 It is quite difficult to imagine the Assyrians staying, after the battle, to blind *14,400* persons — overall, a disadvantageous thing (what to do with *14,400* blind prisoners?). However, the "blinding" probably consisted of gouging out one eye only, as a sign of dishonor, as suggested by Borger in EAK 1, 57 (cf. also ARI 1, n. 177 and Cross, HHI, 149:6f and 157 n. 23 for the O.T. evidence).

15 blind (IGI.NU.DU$_8$) men from Amurru are mentioned in the middle-Assyrian text KAJ 180 (line 13; cf. Harrak, *Hanigalbat*, 223f). The translation of the term, however, is rather disputed (cf. Gelb, JNES 32, 87 and Harrak, *Hanigalbat*, 171 n. 88). Yet the very low number does not indicate that they were actually a consistent group of prisoners and not, say, people whose optical infirmity derived from work accidents or natural causes — another *9* "blind" men are mentioned at line 2, side by side with other people (cf. the remarks of H. Freydank, "Anmerkungen zu mittelassyrischen Texten. 2.," OLZ 80 (1985), 234, suggesting the translation "sehschwach").

A remarkably similar event happened in AD 1014 when, after defeating the Bulgarians in the Struma regions, the Byzantines blinded *14,000* (or *15,000*, depending on the sources) prisoners, save for one in every hundred who was left with one eye to lead his comrades back to the Tsar of Bulgaria (cf. D. Obolensky in *Cambridge Medieval History*, vol. IV/1, Cambridge 1966 [new ed.], p. 518 and n. there).

165 Actually one has to consider that the coincidence between the two numbers is not so singular: they are two grossly "round" numerals, written respectively *4* × 3,600 and *8* × 3,600 — as if they were "40,000" and "80,000" in our system (Borger suggested to read them "15,000" and "30,000," cf. EAK 1, 57 and 82 and cf. also above p. 5). Now, between two one-digit round numbers having a similar magnitude it is very

H.D. Galter in a recent study (JCS 40) dedicated specifically to this episode is rather skeptical of its truthfulness. Actually the texts present the facts in a simplistic and generic form, without the articulation that one would expect for an event of such importance (the number represents a "record" of deportations, for middle-Assyrian times):

> *i-na šur-ru* ^{GIŠ}GU.ZA*(kussî) šárru-ti-ia i-na maḫ-ri-i* BALA*(palê)-ia* 8 ŠÁR ERÍN^{MEŠ}*(ṣabē)* ^{KUR}*Ḫa-at-ti-i iš-tu e-ber-ti* ^{ÍD}*Pu-rat-te as-su-ḫa-ma a-na* ŠÀ*(libbi)* KUR*(māti)-ia ú-ra-a*
>
> "At the beginning of the throne of my sovereignty, at the beginning of my reign, I uprooted *28,800* Hittites from beyond the Euphrates and led (them) into my land" (the other version is even shorter, *ina maḫria palê-a* being missing).[167]

Then the narrative passes to other events not concerning Syria[168] and already known from other texts. There are no exact geographic references. Galter pointed out that the episode, from a narrative point of view, is an extraneous body with respect to the structure of the inscription.[169] He then concludes, in consideration of other factors as well, that the episode had been deliberately invented,[170] showing the intention of Tukulti-Ninurta I to compare himself with his father, using — not by coincidence — twice his number.[171]

---

easy to find some relations: after all there are only nine digits (from *1* to *9*), and most of the combinations which may happen show some fractionary ratio (1:2, 1:3, 1:4 etc.).

166 Weidner hypothesized (*Tn.*, 26 n. to lines 27-30) that the event dates back to the beginning of the reign but had been "forgotten" for some years, since the Assyrians were trying to establish good relations with the Hittites. After the failure of this undertaking, the event was "remembered" and inserted in the inscriptions. One may wonder why an episode so favorable for the Assyrians was ignored in texts intended after all for an "internal" fruition and not for "international" diffusion. It seems at least as likely that the failure of the above mentioned undertakings could have produced feelings of "creative" rage, with the consequent invention of the episode. Borger suggested (in EAK 1, 82ff) that the deportation took place in the final part of the reign, but was antedated.

167 KAH 2, 60:27ff and 61:23ff (RIMA 1, 272 and 275; Weidner, *Tn.*, 26 and 30; ARI 1, 773 and 783). To add the pertinent passage from an inscription recently published in Deller - Fadhil - Ahmad, BaM 25 (p. 471 and 464:24ff), showing some variants on the level of writing. All texts originate from Kar-Tukulti-Ninurta.

168 Note that Ḫanigalbat, in middle-Assyrian sources, corresponds to northern Syria (Harrak, *Hanigalbat*, 1 n. 1) and that Ḫanigalbat had been vassal (or allied) to the kings of Ḫatti before and during the reign of Shalmaneser I (*ibid.*, 77, 164 and *passim*).

169 "Fremdkörper," JCS 40, 224. Cf. also the remarks of Harrak, *Hanigalbat*, 211ff, 237ff.

170 Mostly on the basis of historical considerations. The documentation available barely allows the existence of a war between Assyria and Ḫatti in the late years of Tukulti-Ninurta I — cf. the extensive discussion in Harrak, *Hanigalbat*, 257ff. Note also that in the middle-Assyrian texts "not a single Hittite deportee is ever mentioned" (*ibid.*, 280).

However, it is also (theoretically) possible — even if unlikely — that the number represents a "total" referring to the whole reign, and this would explain its inclusion in the later texts as well as the generality of the passage describing the deportation. It would have been inserted at the beginning of the military narrative to antedate it as much as possible (cf. Borger's suggestion, n. 166 above), or simply for stylistic reasons — consider that the inscriptions of Tukulti-Ninurta I had a particular structure, with the military narrative inserted within the list of epithets, and not related in a clear chronological order.

171 Cf. JCS 40, 227 and 234 — thus also von Soden, *Iraq* 25, 138. One should also consider that the *14,400* Hittites deported is the only quantification of people given by Shalmaneser I's inscriptions, and this is true for the *28,800* of his successor as well (cf. below, p. 194).

Note also the "coincidence" between the *180* cities destroyed by Shalmaneser I during the same episode (3 *šu-ši* URU.DIDLI-*šu a-na* DU_6 *ù kar-me aš-pu-uk*, KAH 1, 13 II 37 = RIMA 1, 184:77; ARI 1, 530) with the *180* cities of Šubaru destroyed by his successor (3 *šu-ši* URU.DIDLI-*šú dan-nu-ti a-pu-u*[*l*] *aq-qur i-na* IZI *aq-lu*, Weidner, *Tn*, Tf. 3:16 = RIMA 1, 236; ARI 1, 693; both appear in the table with the highest quantifications, p. 181). This last passage, however, seems not so artful (at least judging from the context), even if the numeral can be considered "symbolic" (as does Harrak, *Hanigalbat*, 250).

There are some coincidences between Sargon and his successor as well. The *8* bronze lions he had placed in his "palace without rival" in Dur-Sharruken weighed *4,610* talents,[172] while the *8* lions of shining bronze that Sennacherib had set up during the construction of his "palace without rival" at Nineveh weighed *11,400* talents.[173] The same perimeter of Nineveh, measuring *9,300* cubits, was enlarged by *12,515* cubits by Sennacherib, thus measuring *21,815*. Dur-Sharruken had been built by Sargon with a perimeter of *16,280* cubits, according to his inscriptions.[174] Sennacherib opened *18* gates at Nineveh, *versus* the *8* of Dur-Sharruken.[175] One should note that never does any other Assyrian royal inscription specify the weight of some colossi, either of bronze or not, or the perimeter of a city, Assyrian or not, or the number of its gates.

11. IMPROVEMENTS

When speaking of building enterprises, it is possible to compare one's works with those of one's predecessors, as in the case of Sennacherib. If these works concern a building being done anew, or "rebuilt," this comparison may be explicit, and sometimes the quantifications underscore the fact. Thus, speaking of the restoration works at the temple of Ishtar at Nineveh, damaged by an earthquake, two texts of Ashur-resha-ishi report that a wall *15* layers of brick high was demolished, and in its place was erected another with a height of *50* layers, "*35* layers more than before."[176] Tukulti-Ninurta I added *20* layers to the temple of the goddess Nunaittu, that had previously been

---

172 Cf. above p. 56 no. 62 — translations in ARAB 2, 73, 84, 97 and 100. The name of the palace was É.GAL GABA.RI NU TUKU.A (cf. e.g. Winckler, *Sargon*, no. 50:14 = Lie, *Sargon*, 76; Fuchs, *Sargon*, 182 and 340:430).

173 Number written ŠÁR ŠÁR ŠÁR GÍŠ+U, Smith, *First Camp.*, line 83 (Luckenbill, *Senn.*, 97; ARAB 2, 367). Later, as it seems, the lions were augmented to *12* (for architectural reasons or to overtake those of Dur-Sharruken in number as well as in weight?) — cf. above p. 61 no. 75 (ARAB 2, 391 and 413). Note that no Assyrian bronze colossus has survived.

The descriptions of the two palaces show many analogies, besides those mentioned. Sennacherib's was called É.GAL ZAG.DU NU TUKU.A, cf. Smith, *First Camp.*, line 79 — the indication of its name, however, was suppressed in the later inscriptions.

174 The perimeter of Nineveh at the time of Sennacherib is given by CT 26, pl. 30 and p. 27:64 (Luckenbill, *Senn.*, 111): to the original *9,300* cubits Sennacherib had added another *12,515* cubits thus bringing the total to *21,815* "great cubits" (21 *lim* 8 *me* 15 *ina* as_4*-lum* GAL-*ti*), cf. also above, p. 62 (the passage appears in ARAB 2, 396).

The measure of the perimeter of Dur-Sharruken is mentioned above, p. 140.

175 Sennacherib: cf. above, p. 62 no. 77 (there are several versions). Sargon: Weissbach, ZDMG 72, 182:40; Lyon, *Sargon*, 38:65, 45:79f and 50:47 (ARAB 2, 85, 108 and 121).

176 [3]5 *ti-ib-ki*MEŠ *ana maḫ-ri-e lu ut-te*[*r*], RIMA 1, 311:10f (A.0.86.1; cf. Weidner, *Tn.*, 55:10f and 56:6f; ARI 1, 951). The texts have been reconstructed by Weidner and Borger (EAK 1, 103ff).

All the passages relating to enlargements of buildings, including the quantifications, included in the royal inscriptions up to the reign of Tiglath-pileser III are collected in Lackenbacher, RB, 217 (to which add RIMA 2, 180:2′) — cf. also the paragraph on "surpassing his predecessors" at p. 76f (as well as at p. 95). See also Lackenbacher, PSR, 54ff.

For the later periods, Sennacherib: "to *39* bricks I augmented it (*ú-tir*)"; "to *220* bricks I raised it (*e-la-niš*)" (Thompson, *Iraq* 7, 90:6 and 7). Cf. pp. 58ff for the reports concerning the "palace without rival," in some cases specifying both the older and the new measures.

Cf. also Sargon in Lie, *Sargon*, 44:277ff (the Babylonians who "raised the wall more than before" — within a military narrative).

rebuilt by Shalmaneser I with a height of *72* layers, according to the same text.[177] One may find specified only the varied measures: "I made it *15* feet longer and *5* and *½* feet wider," without the new dimensions nor the previous ones being specified.[178]

Either using quantifications or not, the boast of having surpassed his predecessors (or, anyway, the former situation) is quite frequent in relation to military facts as well:

> ŠU.NIGIN*(napḫar)* 2 *lim* 7 *me* 2 ANŠE.KUR.RA[^{MEŠ}*(sīsē) ṣi-im-da-at* ^{GIŠ}]*ni-r*[*i* ... *a-na*] *e-*[*m*]*uq mat-ia eli šá pa-an ú-šá-tir ar-ku-ús* "Altogether *2,702* horses in teams, [and chariots (?)] more than (even) before, for the forces of my land I had in harness" (Tukulti-Ninurta II).[179]

Quite frequently it is boasted of having added people and territory to Assyria: "I added land to Assyria (and) people to its population" (Tiglath-pileser I).[180] After all, it was the god Ashur who ordered Tiglath-pileser I to extend the border of his (the god's) land: "The god Ashur (and) the great gods, who magnify my sovereignty (and) who granted as my lot power and strength, commanded me to extend the border of their land."[181]

In other cases, reference is made to single episodes, such as the augmenting of the royal forces thanks to the inclusion of some new contingents:

> 50 ^{GIŠ}GIGIR*(narkabti)* 2 *me* ^{ANŠE}*pet-ḫal-lu* 3 *lim* ^{LÚ}*zu-uk* GÌR^{2}*(šēpē) i-na lìb-bi-šú-nu ak-ṣur-ma i-na* [*eli*] *ki-ṣir šar-ru-ti-ia ú-rad-di*
>
> "*50* chariots *200* horses *3,000* footmen among them I selected and I added to my forces" (Sargon).[182]

---

177 20 *ti-ib-ki eli-šu-nu ú-ra-di*, cf. RIMA 1, 264:11ff (A.0.78.17; cf. Weidner, *Tn.*, 22:20f; ARI 1, 757). See also Tukulti-Ninurta II: "*300* [layers] to the *20* bricks (existing) I added (*ú-rad-di*)" in Schramm, BiOr 27, pl. VI and p. 154:57f (RIMA 2, 178:139f; ARI 2, 480).

178 Tiglath-pileser I; Weidner, AfO 18, 352:57f (RIMA 2, 44 and ARI 2, 101 end; the measure is the GÌR = *šēpu*, "foot." For other instances cf. RIMA 2, 180:2′ (Tukulti-Ninurta II, very fragmentary; added 35 *tibki*), and the modifications carried out by Shalmaneser I to the shrine of Ashur (*16* cubits higher and *2* strata of bricks wider) — *ibid.*, A.0.77.1 line 145 (ARI 2, n. 188). Cf. also Tadmor, *Tigl. III*, Ann. 28 (width of a palace expanded by *6* cubits) and Streck, *Asb.*, 170:41ff (ARAB 2, 914; fragmentary).

One example of "improvement" of buildings, yet not quantified, comes from Adad-nerari II: *eli maḫ-ri-e ma-di-iš ut-te-ir* (referring to a temple), KAH 2, 84:130 (RIMA 2, 154; ARI 2, 437).

179 Schramm, BiOr 27, pl. V and p. 154:48f (RIMA 2, 178:130f; ARI 2, 477). Cf. also the predecessors — not quantifying the forces — Adad-nerari II in KAH 2, 84:120f (RIMA 2, 154; ARI 2, 435) and Ashur-dan II in Weidner, AfO 3, 155 and 158:21f (cf. EAK 2, 1 Rs. 22; RIMA 2, 135:66f; ARI 2, 368).

Tiglath-pileser I already had affirmed a similar thing: GIŠ.GIGIR.MEŠ *ṣi-im-da-at ni-i-ri a-na e-muq* KUR-*ti-ia eli ša pa-na ú-tir ú-šar-ki-is* "chariots and teams of horses for the forces of my land more than before I had in harness," AKA, 92:28-30 (RIMA 2, 27; ARI 2, 48).

Similar passages, but referring to the accumulation of grain (not quantified; always with *eli ša pān ušatir atbuk*) appear in the same inscriptions: Tukulti-Ninurta II in Schramm, BiOr 27, lines 50f; Adad-nerari II in KAH 2, 84:121; Ashur-dan II in Weidner, AfO 3, line 20.

180 AKA, 35:59f (RIMA 2, 14; ARI 2, 11); the passage follows: "(and) the border of my land I extended (*ú-re-piš*)." Cf. also RIMA 2, 27:31f (= same inscription) and 33:20 (fragmentary). The motif is taken up also by Tukulti-Ninurta II (RIMA 2, 178:133) and Ashurnasirpal II (*ibid.*, 292:100f; his titulary includes this passage: "these lands … [names them] I added to the territory of my land," cf. e.g. *ibid.*, 304:17).

A text of Tiglath-pileser III mentions the addition of land to some Assyrian provinces (cf. Tadmor, *Tigl. III*, Summ. 9 obv.).

181 *mi-ṣir mat-ti-šu-nu ru-up-pu-ša iq-bi-ú-ni*, AKA, 33:46ff (RIMA 2, 13; ARI 2, 11); cf. also, just after, AKA, p. 48 (RIMA 2, 17:97-100).

The *topos* of enlargements will be discussed again in ch. 5 §5.

182 Lie, *Sargon*, 12:75; Fuchs, *Sargon*, 94 and 316 (ARAB 2, 8) — from Karkemish, cf. above p. 52 no. 42, and cf. Table 2, p. 106, for the forces collected by Sargon.

Or a tribute is augmented:

GUN*(bilta) ù ma-da-at-ta eli ša pa-na ut-ter i-na muḫ-ḫi-šu áš-kun* "Tribute and tax more than before I imposed."[183]

12. "UNUSUAL" QUANTIFICATIONS

From some of these last examples we can deduce that the number sometimes expresses the absence of comparison with the predecessors, being a measure without precedents.

Yet sometimes it happens that the king achieves something that nobody before him did, something new. He thus inserts himself in the rank of the "creators."[184] It could be that this "something new" does not represent a true creation, but simply a new record: "a royal residence of *95* great cubits in length and of *31* great cubits in width, as none of the kings my fathers had built, I did build."[185] In this case, since royal palaces had already been built, it is necessary to remark explicitly that it is a *record*, and if a quantification underscores the fact, it is even better. This is a "priority" too. We are not facing a simple enlargement, or augmentation, of something "existing" (tributes, horses, etc.), but we witness the realization of something that surpasses all the predecessors. The number can also serve to "measure" the very priority: "the site of this (new city) that none among the *350* ancient princes who lived before me ... had thought of nor did he know ..." (thus Sargon, upon choosing the site of Dur-Sharruken to found the new city).[186]

13. CONCLUSIONS

It is necessary to stress that the framework just outlined concerning certain editorial processes applied to the argument "how much" is excessively synchronical. Most of the remarks set out above have a modest general value and must be considered as referring to an exact historical phase, if not even to a single campaign or episode of a certain inscription. Furthermore, these considerations cannot be easily used to explain any specific case. In fact, there

---

183 Tiglath-pileser I: AKA, 82:34f (RIMA 2, 25); other analogous examples in Ashurnasirpal II: RIMA 2, 200:95f, 202:10f, 208:78f, 215:47f; Shalmaneser III: Cameron, *Sumer* 6, 18 IV; Sargon: Lie, *Sargon*, 10:71.

Other "improvements":
Sargon: *na-gu-ú šu-a-tu eli šá maḫ-ri par-ga-niš ú-šar-bi-iṣ* "that province more than before I caused to rest in safety," Lie, *Sargon*, 50:14; cf. also there 18:103 (*22* forts: KUR-*ud a-na mi-ṣir [mat Aš-šu]r ú-tir*!-*ra*); 4:16 (deported: *eli šá pa-na ú-še-šib* "more than before I caused to dwell."

184 This is the motif of the "heroic priority," very dear to the Assyrian royal inscriptions: see M. Liverani in S. Moscati (ed.), *L'alba della civiltà*, Torino 1976, 461f; *id.*, in *Power and Propaganda*, 308f. The topic has been the subject of a thesis *ad lauream* by R. Gelio (University of Rome 1976).

185 Borger, *Asarh.*, §27 episode 22 5f (ARAB 2, 698).

186 Lyon, *Sargon*, 34:44ff (Fuchs, *Sargon*, 38f and 293; ARAB 2, 120); cf. also Weissbach, ZDMG 72, 180:31ff (Fuchs, *Sargon*, 78:29f; ARAB 2, 83).

still remains an ample margin of uncertainty concerning which were the "choosing" standards applied to each case, while, on the other hand, it is obvious that a certain degree of casualness is to be accounted for. To explain better the issue, we shall take into consideration the earliest version of the 3rd *palû* of Shalmaneser III as given by edition "A" of his annals, appearing on the "Kurḫ monolith". In two similar instances, two homologous items are once quantified and once not: in a first episode (a), it is reported that "*3,400* of his warriors I cut with the sword," while in reference to another episode of the same campaign (b), it is said that "*many* of his warriors I killed."[187] Is it that in (b) there was no data, either exact or approximate, available? Rather, we should think that it has not been used: in (b) the quoted statement is followed by "*3,000* of their prisoners (together with) their oxen, their sheep, horses, mules, colts *without number* I carried off" — showing that some data had been collected.[188] We may hypothesize that in this case the scribes following the army were occupied in counting the prisoners and the booty, for obvious reasons. In (a), where there were neither prisoners nor booty (they are not mentioned), the scribes had to be content with the counting of the enemy's losses. In (b), the animals looted have remained "without number" perhaps because it has been decided not to use the data collected (a little too low?) replacing them with a less binding expression, *ana lā mēni*. Subsequent versions of the same campaign preserve a report (now much shorter) of the first episode, while the second disappears.[189] The first one, (a), was obviously thought to be more important and this has determined the collecting of the data "enemies killed" immediately after the battle and/or the using of that data during the editing of the text, since no prisoners had been carried off — a number had to be collected for that event. In (b), an infinite number of hypotheses could be pointed out to explain why it has been preferred to use the number of prisoners.[190] Anyway, in these passages we can detect a specific behavior: *a number* (at least) had to quantify the victory, especially if it was an important one.[191] Thus, e.g., they took the "highest" number among those available, and passed over the others in silence — if not that of the prisoners, then that of the killed. In effect we may detect the tendency to suppress, in the later inscriptions, the less relevant numbers, to leave space instead for the "higher" ones.

---

187 Respectively: 3 *lim* 4 *me mun-daḫ-ḫi-ṣi-šú ina* GIŠ.TUKUL.MEŠ *ú-šam-qit* (ICC, 8:49f = Peiser in KB 1, 166; ARAB 1, 605) and GAZ.MEŠ-*šú-nu* ḪÁ *a-duk* (ICC, 8:64 = KB 1, 168; ARAB 1, 607). For the first passage, cf. also p. 49 no. 23.

188 3 *lim šal-la-su-nu* GU_4.MEŠ-*šú-nu ṣi-ni-šú-nu* ANŠE.KUR.RA.MEŠ ANŠE *pa-ri-i a-ga-li a-na la me-ni áš-lu-la* (*ibid.*, line 64f).

189 In edition "B" (cf. ARAB 1, 619) the killed become *3,000*, while of the second episode just the conquest of Ḫubuškia, the main center, is recalled, similarly to the subsequent editions. Edition "C" does not even mention the killing of the warriors within episode (a), still given with some detail (cf. Cameron, *Sumer* 6, 12), while in the subsequent editions the exceptional compression of the whole third campaign allows just a mention of the two episodes (cf. ed. "F" in Michel, WO 2, 146:43f = ARAB 1, 560).

190 The possibility that these numerals represent "totals," including also the quantities pertaining to episodes that are narrated without giving numerals, as we saw above, p. 96, has to be excluded.

191 Cf. how a *summary inscription* of Tiglath-pileser I (RIMA 2, A.0.87.2; cf. p. 24 n. 22 above) contains in each paragraph, even if very short (the average is three lines), the report of a whole campaign, always complete with one or two quantifications that concretely express the issue.

14. PREFERENCE FOR THE HIGHEST NUMBERS

It is not difficult to note how, in some cases, the most relevant quantifications are taken up in later or shorter inscriptions at the expense of other ones. Several examples can be found, some of which have already been mentioned: see for instance the number of people deported from Muṣaṣir by Sargon, originally higher than those of the animals,[192] the sole one to appear on the "display inscriptions," though altered; or the number (*20,500*) of enemies killed at Qarqar by Shalmaneser III (6th campaign), the sole number to appear within the section referring to the first six campaigns in the final edition ("F") of his annals (cf. table at p. 121). In the earliest version (the monolith inscription, ed. "A") the number was *14,000*, and was surpassed by another one (*17,500* prisoners; 4th *palû*), yet it was the highest one in the intermediate editions ("C" and "E"), where both numerals had varied (p. 49 nos. 24 and 25):

	ed."A"	ed."C"	ed."F"
Prisoners (4th *palû*)	*17,500*	*22,000*	not appearing
Killed (6th *palû*)	*14,000*	*25,000*	*20,500*

It has been pointed out that the value of a quantification is a function of the quantified object (p. 6). During the 1st campaign Sennacherib carried off from Babylonia *208,000* people, *7,200* horses (and mules), *11,073* asses, *5,230* camels, *80,050* oxen and *800,100* sheep.[193] Now, the number referring to the people is, without doubt, the most relevant (in a certain sense it is "higher" than that of the sheep), and it is the sole one to appear in the later versions of his annals. The final edition quantifies the prisoners only in two occasions: they are *208,000* and *200,150* (both referring to UN^{MEŠ} = *nišē*).[194] The older editions have only two other quantifications of prisoners comparable in quantity, yet smaller: *80,000* bowmen[195] and *150,000* warriors.[196]

But a case from the reign of Tiglath-pileser III is perhaps of significance, even if the conclusions that can be drawn may appear too speculative. A rather fragmentary *summary inscription* from Nimrud[197] gives us the total of people deported from the cities of [Sarrabā]nu, Tarbaṣu, Yaballu, Dūr-Baliḫāya and Malilatu: *155,000*. According to what is stated in another *summary inscription* that relates the same events in a more detailed manner, Tiglath-pileser had deported *55,000* people from Sarrabānu, *30,000* from Tarbaṣu and Yaballu, *40,500* from Dūr-Balīḫāya, plus, simply, "the people" of Amlilatu.[198]

---

192 Cf. above p. 54 no. 49.

193 There are some variants: cf. p. 58 nos. 67ff.

194 The two quantifications are discussed above, pp. 113ff (cf. also the remarks at p. 122).

195 ERÍN^{MEŠ} GIŠ BAN^{MEŠ} = *ṣābē qašti*, Luckenbill, *Senn.*, 49:9 (ARAB 2, 257); the number appears only in the text with the first campaign.

196 ERÍN^{MEŠ} MÈ / *ta-ḫa-zi* = *ṣābē tāḫāzi*, Luckenbill, *Senn.*, 89:48 and 92:14 (ARAB 2, 352 and 357; cf. the remarks above, p. 88 n. 169). Besides, in the inscriptions of this ruler, also some quantifications referring to Assyrian soldiers appear: cf. p. 89 (total of bows and shields annexed to the army), still, however, much lower than those mentioned.

197 Tadmor, *Tigl. III*, Summ. 11 (photo: Pl. LVIII; copy: Rost, *Tigl. III*, Pl. XXXIV; cf. ARAB 1, 805-807); passage quoted at p. 91 above. Tadmor (*loc. cit.*, 193) suggests it might be earlier than the other parallel text, Summ. 7.

198 Tadmor, *Tigl. III*, Summ. 7:16ff (Rost, *Tigl. III*, 58; ARAB 1, 789ff).

Since *55,000* + *30,000* + *40,500* gives *125,500*, it would seem that the people of Amlilatu were *29,500* (that is *155,000* less *125,500*), or a number smaller than that of the other cities of which the number of the deported has been specified. Naturally, the possibility that the *155,000* of the later text simply derives from the *55,000* relating to the first city, to which 100,000 was added to make up for the other cities, cannot be ruled out.[199]

199 Cf. Tadmor, *Tigl. III*, 161 n. 16. Cf. above, p. 95 for a similar procedure.

5. CONCLUDING REMARKS

The remarks collected here are intended as a general chronological overview of the topic. Some of the facts pointed out above will be recalled, but in some cases I will examine specific arguments that are better dealt with within a general concluding survey, or that simply did not fit elsewhere. In most such cases, however, the remarks refer to a wider period, and not to a single reign (i.e. that appearing in the section titles).

1. Function of numbers

In texts that are not receipts nor commercial documents of any other kind the use of a number has no practical scope, and does not aim at defining an exact quantity, verifiable at any reading of the document. In the Assyrian royal inscriptions, as in other narrative contexts, from a stylistic point of view the number rather represents additional information, or an "amplification." The quantities need not be necessarily exact, but it is sufficient that they are indicated with the level of approximation that each time will be thought necessary. Thus non-numerical quantifications ("many," "several," etc.) may be used as an alternative to the numerical ones, which could be more or less "exact." Comparing the two possibilities, we may observe that, in practice, the number narrows the reader's options, preventing him from falling back on a wider set of possibilities.

2. Inscriptions and numbers

The higher concentration of numerals in the longer inscriptions is obvious, even if some of these are almost completely devoid of them. The shorter inscriptions that do not relate military or building activities do not give any quantifications. As for the building inscriptions or the building reports appended to historiographic texts, a high level of detail is necessary to find some data: one needs to consider the reports on the "palace without rival" and on Sennacherib's works at Nineveh (pp. 59 ff.), referred to by some of his longer inscriptions.

The degree of completeness of the information is important also within the military sections: compare for instance the latest edition of Shalmaneser III's annals with the preceding ones. It contains much less detailed information in relation to the first 20 campaigns, devoid of most of the numerical data reported by the older versions. Overall, it is the annalistic edition having the lowest quantity of numbers, yet not the shortest (cf. pp. 42-43). On the other hand, the "banquet stela" of Ashurnasirpal II and the "letter to the God" of Sargon, texts narrating very specific events, can afford to report very lengthy lists of commodities, fully quantified.

In the "letter" the list is placed within the concluding section of the text (cf. p. 128), corresponding to the *climax* of the military campaign described there: the conquest of the city of Muṣaṣir. This corroborates the typical tendency to concentrate the quantifications in an appendix to each military narrative, to underscore tangibly the extent of the "victory."

3. Tiglath-pileser I and his successors: the "round" numbers

In the period from Tiglath-pileser I to Ashurnasirpal II the use of high numbers having one or, at most, two digits prevailed. These "round" numbers (1 *lim* 2 *me* and the like) were used especially for the quantifications having a higher ideological relevance — e.g. enemies killed. Their frequency is partly due to the presence of some booty or tribute lists, usually containing only "round" numbers. Yet, booty and tribute are, after all, rarely quantified: the annals of the 5th year of Tiglath-pileser I report only the quantities of some objects destined to be donated to the gods of Assyria and of a few other booty items corresponding to particularly important episodes, described in great detail. The other inscriptions of this reign are completely devoid of such quantifications. In the subsequent reigns, too, the quantifications of tribute and booty appear in particular contexts (the so-called "itineraries," cf. p. 97) or within some campaigns not recording particularly important military deeds, and thus leaving space for some tribute lists.

The few "exact" numbers appearing in the inscriptions of the predecessors of Shalmaneser III most of the time do not come from military contexts (consider, e.g., the *Distanzangaben*), or, in any case, are not particularly "high" (cf. p. 188). The frequency of "round" numbers might be explained by the impossibility of having at hand detailed reports of the events, in view of the fact that sometimes the inscriptions were composed immediately after the conclusion of the last narrated campaign. Nothing prohibited, however, making "specific" the numbers in the later inscriptions, but most of the variations certainly are not due to such an aim. Only in one case does the variation involve the replacement of a "round" number with an "exact" one, when the *16,000* soldiers killed by the Assyrians during the 18th campaign of Shalmaneser III become *16,020* in a later inscription (cf. p. 71), yet we are surely not facing an "updating": if this was the case, then with the first approximated evaluation they nearly hit the mark!

In any case, it was certainly possible (except in cases where information was lacking) to use "exact" numbers anywhere, but considering the aims and contents of the royal inscriptions, a writing such as 1 *lim* perfectly fulfilled the function of expressing a "high" quantity, which was its main purpose. The approximation was obviously due to editing choices, and not to vagueness, carelessness or negligence. Indeed, it was facilitated by the writing system, which allowed not specifying the lower order digits. Thus, one has to consider that such writings as 1 *lim* may either express exact quantities or approximate and indefinite realities.[1] Within a royal inscription, however, it is fairly clear that the former solution is very unlikely, at least in many contexts.[2] Therefore, such a number should be translated "a thousand" and not "1,000," and, similarly, the translation of 2 *lim* could be "some two thousand" or even "some thousands." This would be especially the case with those numbers that employ the digits 1 and 2, the most commonly used.[3] Clearly, in such "high" and "round" numbers the exponent (*lim*) is more important than the digit (1 or 2 or whatever). Most of the variations listed in ch. 3 actually concern the digits and only rarely involved a change in the exponent.[4] Most of them are due to copying errors: even if these oversights were quite frequent, still the variants have almost always the same magnitude, a fact stressing the major interest of the writers of these inscriptions, that is the greatness of the number and not the accuracy or the exactness of its digits.

Up to (around) the 9th century, the tendency to use "round" numbers anyhow and anywhere, in any context, seems thus quite clear. What was of interest was to represent "high" quantities, and the "round" numbers could well fulfill this function, as their roundness expressed emphasis and hyperbole.

4. Feasibility and exaggerations

Besides some gross "round - high" numbers, mostly referring to killed or prisoners, such as *10,000* or the like, several "semi-round" numbers (such as *1,200*, etc.) appear in the inscriptions of this epoch (Tiglath-pileser I - Ashurnasirpal II). In these cases, a certain degree of exaggeration might be expected, since these inscriptions show many variations,[5] and since they very frequently use the digits "1" and "2" (and "3" in Ashurnasirpal II's "annals"), which makes these numerals very homogeneous and repetitious. However, we have no good reason to doubt their basic veracity, since they represent quite modest (and feasible) quantifications. This applies even more clearly to such figures of killed as *332*, etc., which possibly derive from authentic accountings made on the spot. Editing habits have to be taken into consideration: thus, for instance, Ashurnasirpal II's inscriptions quite frequently use "exact" (and modest) numerals for the enemies killed during a battle or after a pursuit (such as the just mentioned *332*), while using "round" numbers for

---

1 For the O.T. ambient, it is commonly acknowledged that some numerals may represent conventional solutions that, under a seemingly exact form, actually express indefinite realities: cf. e.g. *Dictionnaire de la Bible*, Paris 1895ff, vol. 4, col. 1683; *Encyclopedia Judaica*, Jerusalem 1971, vol. 12, col. 1255. Fouts, JNES 53, 209 draws attention to a Ugaritic literary text showing a "round" hyperbolic number (*3,000,000*) in synonymous parallelism to the expression "beyond counting".

2 See the remarks above, p. 5.

3 Cf. Table 2 pp. 42-43, *sub* 7. Segal, JSS 10, 3 remarks that in the Semitic langaugues "1 is basically an adjective. 2 is regarded by Semites as the extension of 1; this is clear enough from the morphology of the dual number. For Semites plurality begins with the number 3."

4 See p. 48. This applies especially for the periods in which the "round" numbers were used more frequently: see how the most consistent variations appear almost only in later periods (p. 69, *sub* B: relation between the variants), while in the preceding ones the variations concerning a single digit only are very frequent (p. 68, *sub* A).

5 Cf. var. nos. 5, 8, 9, 12-14, 17 and 18 in ch. 3 §1. Also such figures as *40* and *50* can be doubted on this basis (cf. var. 10f. and 16).

those killed after a siege or the executed prisoners (e.g. *800*, *200* etc.), and "round - high" numbers for the deported (e.g. *3,000*, *2,000* etc.) — these latter being very repetitious numerals.[6]

Even if exaggeration occurred, it is noteworthy that most quantifications remain within the field of feasibility. This threshold is first surpassed by the *120,000* of Shalmaneser III's *armada*, a quantifications that does not appear in the later inscriptions.[7]

5. The "summaries"

Most of the longer royal inscriptions of this same period (from Tiglath-pileser I to Shalmaneser III) included, at the end of the chronological (or geographical-chronological) exposition of the events, one or more passages with summaries of the king's achievements. The annals of Tiglath-pileser I show the widest repertory, with many motifs that will be taken up by his successors: they specified the total number of lands conquered (yet recalling, in the meantime, that some campaigns had not been mentioned),[8] expose the tangible results — numbers — of the sporting deeds of the king, tell of the building of various palaces and temples in the Assyrian cities, boast of having stored more grain than ever before, of having formed herds of animals and gardens of rare plants, of having harnessed more chariots and teams of horses than ever before and of having added land and population to Assyria.

During the later centuries, as we mentioned, many inscriptions included one or more of the just listed paragraphs.[9] Those referring to the hunting exploits, inevitably quantified, appear in many texts, and are clearly inspired by the model of Tiglath-pileser I (cf. ch. 4 §9). Very fashionable also were the references to the palaces and to the grain (Ashur-dan II, Adad-nerari II, Tukulti-Ninurta II, Shalmaneser III), to the herds (Ashur-bel-kala, Adad-nerari II, Ashurnasirpal II), to the chariots and teams of horses (Ashur-dan II, Adad-nerari II, Tukulti-Ninurta II, Shalmaneser III)[10] and to the addition of land and people to Assyria (Tukulti-Ninurta II, Ashurnasirpal II). Some original motifs could appear here and there, as e.g. the re-establishment of order in Assyria (Ashur-dan II).

---

6 Cf. below, p. 195. Note how Ashurnasirpal's "annals" often specify that "a pile of heads I built," which may reflect the act of accounting the killed (see also p. 118 n. 8), but never specify, e.g., "I counted as spoil," in reference to prisoners or booty.

Piles of heads or corpses are often mentioned in the text of the campaigns of the 2nd and 3rd year, reporting many quantifications of killed enemies, which include the only three "exact" numerals (see above, p. 29; text in RIMA 2, A.0.101.1 or ARI 2, CI, 1, lines I 99 to II 49). Cf. in particular line II 36, where the pile of heads included *326* killed, and cf. also lines 41ff just below: *172* were killed and, apparently, their heads were hung on trees. In another instance the killed were *3,000*, while in two cases no numeral is given. Outside these campaigns, the erection of a pile of heads is mentioned only twice, at line I 64 (in this case the killed were *260*), and at III 108 (referring to *400* prisoners). Three times it is specified that "I cut off their heads" (no "pile" is mentioned), and "round" numerals are given: cf. lines II 71 (*50* killed), 107 (*800* killed) and III 106 (*600* killed). Once the hands of *200* captives were cut off (II 115), while in two instances the enemies were impaled: cf. III 84 (no numeral) and III 111 (*780* prisoners).

7 They will be mentioned again below. For the *50,000 ummāni* of Ashurnasirpal II, also appearing in one text only, cf. n. 47 below.

8 *42* lands and their rulers, cf. above p. 88. In two later *summary* inscriptions (A.0.87.3 and 4), it is replaced by the *28* crossings of the Euphrates.

9 Some examples have been cited at p. 153 ("improvements"). Cf. also, in general, RIMA 2, p. 8.

10 The last two give the numbers as well, as do several cylinders and prisms of Sennacherib within the *Abschlusspassus* appended at the end of the military narrative: cf. below n. 45 (and §§8 and 9 in general).

6. Tiglath-pileser I and his successors: general overview

The *summary* inscriptions of Tiglath-pileser I contain a series of very short reports of the various military campaigns, chronologically ordered according to the annals, and separated by horizontal lines. The earliest of these inscriptions contains one or two quantifications for each paragraph.[11] Each military action, with the inevitable victory, is thus concretely underscored. One of the latest inscriptions of this king, though continuing to report some number for the (very abbreviated) first campaigns, instead uses only non-numerical quantifications for the later ones (*ana lā mīna, ma'attu* etc.).[12] It is the sole example of the tendency to replace the use of numbers with that of non-numerical quantifications within a single reign, a tendency that will be manifested, on a more ample scale, across the reigns of the later Sargonids (cf. below). Both within the annals and the mentioned *summary* inscriptions, we may note also the tendency to make use of as many *different* non-numerical quantifications as possible. Both categories of texts, in fact, show a fairly assorted series of "samples."[13]

In the inscriptions of Adad-nerari II, the use of non-numerical quantifications is frequent, apart from the passage that includes the "itinerary," which has neither numbers nor any other quantification.[14]

In contrast, Tukulti-Ninurta II's inscriptions do not use non-numerical expressions, and the lengthy "itinerary" passage includes several quantifications of received tribute. These figures are quite realistic, considering their modesty and the kind of narrative (cf. p. 97). Outside this very detailed passage — in the annals the other campaigns are presented in a very brief fashion — his inscriptions never quantify tribute or booty: it is thus evident how the detail of the narrative determines the possibility of including tribute lists and the relevant quantifications.

Ashurnasirpal II's "annals" show a rather unusual structure (cf. ch. 2 §8), and this may explain their inconsistent behavior in relation to "quantity" (for instance, in some sections the tribute is often quantified, in others it never is).

7. Shalmaneser III: years of reign and palê

The dating formulae used in the Assyrian royal inscriptions are not necessarily a punctual and "objective" chronologic reference, and might be subject to a "distorted" usage. Thus, the formulae with which the annalistic inscriptions were subdivided into campaigns or years of reign (*ina X palê-ia*, "in my Xth year of reign" and the like), may conceal fictitious chronologies, especially in the later periods.[15] To concentrate as many events as possible within

---

11 Cf. p. 24 n. 22 (text RIMA 2, A.0.87.2).

12 Cf. A.0.87.4, lines 22ff (cf. p. 25). The inscription dates to about the 20th year.

13 Cf. p. 25 and n. 25 (keep in mind the brevity of the military narrative of this inscription). The same thing may be observed also in the annalistic inscriptions of Sennacherib: cf. p. 39, nn. 83 and 85 for Esarhaddon.

14 RIMA 2, A.0.99.2: ll. 105-19; the only quantification to be found in the passage is *ana paṭ gimriša*, "in its entirety" (referring to the land of Laqû), while no numerals appear.

15 On the BALA = *palû* cf. Tadmor, JCS 12, 26ff and Ford, JCS 22; in general, on the dating formulae used by the Assyrian royal inscriptions: Tadmor, ARINH, 14-21; Levine, HHI, 65. Weidner (AfO 14, 52f) thought that a *palû* could cover, at times, even 2 or 3 calendar years (cf. the discussion in Tadmor, JCS 12, n. 26).

the shortest time span is certainly not a device foreign to the neo-Assyrian ambience. The first year of reign was invariably characterized by outstanding military events according to the royal inscriptions.[16] In every subsequent year the king faced anew the "enemy." Thus, while some of the older annalistic inscriptions of Shalmaneser III "skip" some years, the later editions, in particular the black obelisk, present a military report, though short, for each year of his reign.

In the annalistic inscriptions of his predecessors each year was dated through the *līmu* (i.e. "eponym"), which allowed some of them to be ignored since this system did not make it explicit whether they followed each other directly or not. To know this, one had to remember their order by heart or to have a *līmu*-list at hand. Furthermore, most of his predecessors actually reigned for a much shorter period than Shalmaneser III, or, in any case, no complete annalistic inscription of theirs has survived. As a result, the report of each campaign was given in greater length, which did not allow a clear perception of their number.[17]

In the earliest annals of Shalmaneser III the events are still dated by eponyms. Thus the monolith, edition "A,"[18] though extending up to the 6th year, could ignore the 5th one, while edition "B"[19] reports only the events of the 3rd, 4th, 8th and 9th year, the last two dated by eponyms. In edition "C,"[20] instead, the *palû* = year of reign system is employed,[21] and it contains a report for *each* of the first 16 *palê*. Ed. "D" simply adds the 18th *palû* (thus skipping the 17th),[22] while edition "E" again presents all the first 20 *palê*, including the 17th. In this text the paragraph on the hunting exploits reported in appendix to edition "C" no longer appears, its place being taken by the "total" booty referring to the first 20 years.[23] However, the hunting exploits of the king appear within the 17th and the 19th *palê*, with two virtually identical passages.[24] The objective of completing the narrative of every *palû*, thus, had been achieved by inventing one (at least)[25] of these "hunting campaigns" and leaving out, to avoid repetitions, the "total" of animals killed and captured, which, in turn, is replaced by the "total" of the booty. The latter, with its very

---

16 This is one of the main arguments dealt with by Tadmor's article in ARINH. Cf. also *id.*, JCS 12, 31.

17 This is the case in the "annals" of Ashurnasirpal II, which reach up to the 18th year, though skipping some years (cf. pp. 29-30; the number of campaigns actually represented is debatable). The latest inscriptions of Tiglath-pileser I, as pointed out above, relate very short reports for some campaigns: the fact that they are not dated, however, makes it possible to "skip" some of them — thus the campaigns of the annals of the 5th year, A.0.87.1, in the inscription 87.4 (*circa* 20th year) are reduced to 4 paragraphs.

18 Cf. p. 30 n. 52.

19 Inscribed on the bronze bands of the Balawat gates, (p. 31 n. 56).

20 Cameron tablet (p. 32 n. 57).

21 Possibly under the influence of the habit of counting the Euphrates crossings, which from this edition onwards are consistently referred to and numbered (cf. p. 138). The reasons for using the term BALA = *palû*, "term of office," instead of the typically Babylonian *šattu*, with the proper meaning of "year," do not concern us here: cf., however, Tadmor, JCS 12, 24ff (and Schneider, *Shalm.*, 76ff).

22 Cf. p. 32 n. 60.

23 Cf. p. 112.

24 Cf. pp. 148f. This fact may also have induced an error in the number of the Euphrates crossing of the 19th *palû*, cf. above, p. 50 no. 28 (and nn. there).

25 The entry for year 17 of Shalmaneser III in the eponym lists is broken, while year 19 (= 840 BC) reads "[to the] cedar [mountain]," thus confirming the annals: cf. Ungnad in RlA 2, 433 and, recently, Millard, *Eponyms*, 29 and 56 (cf. also Reade, ZA 68, 251). Cf. also p. 32 n. 59 above.

high numerals, is indeed a very good replacement for the "total" numbers referring to the hunted animals.[26]

In the latest annalistic edition from this reign (edition "F," given by the black obelisk) — obviously — each year is represented, up to the 31st. However, both the hunting exploits of the 17th and 19th *palê* as well as the "totals" appearing at the end of edition "E" have disappeared. The former do not appear simply because of the notable abbreviation of the account of campaigns 1-23, a procedure intended to leave more room for the more detailed accounts of the later years.[27] It did not seem important anymore to record the killing of some beast during the sporting relaxations of the king, some 12-14 years before in the king's youth. The disappearance of the "totals" is probably due to the abrupt conclusion of the document, without a building report or an epilogue of any other kind.

Thus, as time went on, and as the length of the reign increased, the narrative of each year (actually, of the early years) had necessarily to be shortened, and this resulted in producing annalistic inscriptions composed of a series of short or very short paragraphs, each preceded by the *palû* number, which therefore assumed a particular relevance. This editorial practice demanded recording a report for each regnal year, so that the readers would not have any gap. Such a period of "vacancy" (i.e. lack of military activity) was commonly recorded by the contemporary chronicles.[28] Evidently, this was not admissable by the annalistic inscriptions: the king had to confront and defeat the enemy each and every year.[29]

In relation to the later campaigns, Shalmaneser's last annalistic text shows an interesting phenomenon, implying a diverging policy, since in this case the editor manipulated the chronology so as to shorten the length of the reign.

---

26 The "totals" of beasts killed/captured across the entire reign, whose numerals in most cases represent a *record* within their category, are replaced in the new text by the (obviously) much lower number of beasts killed/captured during one single hunting campaign.

27 Cf. the remarks of Olmstead, *Historiography,* 8, 21 and 27. One may suppose that between edition "E" and "F" another intermediate version had existed, coming up to the 23th campaign and containing the detailed narrative of campaigns 21-23 (cf. *ibid.*, 24f). Thus another edition could also a find place between ed. "B" (up to the 9th camp.) and "C" (16 campaigns, very synthetic up to the 13th).

28 The Assyrian "eponym chronicle" records, side by side with the names of the functionaries, also the most important events of each year — usually the main objective of the military campaign. If no campaign took place, often the expression *šarru ina māti*, "the king (stayed) in the land" appears: cf. Ungnad in RlA 2, 428ff or Millard, *Eponyms*, 40ff — sometimes the exact place is indicated, e.g. "the king stayed in Kish," cf. *ibid.*, entry for year 710.

The Babylonian chronicles usually allocate one paragraph for each regnal year, cf. e.g. Grayson, ABC, chronicles no. 2-5, 7 and 11-12, and in this case too no year is skipped, with very few exceptions (e.g. Esarhaddon's 9th year). Thus, in the "Nabonidus chronicle" a blank space is left after the entry for the 8th year, lacking any event to narrate (*ibid.*, 107:9). According to Grayson the scribe did not have information at hand for this year and thought of leaving a space to be filled in later on. In this case one may ask why he left only one single line of room, for information of which possibly he did not even know the length, a space almost certainly destined to be inadequate — and to be filled within a short time, before the clay dried.

Other chronicles, on the contrary, record information relating to selected years only (e.g. Grayson, ABC, nos. 1 and 14-15), but also may employ the expression *šarru ina māti*, in the Assyrian manner, to indicate the lack of military campaigns (e.g. *ibid.*, 75:3′).

29 Cf. also the remarks of Schneider, *Shalm.*, 87. Similar problems were encountered by Sargon's scribes, who had to face the lack of campaigning in the first year of their monarch. In the "Nineveh prisms" they placed in *palû* one the campaign of *palû* two and so on, while in the later Khorsabad annals they returned the campaign of *palû* two to its place, transposing a number of details into a "new" *palû* one, purposely created (cf. Tadmor, JCS 12, 93 and Ford, JCS 22, 84b). Tiglath-pileser III's annals, also dating the events by *palû*, do not show similar editorial problems. In this case, they started to count the *palê* not with the first full year of reign, but with the *rēš šarrūti*, which was 11 months long (cf. Tadmor, *Tigl. III*, 232).

If we compare the black obelisk with the Assyrian eponym canon, we would notice that the year-numbers of the later campaigns are two too low.[30] It has been suggested that the campaign to Damascus and Danabi of the 21st *palû* amalgamates the two campaigns of the 21st and 22nd actual years.[31] In addition, since the black obelisk records three campaigns to Que (Cilicia) instead of four, yet specifying that the last one was the "fourth,"[32] it appears that its editor either left out altogether or amalgamated one such campaign with another, thus creating an artificial chronology.

Shamshi-Adad V[33] and Sennacherib[34] make use of the term *girru*, "campaign": the annalistic inscriptions are thus subdivided into campaigns, not necessarily lasting one year, and consecutively numbered. This system thus definitively solved the problem of a possible lack of campaigning for certain years. The *apex* of this trend is represented by Ashurbanipal's prisms, in which the material is only seemingly chronologically related.[35]

8. "Exact" numbers

The above mentioned "total" booty of edition "E" represents the first clear example of "high" and "exact" quantifications (cf. p. 93). In the earlier periods the "exact" numbers were quite rare and mostly reserved for particular contexts.[36] The "round" numbers are still very common in edition "A" of Shalmaneser's annals, but also the later inscriptions bear some examples of

---

30 Here I follow Reade, ZA 68, 251-255. Grayson, BiOr 33, 140f suggested a different reconstruction of the chronology, implying that the eponym canon is in error, which is unlikely, yet being unable to explain many of the discrepancies between the two sources still resulting.

31 E. Forrer, *Zur Chronologie der neuassyrischen Zeit* (MVAG 20/3), Leipzig 1916, 9ff; Reade, ZA 68, 260.

32 Cf. Michel, WO 2, 222: 132 (ARAB 1, 583; 26th *palû*).

33 For the chronology of his reign cf. (recently) Reade, ZA 68, 257ff and Grayson, BiOr 33, 141ff.

34 Cf. Levine, HHI, 73. For Esarhaddon's prism inscriptions cf. the remarks at p. 39 n. 84.

35 The rearrangements introduced during the editing of the new editions of the annals of this king could involve even the leaving out of entire campaigns and the addition of new ones, perhaps taking them over from those of his father Esarhaddon, as is the case with the Egyptian one and with that against Tyre (cf. Olmstead, *Historiography*, 54f; Piepkorn, *Asb.*, 8 n. 4; several authors, however, think that these campaigns are to be credited to Ashurbanipal, cf. e.g. recently Grayson in ZA 70, 244; a synthesis of the problem appears in Aynard, *Prisme*, 18f). On the characteristics of the inscriptions of Ashurbanipal cf. Grayson, ARINH, 43; *id.*, ZA 70, 228 and 245; Tadmor, ARINH, 21; Cogan - Tadmor, Or 46, 82f; Aynard, *Prisme*, 1ff; Olmstead, *Historiography*, 53ff. The king's deeds are given according to a geographic order, and the chronologic references are only given to respect the annalistic "formula." It is obvious that the historical value of such texts is relative, as to define them "annals" is inexact (Grayson thinks they may be compared to "historical novels").

36 See the *47,074* and *69,574* appearing in the "banquet stela" of Ashurnasirpal and connected to the numeral "74" which is mentioned on pp. 141-142. For Tukulti-Ninurta's II "totals," relating to teams of horses, cf. below.

For a general framework cf. the list of "exact" numbers on pp. 188f (including the notes at its end). Other numbers that may be defined "exact," as far as the predecessors of Shalmaneser III are concerned, are lower than "10,000":

Tiglath-pileser I: *641* years (RIMA 2, 28:64);

Tukulti-Ninurta II: *2,702* (!) horses (*ibid.*, 178:130);

Ashurnasirpal II: soldiers: *332* ERÍN.MEŠ (*ibid.*, 201:111); *1,460 mundaḫṣu* (*ibid.*, 203:28); *326 muqtablu* (*ibid.*, 204:36); *172* ERÍN.MEŠ GAZ (*ibid.*, 204:41).

To these add some numbers referring to the hunting exploits (cf. ch. 4 §9).

Millard (*Fs Tadmor*, 220) states that "figures of all sorts are present throughout the texts, and «round» ones mix with «exact» ones": this is true, since both "categories" can be represented within a single inscription, but it is as much true that the "exact" numbers are typical of a later epoch, when the "round" ones were less frequent. Tiglath-pileser III's annals possibly represent the watershed, having a similar number of instances of both "categories" (cf. Table 1 on pp. 42-43).

them.[37] However, all these quantifications are included within the narrative of the military campaigns: thus, with a few exceptions, the round numbers quantify specific events — as happened in earlier times — such as the killing of enemies during a battle, the capture of prisoners after a siege, etc. We are still dealing with numerals having, in many cases, a purely indicative value, not to be taken literally: as pointed out above, a translation as "a thousand" would be preferable to "1,000," and "a myriad" (in the modern sense) to "10,000." The use of these numbers certainly does not seem so strange: in modern European languages too, if one wanted to narrate kingly deeds in a pompous and emphatic manner, "I took a myriad of prisoners" could be more effective than, say, a prosaic "I took 3,617 prisoners," and, within a narrative context, it would improve readability. But, when affixing the "total" booty round numbers could not be used: for instance, a statement such as "during my first *20* campaigns I took 100,000 prisoners, killed 80,000 enemies, carried off 50,000 oxen, etc." would not fit the context of "total." The "total" is the result of a calculation, and the calculation must be exact: what is the sense in relating such a sum if one does not know how much it amounts to, save in very approximated terms? The sole sense of a "total" is that of giving some data, pure and simple. No battle, no victory or looting is narrated. Moreover, the numbers of the "totals" will necessarily be very high, and a very high number can hardly be trusted if it is so roughly rounded and approximated.

The old-style "summaries" (cf. above) appended at the end of the texts by Tiglath-pileser I and his successors, boasting of having piled up more grain or of having harnessed more teams of horses than their predecessors were no longer suitable: some "data" had to be given. The grain is not easy to quantify, but the horses are, and thus with the *2,702* horses of Tukulti-Ninurta II the "exact totals" first make their appearance.[38] Actually, the round numerals were also not very frequent within the hunting exploits, the sole "totals" to be quantified up to this time — yet these digits were much smaller in comparison with the total booty appearing in Shalmaneser's annals ed. "E."

The use of "round" or "exact" numbers, thus, is not completely casual; the former were preferred for quantifications within the narrative of the campaigns, as in earlier times, and the latter for the "totals." In Shalmaneser's annals ed. "D," however, we have the first examples of "exact" numbers enclosed within the annalistic narrative, and this will become more frequent later on.[39]

---

37 Actually, the round numbers appearing in the later annalistic inscriptions (from ed. "C" on) are included within the text of the first campaigns, and obviously had been recopied from the older texts: a single "correction" had been made, in edition "E" (*16,000* > *16,020*, cf. p. 50 no. 27). This case, however, is quite significant, since it represents the sole example of the "category" shift (round > exact), as we have repeatedly remarked.

38 Cf. p. 153 and below n. 48. They appear within the "summary" passage (for which see above §5). Similar quantifications will appear also in Shalmaneser III's annals, ed. "C": *2,002* chariots and *5,542* *pētḫallu* (cf. p. 50 nos. 29f for the variants). In later times, Sennacherib's cylinders and prisms will prefer "round" numbers for the "totals" relating to the Assyrian army, yet will replace them with "exact" ones in the later versions (cf. p. 89).

Ashurnasirpal II's inscriptions do not employ any such "total." For instance, his "annals," which have a quite unusual structure, as we have frequently observed, do not report any of the usual "summaries" (cf. above) at the end of the military narrative.

39 Page 50 n. 24; Shalmaneser's annals ed. "E" added several exact numbers within the text of the new campaigns (cf. also preceding note and ch. 2 §13; edition "F" does not report any new "high" quantification).

On the other hand, we are unable to explain in each case the alternation between "round" or "exact" numbers — indubitably an ingredient of fortuitousness is involved. Yet, in some cases stylistic or editorial habits may explain this alternation. Thus, for instance, the "letter" to Shamshi-Adad V from the god Ashur, in connection with the capture of the city of Bēru, quantifies "*30,000* prisoners, their oxen [and sheep without number I took as spoil (?)], *476* cities of the neighborhood [I devastated, destroyed, burned with fire]."[40] These are not "totals," and therefore the "round" number of prisoners is in line with what was used some years before. As for the cities, it has to be remarked that the inscriptions of this ruler always quantify them with "exact" numbers,[41] just as in the preceding reign (Shalmaneser III) and in the following ones. Thus, some clear attitudes are involved: this could have depended either on the nature of the data available (i.e. exact numbers for the cities and round ones for prisoners had been collected) or on the editing procedures adopted (i.e. exact numbers for the cities and round ones for prisoners had been used) — I would, however, incline towards the latter. We remarked on how the narrative may demand a "high - round" number for livestock or prisoners to get a roaring "hyperbolic" effect. A similarly approximated number, though necessarily smaller, would not be so effective if referring to cities — just as, in our languages, referring to a specific event (or even to a campaign), "I conquered some hundred cities" would not be as impressive as "I carried off some myriads of prisoners" (which could be the translation of the *30,000* mentioned here above). The cities represent entire communities, they represent the result of several repeated and successive acts of conquest (or destruction): how can they be measured by roughly approximated figures? Such figures, necessarily on the order of a few hundred, could never be so roaringly bombastic — it is definitely better to use an "exact" number for the cities.

9. Some remarks on the "totals"

In some cases the "total" appears at the end of the introductory section, containing the royal "titulary," following the listing of the territorial expansions of the conquests[42] — thus in a proper place. These "totals" are mutually exclusive, since there are no examples of inscriptions reporting, e.g., the "total" of the lands conquered at the end of the titulary[43] — before the narrative of the campaigns — and the "total" booty after them. One or more "total" quantifications could be reported, inserted either before *or* after the narrative of the campaigns.

This rule is generally valid for the annalistic texts, but not necessarily for the *summary* inscriptions, which, following an associative principle in relat-

---

40 Weidner, AfO 9, 102:11f and 18f (re-edited as SAA 3, 41). The text actually does not represent a "royal" inscription, but a reply letter from the god Ashur to Shamshi-Adad V.

41 Cf. one instance at p. 50, no. 32. Cf. also p. 34 n. 67.

42 One instance is given by the *40,400* soldiers (this is also a relatively "exact" number) deported, according to edition "B" of the annals of Shalmaneser III, from some regions mentioned just before: cf. p. 94 and n. 196.

43 A "total" has to be placed at the end of a document, thus it is preferable to consider it as appearing at the end of the titulary rather than before the campaigns.

ing the events, could (theoretically) present some "totals" inserted at any point. This interpretation could be applied to the number of prisoners carried off from Ḫatti as given by a *summary* inscription of Shalmaneser III, *87,500* (cf. pp. 94f). This quantification is not immediately perceivable as a "total," since it appears within the military narrative, between two passages concerning, respectively, events of the 1st and the 4th *palê*. In view of the relevance of the numeral (compare the *110,610* prisoners carried off during the first 20 years of that king, p. 93) and of the vagueness of the geographical references we may consider it as the "total" of prisoners deported from Ḫatti (that is from the West) during the whole reign. The not strictly chronological presentation of the material allows the "totals" to be inserted in any place, and thus the *87,500* "Hittites" were reported where Ḫatti is mentioned for the first time. Actually, this is the *only* quantification of prisoners in the text under discussion, which has also only one quantification of killed enemies (*13,500* soldiers of Arramu, cf. p. 49 no. 23) — this could mean that both these numerals represent general totals. However, if the characteristic of "total" — even if "hidden" and not explicitly identified as such — is quite obvious, as far as the narrative context is concerned, this does not allow us to take these numbers as genuine, as both of them show singular analogies with two similar quantifications given by an older annalistic inscription (*17,500* and *3,400* / *3,000* respectively). Actually, it seems rather clear to me that the scribe picked up *those* numerals, referring to specific events, and, since now they had to appear within a "summary" context, and thus would have been intended as "totals," he thought that a higher value was needed, and consequently he "improved" them.[44]

In any case, some quantifications may actually represent partial "totals," relating to wider accountings (e.g. a whole campaign), though referring to a single episode. This is the case with the report of the 1st campaign of Sennacherib, relating the total booty of the campaign at the very end of the military narrative, thus making it easy to identify it as a total (cf. p. 90). In all the subsequent annalistic texts the passage is embodied within the narrative of the campaign itself, displacing after it a couple of other episodes, and is subjected to some textual modifications in order to make it appear as referring to a specific event. This operation, in my view, was not carried out for the purpose of concealing a "total" (= higher) quantification within a specific context, thus making it seem a more relevant one. Rather, the editor of cylinder "Bellino," the second edition of Sennacherib's annals, having two campaigns, felt that such a "total" represented by a detailed list including several "exact" numbers was not acceptable within the military narrative (even if at the end of the first campaign), since, as we saw, as a rule it had to be appended either at the close of the whole military account or at the end of the "titulary."[45] Thus, he "transformed" it in order to make it appear a "partial" quantification. Such a procedure was probably quite common within the annalistic texts, at least in this epoch, and therefore we may suspect that

---

44 Cf. p. 157 for another (possible) similar instance.

45 The latter habit, actually, was given up in later times. Note that, starting from cylinder "Rassam," including three campaigns, the military account was concluded by the so-called *Abschlusspassus*, reporting the consistence of the Assyrian army (cf. p. 89), which actually replaces the "totals" of the one campaign cylinder. The later eight-campaign prisms, however, will again exclude the *Abschlusspassus*.

some quantifications, especially if relative to the first years, may apply to a whole campaign instead of a single episode. It is very unlikely, however, that any such quantification could represent a "general" total (i.e. referring to the entire reign), since most of the annalistic texts were made up from the older versions, expanded by adding the new campaigns (this, at least, is the case with Sennacherib). Consider also that there are very few examples of consistent "improvements" of numerals in the course of time that might suggest that they included subsequent additional homologous quantifications.

10. Shalmaneser III: general overview

To this reign belong some of the just mentioned "improvements": the most striking case is that of the Syrians killed at the battle of Qarqar (6th year), a variation that only with difficulty could be interpreted as inclusive of further groups of enemies killed in subsequent campaigns to the West.[46] To the reign of Shalmaneser III belong also the most striking examples of the use of numbers with purely literary functions: see the *12* kings of the Syrian coalition (pp. 134ff), or the motif of the Euphrates crossing (pp. 136ff).

This ruler once fielded, according to some of his annalistic inscriptions, an army of *120,000*, a very impressive number indeed (cf. discussion at pp. 107ff). It may be compared to the army of *50,000* mentioned in an inscription of his father, Ashurnasirpal II.[47] In both cases, however, the quantifications, in spite of their numerical relevance, did not become very popular, since Shalmaneser's disappeared from his later annals, while Ashurnasirpal's appears in one text only (though represented by some duplicates) — evidently, it was preferable to exaggerate the figures of the enemies. Anyway, the use of large "round" numbers for their own armies is easily explainable since they were very adept at this type of showy boasting.[48]

11. Tiglath-pileser III and Sargon: general overview

The annals of these two rulers never quantify the enemies killed (cf. pp. 35ff), but frequently, instead, quantify prisoners and deportees, and this is in accordance with the fact that Tiglath-pileser III and Sargon were responsible for the most intensive mass deportations of the Assyrian empire.[49] Here it is of interest, however, to remark how willingly their inscriptions report the quantities, and that these quantities are rarely given in round numbers. Tiglath-pileser's annals report also modest numbers of prisoners, but, in this case, they list several cities one after another, each followed by the pertinent

---

46 Cf. p. 49 no. 25 and p. 107. For other such variations cf. discussion in ch. 3 §4.

47 50 *lim um-ma-ni ad-ki a-na mat me-e[ḫ]-ri al-lik*, "I mustered 50,000 troops (and) I marched to the land of Meḫru" — later, these troops (here *ummāni* might also be translated as "workmen") were employed to cut wood for the Ishtar temple — Thompson, AAA 19, 109:28 (RIMA 2, 309 and ARI 2, 717; for the duplicates, cf. Thompson, AAA 19, 112). Cf. also the *150,000* enemies killed by Sennacherib's army, a numeral that appears in a few inscriptions only (see above, p. 116).

48 The above mentioned *2,702* horses of Tukulti-Ninurta II represent an exception since they appear within the context of "totals" at the end of the military narrative. This also applies to the similar "totals" given by some inscriptions of Sennacherib (cf. n. 45 above).

49 Cf. Oded, *Deportations*, esp. p. 20.

number of deported.[50] The lists thus obtained have an "amplifying" effect, stressed by the presence of numbers, though not high, but preferably not "round" either. It would not be very effective, actually, to use "round" numbers in such a context since they would be too repetitive, especially if only low digits such as "1" or "2" were used, as in the preceding periods — the result would be something like "1,000 deported from A, 1,000 deported from B, 2,000 deported from C, 1,000 deported from D, etc., etc." It is possible that some of these lists report official data, therefore genuine; in effect, the numerals appearing in the mentioned passages seem quite varied, and not artificially "created."

If we enlarge the picture to a whole text of the annals of Tiglath-pileser III, we would notice that the numerals are differentiated in many respects, not only alternating round and exact numbers, but also reporting numbers of different magnitude and of as assorted digits as possible.[51] In inscriptions bearing many quantifications, most of which are referring to a single kind of object (prisoners, in this case), such a policy is easily conceivable. Thus, in the inscriptions of this reign and that of Sargon, it would be easy to find "high" numbers, on the order of some thousands or tens of thousands, formed by two, three and even more digits, often high in their turn (such as, e.g., *9,400* or *40,500*). Not always can these numbers be defined as "exact," nor are they "totals," as in the case of Shalmaneser III. At the same time, the "round" numbers with low digits ("1" or "2") have (almost) completely disappeared.[52] The texts of Sargon, in addition to the prisoners, show a certain inclination to quantify the size of Assyrian enemy contingents (cf. p. 106). It often happens, also, to find lists of names of conquered (or destroyed) cities, which are concluded by recalling their number (cf. p. 190).

12. "Exact" numbers (Sargon and Sennacherib)

The tendency to use higher numbers shown by Tiglath-pileser's inscriptions will be fully developed in Sargon and Sennacherib's times. Their inscriptions are quite inclined to use "very high - exact" numbers deriving from the sum of a very high "round" numeral with an "exact" one on the order of some hundreds and some tens (as, e.g., *100,225*, or *200,150*).[53] These numbers in some cases appear very artificial, owing to their form of "round numbers exacted," and are subject to numerous variations. It is therefore rather unlikely that they could be the sum of partial accountings represented by "round" and "exact" quantities (cf. pp. 86f). On the contrary, there is not one single example of a very high number formed by a single digit, as would be, e.g., "100,000."[54] The nearest to it is perhaps represented by the above

---

50 Cf. above pp. 100ff for such lists (note also n. 239 there, to which add the instances appearing in Tadmor, *Tigl. III*, Ann. 9:10 and Ann. 23:14f = Rost, *Tigl. III*, 34:206ff; ARAB 1, 777).

51 See Table 1 on pp. 42-43 *sub* 7: the annals of Tiglath-pileser III use all the digits from "1" to "9" ("5" being the most frequent) — the sole case among the texts examined.

52 For some instances of round numbers, cf. the index (pp. 193 ff.), or, for the reign of Tiglath-pileser III, cf. just below, n. 54.

53 Mentioned in chapter 3 §11; cf. also p. 188.

54 The highest number formed by a single digit is *80,000*, relating to a contingent of bowmen (therefore it may represent a conventional quantity). It appears in the report of the first campaign of Sennacherib (Smith, *First Camp.*, 31:9 = Luckenbill, *Senn.*, 49), but not in any later inscription. This applies also to

mentioned *120,000* men of Shalmaneser III's *armada*, a number that was not taken up by his later inscriptions. Thus, apparently, while in earlier periods the use of "round" numbers to represent quantities on the order of thousands or of some tens of thousands seemed proper or sufficient, when passing to higher magnitudes (hundreds of thousands), a change in this "policy" was necessary. Numbers on the order of hundreds of thousands with one single digit, as "100,000" or "200,000," would have resulted in abnormally inflated figures, and therefore not realistic at all. The new imperial dimensions and aspirations towards a (truly) universal dominion would explain the tendency to increase the numerals, both in relation to the dimensions of the palaces and to the deeds narrated in the foundation deposits or on the inscribed bas-reliefs. This *grandeur* implied a different strategy in using quantifications: to use figures on the order of hundreds of thousands, yet "exact," and also "complex" (more digits, higher digits).

These "high - exact" numbers, thus, represent the attempt to associate the sense of emphasis given by the "high - round" number ("100,000," "200,000" etc.) with the (appearance of) truthfulness given by their "exact" form. This mode of quantification is observable in counting sheep and oxen taken as booty, which were more easily quantified in very high measure. When other objects or other animals were mentioned (and quantified) within the same context (which often happened), here, too, "exact" numbers were employed (cf. for instance variations no. 49ff and 67ff in ch. 3). In short, the habit was to quantify a given booty, formed by several objects, using either only "exact" or (in an earlier period) only "round" numbers. Anyway, each category retains its proportion: to very high numbers of sheep correspond smaller numbers of oxen, people, etc. — each of them, however, is equally "high" within its category. The increase in the number of digits (and of their individual value) allows the underscoring of the exceptional quality of the number in each case, even when its absolute value is lower.[55]

13. Alterations

In at least one instance, one of these "high - exact" numbers was produced by multiplying the exponent of the first digit; this is the passage *1,235 / 1,285* > *100,225* which has been discussed above (at p. 73) and referring to the sheep taken at Muṣaṣir. This fact, evident in itself, does not allow, however, the assumption that the method of multiplying the first digit by *100* had been systematically employed (cf. ch. 3 §11). It seems clear that the authors of these texts mostly worried about the stylistic and editorial conventions, which called for the use of a certain number (that is a number of a certain type, "high," "exact," etc.) in a given place. Thus, the "total" booty of Sen-

---

the *50,000* troops mentioned by a text of Ashurnasirpal II (cf. here above). In addition, a couple of *30,000*'s from Tiglath-pileser III's times have to be considered: 30 *lim* UN.MEŠ (cf. Rost, *Tigl. III*, 58:18 = Tadmor, *Tigl. III*, Summ. 7) and the *30,000* camels mentioned at p. 85. Cf. also the *30,000* captives mentioned here above in n. 40 (Shamshi-Adad V).

55 A number with many digits, in the cuneiform numerical notation, is actually *longer*. The example representing the *climax* in this direction is that reported at p. 58 (nos. 67ff, Sennacherib) — to be noted also is the employment of high value digits such as "6," "7," and "8," much more showy (and "filling") from a graphical point of view.

nacherib's 1st campaign, including *208,000* people, *800,100* sheep, etc.,[56] originally appearing within a "total" context (appendix to the cylinder with one campaign), had to be given by "exact" numbers on the order of some hundreds of thousands, in accordance with the compositional habits pointed out above (cf. §8).

Another striking example of "high - exact" quantification is represented by the people deported from Judah by Sennacherib during his 3rd campaign: *200,150* people (cf. above, pp. 114f). It is worth noting that this is the only other quantified deportation of this reign: in this period (from Sennacherib on) the quantifications are much rarer than earlier, and their size is in any case very considerable (see also below, §14). Thus, it is not surprising that this deportation is represented by a number of the size of that of the deported from Babylon of the 1st campaign, which represented a sort of "standard" to which to refer. It may be that this number originates from a "2,150," as indirectly suggested by Ungnad (cf. p. 73). In any case, since it is very hard (if not impossible) to believe that over 200,000 people could have been taken from a country the size of Judah, we have to conclude either that an "official" number had been altered so as to reach a magnitude of a couple of hundred thousands, or that such a number had been plainly invented. Within the context in which it appears, "high" and "exact" quantities had to be represented anyway, regardless of the availablity of exact and/or genuine data.

At this point, a word needs to be said on the subject of "original accountings," at least in relation to deportees.[57] In the case of the just mentioned deportations of Sennacherib, the prisoners were captured in different places (and at different moments), probably by different detachments of the army, and deported in small units to their destination.[58] Most likely, they never were gathered in a single place, which would have been a complicated and, after all, useless operation. Thus, only partial accountings were carried out, which eventually reached the Assyrian capital.[59] Supposing that this information was accurate[60] and comprehensive, one still may question whether general statistics had been calculated, and if the editors of the royal inscriptions looked back at these records to check out the evidence. After all, would any possible reader of the royal inscriptions have enough familiarity with such figures of people and animals to understand if they were genuine or exaggerated? Who ever saw 200,000 prisoners, or 800,000 head of sheep? Counting deportees (and livestock) is a rather elusive subject: we may safely assume that some of these figures are improbable if not impossible (cf. above pp.

---

56 Cf. p. 58 no. 67, p. 90, p. 113.

57 See also above pp. 117ff.

58 Cf. above, p. 76 n. 124. This applies also, to some extent, to any quantification of deportees or livestock referring to entire lands and not to single cities.

59 See the administrative texts, some of which are mentioned in the notes to pp. 77ff. As we noted, possibly some of these records were employed in Tiglath-pileser III's annals (cf. above p. 171 and pp. 100ff).

60 Indeed, one has to take into account the possibility that inaccurate, exaggerated, false or simply wrong figures were known in Nineveh. Falsehood is theoretically possible either on the side of the people responsible for the accountings, who may have overestimated the deportees for a variety of reasons ranging from simple laziness to personal advantage considerations, or of high-ranking officials, who may have changed statistics to fit what they believed their king in Nineveh wanted to see. How could he countercheck the veracity of the digits relating to prisoners taken in Palestine or southern Babylonia? Obviously, it was not possible to lie about the amount of tribute or booty of precious metals which had to reach Assyria in any case (cf. here below).

113ff) judging on the basis of our knowledge of historical and modern demographic statistics, censuses, etc. to which we have been accustomed since our school days — still, we have some doubts on the veracity of these numerals.

Going back to the case of the *200,150* deportees from Judah, we may note that the amount of the tribute imposed on the king of Judah on the same occasion, as has been noted (above p. 80), is partially corroborated by the 2nd book of Kings: *30* talents of gold are recorded both by Sennacherib's prisms and the O.T., whereas for silver Sennacherib records *800* as against the *300* talents of the book of Kings. However, it is not necessary to think that the Assyrian sources are exaggerating this figure: as was suggested, the *800* in the Assyrian count of silver may incorporate the tribute from the royal treasury as well as the silver stripped from the temple doors.[61] If so, then the amount of tribute taken from Hezekiah would represent genuine figures. In any case, the authenticity of these numerals should not be extended to the figures of people deported. One should not underestimate the complexity of our sources: archival data may have appeared side by side with unreal figures expressing enormous quantities by "exact" figures.[62]

14. Abuse of numbers

The fact that in the neo-Assyrian period certain numerals on the order of tens and hundreds of thousands had become more and more frequent must have provoked a sort of "devaluation" of the objects quantified in such a manner. Certainly, when the annals of Tiglath-pileser III refer to the deportation of *900* people from a group of cities, such a statement would not have impressed any potential reader, especially as elsewhere (before) in the same text *83,000* people are said to be deported in a single instance.[63]

But it is especially in the later period that an attempt was made to avoid the use of numbers whose value was not sufficiently high. The annals of Sargon never go under *2,530* people, except when they speak of nobles, prefects or warriors, who obviously have a higher intrinsic value.[64] Rarely do they quantify animals, and in some such instances the exaggerations are obvious (cf. e.g. the sheep of Muṣaṣir). The last edition of the annals of Sennacherib never quantifies livestock, not even the horses, even though such quantifications were known in the earlier editions (see the repeatedly men-

---

[61] Cogan - Tadmor, *2 Kings*, 229 no. 14.

[62] H. Tadmor suggested that the passage from "As for Hezekiah ..." on (lines 37ff.), including the tribute list, is a kind of postscriptum to the narrative, added as a compensation for the non-conquest of Jerusalem ("Sennacherib's Campaign to Judah: Historical and Historiographical Considerations," *Zion* 50 (1985), 76ff [Hebrew, English summary: p. X]; cf. also Cogan - Tadmor, *2 Kings*, 248).

[63] Tadmor, *Tigl. III*, Ann. 5:6 and Ann. 19:11 (Rost, *Tigl. III*, 22:132 and 32:182; ARAB 2, 770 and 775) respectively The passage relative to the *900* deportees is quoted at p. 9.

Already in the inscriptions of Adad-nerari III one may observe that the sole numbers that appear on the Kalah slab and on the Saba'a stela concern the remarkable tribute from Damascus, quantified in the order of thousands of talents of metals (cf. p. 51 nos. 33ff).

[64] 2 *lim* 5 *me* 30 U[N.M]EŠ*(nišē)* in Lie, *Sargon*, 20:116 (= ARAB 2, 15). The smallest numbers relative to soldiers are slightly lower: cf. the *2,200 zi-im pa-ni* of 28:168 (ARAB 2, 23) or the *1,000* ANŠE *pet-ḫal* GÌR^{2}-*ia šit-mur-ti* LÚ *zu-uk* GÌR^{2}-*ia*, horses and infantry of the royal guard, in 26:150 (ARAB 2, 22). Concerning the nobles, cf. the *260* NUMUN LUGAL-*ti* (*zēr šarrūti*, "princes") in 24:134 (ARAB 2, 20), while the *28* EN URU.MEŠ-*ni* (*bēl ālāni*) of 16:100 (= ARAB 2, 11) represent cities rather than single persons. In this passage, Sargon states that he received tribute from *28* city prefects of the land of the Medes, thus the quantity of items (quantification of a single tribute) is replaced by the quantity of tribute (number of instances of tribute) — and this happens quite often in later times.

tioned booty from Babylon), while, as just remarked, it quantifies the prisoners only in two instances (*208,000* and *200,150*). In line with this magnitude, a couple of his late texts tell us of *150,000* enemy warriors killed during the battle of Ḫalulē, a quantification that, I would suggest, should be regarded as the extreme peak in this rush to ultra-high numerals.[65]

15. Later rulers

It is possible that this "inflationary spiral" had some influence on the reduction of the number of quantifications present in the inscriptions of Esarhaddon and Ashurbanipal (cf. pp. 39f and here below). Esarhaddon's prisms from Nineveh never use numbers in relation to prisoners, soldiers, horses, oxen, etc., which are the most common quantifications in the annals of Shalmaneser III (cf. pp. 30f). The only exceptions are represented by two augmentations of an earlier (unquantified) yearly tribute, "*65* camels and *10* asses"[66] and "*10* minas of gold, *100 / 1,000* choice gems, *50* camels and *100 / 1,000* leather bags of spices"[67] respectively. A fragment of a prism from that reign records another quantified tribute, expressed by "exact" numerals, probably representing the sacrifical offerings imposed on a series of cities,[68] while another tribute (?) appears in a fragmentary context.[69] In the texts of Ashurbanipal only *2* obelisks from Thebes (cf. p. 124) and *30* horses added to an earlier tribute (again not specified) of the Manneans[70] are quantified. Thus, only tribute was quantified, but specifying only the augmented quantity, while booty rarely is,[71] and prisoners or killed never are. As tribute is a

---

65 Cf. above, p. 116. The "*quasi*-round" nature of this numeral deserves comment: it is possible that the *80,000* Elamite bowmen that were sent to help Merodach-baladan during the 1st campaign of Sennacherib (cf. above n. 54) had represented some sort of guide both from a quantitative and from a formal point of view. As for this *80,000*, note that the quantifications of military contingents are often given in "round" numbers, not just in Sennacherib's time, which, as we observed, may reflect the fact that the army was structured on a decimal basis (cf. p. 82).

66 Borger, *Asarh.*, §27 episode 14:17f. Note that *Nin. B* and *C* do not have the *10* asses, mentioning only the *65* camels (cf. also "fragment B," §72 Rs. 4).

67 Cf. p. 63 nos. 81f (= ARAB 2, 518a, 536 and 551).

68 Winckler, AOF 2, 22 II 14ff (= Borger, *Asarh.*, §80; ANET 294a). The tribute ([*mand*]*attu*) includes "*6* talents *19* minas of gold, *300* [+ *X* minas of silver?], *1,586* garments, [X +] *7* homers? [..., *X* +] *24* ebony [trunks ...], *199* leather [bags?], *1,000 + ? + 40* horses [...], *30,418* sheep [...], *19,323* asses, *1*? barley [...]."

69 Bauer, IWA, Tf. 36 (K 3127+) Rs. 29 (= Borger, *Asarh.*, §79 - to be assigned to Esarhaddon), mentioning *8* oxen and *370* (+ X?) of something else. Cf. also n. 71 below, and cf. the fragmentary prayer from Uruk, mentioning Esarhaddon's grandfather Sargon, recording *60,000* sheep and *6,000* oxen, herds belonging to the temple of Ishtar, published in A. Falkenstein, *Literarische Keilschrifttexte aus Uruk*, Berlin 1931 (No. 46:4f; cf. Borger, AfO 18, 117a — the two numbers are written repectively 1 ŠU *lim*, "one *šuššu* thousands," and 6 *lim*).

70 Prism "A": Streck, *Asb.*, 26:25 (= ARAB 2, 786); pr. "C": Freedman, *St. Louis*, 92:127; pr. "F": Aynard, *Prisme*, 38:52; pr. "B": Piepkorn, *Asb.*, 56:1; cf. also Nassouhi, AfO 2, 102 III 8.

71 Cf. Esarhaddon's Dog River (*Nahr al-Kalb*) stela, mentioning *15* and *30* crowns in connection with the conquest of Memphis (Borger, *Asarh.*, §67:27ff; ARAB 2, 585; ANET, 293b). A fragmentary tablet (K 8692) published in W.G. Lambert, "Booty from Egypt?," *Journal of Jewish Studies* 33 (1982), 61-70 records a booty list including as much as *8,000* talents of silver, [*1*?]*20* gold headdresses, *50,000* strong horses and *60,000* fattened choice oxen, apart from countless (*lā mīna*) sheep, innumerable (*lā nībi*) linen robes, etc. Lambert suggests the text is a royal inscription written in the time of Esarhaddon or Ashurbanipal and describing the looting of an Egyptian capital (possibly Memphis). Since the script is neo-Babylonian, Lambert also suggests that the tablet represents some kind of dedicatory inscription originally accompanying a donation to a Babylonian temple, which may well be the case. The numeral referring to the sheep is, apparently, 1?+ŠU *lim* (but the copy is unclear on this point, cf. p. 70:32). The number referring to horses is really extraordinary: for similar quantifications one might recall certain literary texts, such as, e.g., the "letter of Gilgamesh" from Sultantepe (STT 1, 41:15f; cf. O.R. Gurney, "The Sultantepe Tablets: VI. A Letter of Gilgamesh," AnSt 7 [1957], 127-36).

regular payment, received yearly, its quantification is a "repeating" figure, not comparable to an *una tantum* quantification such as booty or a deportation might be. To report only the increased quantity is a way to underscore this characteristic.

In the Nineveh prisms of Esarhaddon and Ashurbanipal, especially the latter's "Rassam prism,"[72] the quantifications mostly concern cities, which were subject to a modest "devaluation," subdued foreign rulers, and distances between cities, regions and mountains. In some cases the numbers of cities or rulers are preceded by the listing of their names (these lists in substance take up the "amplifying" effect of high numbers).[73] Thus, Ashurbanipal's inscriptions give several quantifications within the building sections or in narrative contexts where they are needed, for instance when it is spoken of the fury of the Assyrian army ("of Ashur and of Ishtar"), used against Elam "once, twice, three times,"[74] or in such expressions as "I did not delay one day nor two."[75] In short, the last two important rulers of Assyria did not participate in the "rush" for the highest numbers, which already had given noteworthy results. Some quantifications appearing in their inscriptions, however, represent an exceptional figure, as in the case of the just mentioned *1,000* choice gems of Esarhaddon and the weight (*2,500* talents) of the *2* electrum obelisks carried off from Egypt by Ashurbanipal.[76]

Another aspect is the shift in style, fully manifested in the inscriptions of Ashurbanipal,[77] which is mainly responsible for the paucity of quantifications to be found in his inscriptions. They are characterized by a very free-flowing style, rich with entertaining elements and assorted curiosities:[78] it is therefore natural that the inclusion of "cold" numerical data in such a context would be out of line. Thus, even in his longer prisms, in which the events are related in greater detail,[79] the argument "quantity" is avoided. However, the numerals which occasionally do appear avoid direct comparison with his predecessors (and this may be intentional),[80] not quantifying those objects most frequently (and immeasurably) quantified by them.

---

72 Streck, *Asb.*, 2ff = ARAB 2, 765ff.

73 These instances are collected at pp. 190f. Most of them actually come from the reign of Sargon, and, if we except his annals, they are characteristic (in the Sargonid period) of the earliest inscriptions of each reign, or anyway of those containing detailed reports of the events: "letter to the God" of Sargon; report of the 1st campaign and "Rassam" cylinder of Sennacherib; prisms "B" and "C" of Ashurbanipal.

74 Streck, *Asb.*, 62:54 (ARAB 2, 816).

75 Borger, *Asarh.*, 43:63 (ARAB 2, 504); for a similar passage cf. Aynard, *Prisme*, 50:46 (ARAB 2, 942); Bauer, IWA, 57:10 (K 3404).

76 Cf. also here above, nn. 69 and 71, and the *55* statues of Egyptian kings or the *32* statues of kings of Elam mentioned in Streck, *Asb.*, 216:6 and 54:48 (ARAB 2, 939 and 810 resp.), or the *1* hundred *bēru* which had been covered pursuing Uaite', king of Arabia according to Streck, *Asb.*, 70:91 (ARAB 2, 823; an incantation written on the reverse of text Rm. 40 even mentions *3,600 bēru*, cf. Bauer, IWA, 92 Vs. 10). Another noteworthy numeral is the *16,000* bows (GIŠ.PAN.MEŠ) readable in a fragmentary passage of the text K 7596 (*ibid.*, 95:3 = ARAB 2, 1113; cf. above p. 106 n. 251 end).

77 Note how also the inscriptions of both Sargon and Sennacherib show a slightly different narrative style within the text of the later campaigns, which becomes less prosaic — a fact stressed by a lower propensity to give numbers in relation to booty, to killed, etc.: cf. n. 77 on p. 37 and n. 83 on p. 39.

78 Cf. n. 35 above. According to Grayson, ARINH, 43, the "annals" of this king recall some aspects of the "Histories" of Herodotus.

79 In the case of Ashurbanipal, the various prism "editions" differ greatly in length, having different sizes and 6, 8 or even 10 sides, this fact not being in relation to the increase in the number of campaigns (e.g., prism "F," one of the latest, has only 6 sides, and has a much abbreviated account for each campaign).

80 Some of his predecessor's tablets have been found in Ashurbanipal's "library" at Nineveh (cf. e.g. *Cat. Suppl.*, XVIII).

16. Non-numerical quantifications

Very frequently, in all periods, the quantity was roughly indicated with such terms as *ma'du* / *ma'dūti* (often written ḪÁ) or *rapšu* (DAGAL),[81] with the meaning of "many." In this case the text editor takes the place of the reader in evaluating the quantity, while the role of "proof," proper to the numbers, is given up.[82]

Some expressions are able to convey the concept of "innumerable": the most common are *ana lā māni* / *mīni* (or *ina lā mīni*), "without number,"[83] and, typical of Sargonid times, *ša nība lā išû*, "without measure," something that cannot be designated.[84] In spite of this, it is possible, in some cases, to establish the actual quantity represented by such expressions: thus Shalmaneser III's army of *120,000* troops is explicitly defined as *ana lā māni*,[85] while, according to a passage of Sargon's annals, his numerous (*gapšate*) troops for *3* days carried off an immeasurable booty (*šallat la nībi*), which, we are told, actually included *90,580* captives, *2,080* / *2,500* horses, *610* / *700* / *710* mules, *6,054* camels and X + *40* sheep.[86]

In some other cases it is possible to have recourse to parallel passages recording a numerical quantification: for instance, the cities "without number" (*ana lā māni*) conquered by Shalmaneser III during his 11th year,

---

81 Both terms are quite common, *rapšu* = DAGAL(.MEŠ) from Shalmaneser I (RIMA 1, 184:102 and 104) on. Very often the booty is defined *kabittu*, "heavy." For *ana mu'de*, "in abundance," cf. e.g. Thureau-Dangin, TCL 3, 22:131; Grayson, AfO 20, 88:13.

82 In some cases one such expression may be used to underscore some exceptional quantification: cf. above, p. 6.

83 These expressions indicate something that cannot (or could not) be counted (*manû* = "to count"). The expression (*ana*) *lā māni/mēni* (mostly referring to booty) first appears, in the *corpus*, with Shalmaneser I (RIMA 1, 184:69), and is used through Sennacherib (e.g. Luckenbill, *Senn.*, 28:21). For *ina lā māni* cf. Borger, *Asarh.*, 99:45 (only occurrence, cf. CAD M/1, 223a); more frequent *ina lā mī/ēni*, but appearing only in Esarhaddon and Ashurbanipal (cf. e.g. Borger, *Asarh.*, §80 I 4; Gerardi, JCS 40, 34:4′).
For an interesting example of *ša menūta lā išû* (still from *manû*) cf. Ashurnasirpal II's passage in RIMA 2, 199:88, "(tribute/booty) which, like the stars of heaven, had no number" (*ša* GIM MUL.MEŠ AN-*e me-nu-ta la i-šú-ú*; cf. also 215:43 and Ashur-bel-kala, *ibid.*, 88:5′). Cf. also the comparison between the number of cities of Uišdiš and the stars of heaven (*mi-i-na la i-šu-ú*) in Thureau-Dangin, TCL 3, 28:164 (Sargon). For *ša lā išû midita* (*middatu*, "measure") cf. Lie, *Sargon*, 78:10 (food offered for the inauguration of the royal palace).

84 *nību* = "mention," "designation" (cf. AHw, 785b). Earliest instances of *ša nība lā išû*: Rost, *Tigl. III*, 14:70 and 34:204. One noteworthy occurrence in Winckler, *Sargon*, 124:141, "*154* talents *26* minas *10* shekels of red gold, *1,604* minas of white silver, copper, iron without measure." Other expressions from *nību*: *ša lā nībi* (e.g. Luckenbill, *Senn.*, 87:33); *ana lā nība* (Rost, *Tigl. III*, 12:65); *lā nībi* (e.g. Lie, *Sargon*, 38:232).

85 Cf. p. 107. Another interesting example is given by Shamshi-Adad V's monolith inscription. It describes how the Babylonian king Marduk-balassu-iqbi, trusting in the multitudes of his armies, which were countless (*ma'di ana lā māni*), offered battle to the Assyrians: during the consequent engagement *5,000* of them were killed and *2,000* taken prisoners (KB 1, 186:37ff = ARAB 1, 726).
Thus, according to Ashurnasirpal II's annals, once the people of Suḫu, Laqû and Ḫindānu, "trusting in the massiveness of their chariotry, their troops, (and in) their might, mustered *6,000* of their troops" and did battle. Of these, *6,500* were then felled by the Assyrians (and this does not even include some fugitives! — cf. RIMA 2, 214:34ff; ARI 2, 579). Similarly, also the people of the land Dagara, "trusting in the massiveness of their troops," offered battle to the Assyrians. The contest involved the killing of *1,460* men (RIMA 2, 244:88 and 203:27 = ARI 2, 554), while another *1,200* of the fugitives were, later on, uprooted (RIMA 2, 204:31, cf. 244:88). Cf. another two instances in *ibid.*, 201:114ff (*3,000* killed) and 213:17ff (*3,000* captured).

86 Cf. p. 55 nos. 58f and n. 48. As we have already noted, in Tukulti-Ninurta II's annals a quite modest tribute, including *3* talents of silver and *20* minas of gold and some other commodities, is indicated as ḪI.A.MEŠ (*mādu*), "numerous" (cf. p. 97 n. 209 — passage reported at p. 122).

according to the "black obelisk," turn out to be *97* cities of Sangar and *100* cities of Arame, if compared with his older annals.[87] Similarly, the repeatedly mentioned "high - exact" numbers representing Sennacherib's booty from Babylonia in his first campaign were replaced by the expression *ša lā nībi* in his later prisms (cf. above p. 58 or pp. 90f).

In any case, in later times the expressions (*ša*) *lā nība* and the like are frequently used, thus making it clear that the quantities had become so high as to be immeasurable. They are very suitable for a "discursive" context, such as Ashurbanipal's prism inscriptions are. And they cannot be disputed, nor taken as untrustworthy, as would be the case of a particularly showy number — obviously, nobody can contest that on a certain occasion they had not been able to count the prisoners or the booty.

Alternatively, the quantity may be indicated in a relative measure, e.g. specifying that *all* the enemies in a given area had been taken prisoner, or killed, etc. — the quantity thus corresponds to the "maximum possible" (or the "minimum possible"). The most common expression is, in this case, *mala bašû*, "as many as they were," also typical of the later reigns.[88] It may be stated that "all (*mala*) the people of Arabia who had gone forth with him, I cut down with the sword."[89] Or it is specified that "not one I left" (*ištēn ul ēzib*)[90] or that "nobody escaped" (*anumma ul ūṣī*).[91]

It is also possible to have recourse to more complex and effective constructions, for instance speaking of "people ... horses, mules ..., who were more numerous than locusts,"[92] or stating that so many (*ina lā mīni*) were the

---

87 Black obelisk: cf. Michel, WO 2, 150:87; older texts: Cameron, *Sumer* 6, 14:69f (= Michel, WO 1, 466); Delitzsch, BA 6, 147:90; Safar, *Sumer* 7, 9:52f (= Michel, WO 2, 34).

Note also that the *130* garments mentioned in the list of booty from Muṣaṣir in Sargon's "letter to the God" (Thureau-Dangin, TCL 3, ll. 366ff = ARAB 2, 172 end), in the annals are given as *ana lā māni* (Lie, *Sargon*, 26:157 = ARAB 2, 22 - fragm. passage, cf. p. 129 above). For a "quantification" of the term GAL ("great") in Sargon's times cf. above, p. 52 n. 36, and for a group of HÁ ("numerous") warriors, later given as *2,800* by Shalmaneser III's annals, cf. p. 48 n. 17.

88 *mala bašû*, earliest occourrence: Rost, *Tigl. III*, 56:10 (Tadmor, *Tigl. III*, Summ. 7).

89 Streck, *Asb.*, 66:117f (ARAB 2, 818). The term *mala* obviously occurs with a certain frequency. Within the titulary often appears *kâlu* ("all," written also DÙ), but also outside it — cf. e.g. Adad-nerari III: "the kings of Kaldu, all of them (DÙ-*šú-nu*), I made my vassals" (Tadmor, *Iraq* 35, 149:22). With *gabbu*, "all," cf. an interesting instance in Wiseman, *Iraq* 14, 34:105 ("all the people of Assyria" invited to the banquet; Ashurnasirpal II).

Also frequently employed are the terms *napḫaru*, "the total of," and *gimru*, "totality," "entirety," mostly in reference to lands or people. Quite often the expression *ana pāṭ gimrišu*, "in its entirety," is used, from Tukulti-Ninurta I on (RIMA 1, 268:9′, 273:68; cf. also *a-na* ZAG *gim-ri-šá*, 245:66).

Other expressions are *ana gimirtišu* (e.g. in Rost, *Tigl. III*, 22:129) and *ana siḫirtišu*, "in its totality," very frequent from Tiglath-pileser I on (cf. e.g. RIMA 2, 14:92).

As for *siḫirti*, cf. Shalmaneser I in RIMA 1, 184:82 ("the totality of the Kašiyari mountains") and his successor Tukulti-Ninurta I in *ibid.*, 240:26 (similar passage, still relating to the Kašiyari mountains; *siḫirti* frequently appears in the inscriptions of this king).

90 *ištēn* is usually written 1-*en*: cf. 1-*en ina* ŠÀ-*šú-nu* TI.LA *ul e-zib*, "not one alive I left" in RIMA 2, 201:108 (similar instance in Tadmor, JCS 12, 88:8). *ištēn* may be replaced by some other noun: *aiumma*, "nobody" (e.g. in Michel, WO 4, 30 V 3); *edu* (e.g. Luckenbill, *Senn.*, 67:8; cf. *ištēn ul akla edu ul ēzib* in Borger, *Asarh.*, 106:34); *multaḫtu*, "fugitive" (e.g. Lie, *Sargon*, 66:448); *napištu* (Saggs, *Iraq* 37, 14:21, Sargon); *peri'šun*, "descendant" (Luckenbill, *Senn.*, 77:15).

91 Borger, *Asarh.*, 58:18. With *multaḫtu* cf. e.g. Streck, *Asb.*, 36:63; with *napištu* Borger, *Asarh.*, 58:11. Also employed is the verb *naparšudu*, "to escape," e.g. in *ištēn ina libbišunu ul ipparšid* (Tadmor, *Tigl. III*, Ann. 15: 9 = Rost, *Tigl. III*, 30: 173). Cf. *edu ul ipparšid multaḫtu ul ūṣi* in Streck, *Asb.*, 74:40. With *sâtu* cf. *multaḫtu lā isīta* in Thureau-Dangin, TCL 3, 28:176.

92 *ša eli* BURU$_5$.ḪÁ(*eribi*) *ma'du*, Streck, *Asb.*, 56:94; Bauer, IWA, 52:13 (cf. ARAB 2, 811 and 920; cf. also Piepkorn, *Asb.*, 58:46). One similar example comes from Shamshi-Adad V's inscriptions: "his captive warriors were given to the soldiers of my land like locusts" (*ki-ma* NAM.ḪÁ^{MEŠ}; KB 1, 186:34f = ARAB 1, 725 end) — actually these captives were *3,000* (cf. just above, line 31). Cf. another example above, p. 116 n. 308.

Other quantity similes may concern booty, compared to the stars of heaven or to flocks of sheep: they are collected and discussed in Schott, MVAG 30/2 (cf. p. 102 there).

camels brought back from Arabia that "throughout my land were sold for *1½* shekels of silver in the markets."[93]

17. Conclusion

I think that we have adduced sufficient evidence to show that the quantifications are not used in a whimsical or impromptu manner. Rather, narrative or editorial practices may explain the use of certain numbers in a given specific context, though these practices can only partly be identified.[94] The authors of these texts show a remarkable variety of behavior in approaching the problem of expressing "how much," even seemingly contradicting themselves within a single inscription, despite its apparent uniformity. Consequently, each quantification should be independently evaluated in the light of its context, and no wide-ranging statement along the lines of "the Assyrian annals report exaggerated numbers of prisoners captured" can be automatically applied to any specific case, nor it can have a general value.

---

93 Streck, *Asb.*, 76:48f and *passim* (the passage appears in several inscriptions: ARAB 2, 948, 827 and 869; cf. also the translation and the remarks of Weippert, WO 7, 46 and 82 n. 30).

Cf. also the passage in Lie, *Sargon*, 38:233f: *ma-ḫi-ri* KÙ.BABBAR *ki-ma si-par-ri i-na ki rib mat* *ᵈAš-šur*ᴷᴵ *i-šim-mu*, "the price of silver equal to that of bronze in Assyria they valued," thanks to the immense (*la ni-bi*) treasures heaped up in the city of Dur-Sharruken. It is, as can be easily understood, a gross exaggeration (cf. Elat, AfO *Bh.* 19, n. 8), even if it is known that during the 7th century silver replaced copper as standard currency (cf. e.g. Postgate, GPA, 25).

94 One thing has to be stressed: when calling attention to these editorial practices, thus questioning the validity of the numbers, as in the case of Sennacherib's deportations, I am not claiming that we should diminish the extent of neo-Assyrian deportations as a whole. It is obvious that many regions were depopulated. Yet, one ought not to take for granted that the royal inscriptions were intended to provide detailed and "objective" information about these events.

APPENDIX

HIGHEST QUANTIFICATIONS

In some cases, different numerals are reported for a single reign, taking into consideration the kind of quantification (if it is a "total," if it refers to Assyrians or to "enemies," etc.) and the kind of inscription in which it appears.

A small number of very modest "highest" quantifications have been ignored. References are given to copies (with transliterations in parentheses).

CITIES

Shalmaneser I	***180***	(3 *šu-ši*); cf. p. 151 n. 171
Tukulti-Ninurta I	***180***	(3 *šu-ši*); cf. *ibid.*
Tiglath-pileser I	***25***	AKA, 54:58 (RIMA 2, 18)
Adad-nerari II	***40***	(?), KAH 2, 83 r. 1 (RIMA 2, 144)
Tukulti-Ninurta II	***30***	Schramm, BiOr 27, pl. I:35 (p. 149; RIMA 2, 173)
Ashurnasirpal II	***250***	AKA, 340:117 and 237:35 = Le Gac, *Asn.*, 86 and 147 (RIMA 2, 211 and 261:84 - same quantification)
Shalmaneser III	***250***	ICC, 97:189 (Michel, WO 2, 232)
Shamshi-Adad V	***1,200***	1 R, 30 and 33:36 (KB 1, 182)
(Adad-nerari III)	***331***	total of towns built by Palil-ereš, governor of Raṣappa - Page, *Iraq* 30, pl. 39:20 (p. 142; copy also in Cazelles, CRAIBL 1969, 109)
Tiglath-pileser III	***591***	ICC, 73a:14 = Tadmor, *Tigl. III*, Pl. 20:16 (Ann. 23, but cf. n. there)

Sargon II	***140***	annals: Botta, 76 = Winckler, *Sargon*, no. 9:10′ (Lie, *Sargon*, 24:140)
	430	letter to the God: TCL 3, line 422
Sennacherib	***820***	cf. p. 57 no. 66, *sub* (1) and (2)
Ashurbanipal	***75***	CT 35, 49 III 18 (Piepkorn, *Asb.*, 56:6) and *passim*

PEOPLE

Shalmaneser I	***14,400***	*gunni*, elite troops, cf. p. 150 n. 162
Tukulti-Ninurta I	***28,800***	ERÍN.MEŠ, soldiers, cf. p. 151 n. 167
Tiglath-pileser I	***20,000***	two quantifications: cf. p. 46 no. 5 (Muški men/troops) and AKA 77:87 and 120:18 (RIMA 2, 24 and 34; Qumānu hordes, *ummāni*)
	2,000	*šallatu*, prisoners, AKA, 120:17 (RIMA 2, 34:34)
"broken obelisk"	***1,000***	ERÍN.MEŠ, AKA, 129:1 (RIMA 2, 101)
	4,000	*šallatu*, AKA, 129:4 (RIMA 2, 101)
Ashurnasirpal II	***69,574***	"banquet stela," *Iraq* 14, 44:149 (RIMA 2, 293)
	50,000	*ummāni* (Assyrians), AAA 19, 109:28 (RIMA 2, 309)
	6,500	*muqtablu*, (enemy) fighters, AKA, 356:37 = Le Gac, *Asn.*, 98 (RIMA 2, 215); cf. p. 177 n. 85
Shalmaneser III	***120,000***	ERÍN.ḪÁ.MEŠ (Assyrians), cf. p. 107 n. 256
	110,610	*šallutu*, plus [next line]
	82,600	*dīktu* (killed), "totals" relating to the first 20 years, cf. p. 93 n. 193
	87,500	ERÍN.ḪÁ.MEŠ captured (*nasāḫu*), apparently a "total," cf. p. 94
	22,000	ERÍN.ḪÁ captured (*nasāḫu*), cf. p. 49 no. 24
Shamshi-Adad V	***30,000***	*šallatu*, KAH 2, 142:11 and 18 (Weidner, AfO 9, 102; SAA 3, p. 108)
	13,000	*mundaḫṣu*, fighters killed (*maqātu* Š), 1 R, 31 + 34:27 (KB 1, 186)

Tiglath-pileser III	***155,000***	UN.MEŠ, people carried off (*šalālu*; fragm. passage), Rost, *Tigl. III*, Pl. XXXIV:14′ (Tadmor, *Tigl. III*, Summ. 11)
	72,950	people (?) carried off (?) (annals), Tadmor, *Tigl. III*, Pl. XIV:9′ (Ann. 17).[1]
Sargon II	***90,580***	ERÍN.MEŠ (*šalālu*), Botta, 163:8′ = Winckler, *Sargon*, no. 55 (Lie, *Sargon*, 62:6)
	350	*malkē*, (Assyrian) princes, Lyon, *Sargon*, 7:44 (p. 34); Botta, 160b:2′ = Winckler, *Sargon*, no. 61 (Weissbach, ZDMG 72, 180:29)
Sennacherib	***208,000***	UN, "total" prisoners of 1st campaign, cf. p. 58
	200,150	UN, counted as booty, cf. p. 114 n. 296

WEAPONS AND WAR IMPLEMENTS

Tiglath-pileser I	***120***	chariots, two quantif., both 2 *šu-ši*: AKA, 49:3 and 68:94 (RIMA 2, 17 and 21)
Tukulti-Ninurta II	***2,702***	horses (Assyrian forces), *eli ša pān ušatir*, Schramm, BiOr 27, pl. V:48 (RIMA 2, 178:130)
Ashurnasirpal II	***40***	chariots, received as tribute, cf. p. 128 n. 61
Shalmaneser III	***5,542***	cavalry, Assyrian forces, cf. p. 50 no. 30
	2,002	chariots, cf. *ibid.*, no. 29
	470	cavalry (enemies), cf. p. 50 n. 24
	1,121	chariots, cf. *ibid.*
Shamshi-Adad V	***100***	chariots, 1 R, 31 + 34:44 (KB 1, 186)
	200	cavalry, *ibid.*
Sargon II	***300***	chariots (Assyrian), cf. p. 56 no. 61; in the annals: ***150*** (Assyrian), cf. p. 56 (no. 60)
	1,500	cavalry (Assyrian), in the annals, cf. *ibid.*

1 Cf. also Rost, *Tigl. III*, Pl. XIX; ICC, 71a:7 and 72a:9. The quantified term is unclear. Another *83,000* [+ X?] appears in a very fragmentary context (Tadmor, *Tigl. III*, Pl. XVI:11 - Ann. 19), possibly it refers to resettled deportees (cf. *ibid.*, p. 62 n. to line 11 and here above, p. 101 n. 225).

ANIMALS (BOOTY AND TRIBUTE)[2]

Tiglath-pileser I	***1,200***	horses	imp.tr.	AKA, 70: 19 (RIMA 2, 22)
	2,000	oxen	imp.tr.	*ibid.*
Tukulti-Nin. II	[*X* +] ***40***	oxen	rec.tr.	Schramm, BiOr 27, pl. IV:9; or ***100*** [+ *X*] (*ibid.*, l. 12; RIMA 2 176, lines 91 and 94 resp.)
	1,200	sheep	rec.tr.	*ibid.*, l. 18 (RIMA l. 100)
Ashurnasirpal II	***460***	horses	rec.tr.	cf. p. 128 n. 61
	2,000	oxen	rec.tr.	cf. *ibid.*
	10,000	sheep	rec.tr.	AKA, 369:74 = Le Gac, *Asn.*, 108 (RIMA 2, 218) and ICC, 44:32 (RIMA 2, 227:50)
Shalmaneser III	***9,920***	horses	boot.	"total," incl. mules? cf. p. 93 n. 193
	35,565	oxen	boot.	"total," cf. *ibid.*
	184,755	sheep	boot.	"total," cf. *ibid.*
	14,000			(UDU.DAM.QAR.MEŠ), Wiseman, *Iraq* 14, 42 and 35:107 ("banquet stela," cf. above)
Tiglath-pileser III	***2,000***	horses	tr.?	2 R, 67:65 (Tadmor, *Tigl. III*, Summ. 7 rev. 15)[3]
	5,000		boot.	cf. p. 83 n. 158
	20,000	oxen	boot.	cf. p. 85 n. 165
	19,000	sheep	boot.	Rost, *Tigl. III*, Pl. VIIIa:2′ (Tadmor, *Tigl. III*, Ann. 5)
	30,000	camels	boot.	cf. p. 85 n. 165
Sargon II	***4,609***	horses	rec.tr.	fragmentary variant ***8,609***, cf. p. 55 no. 56. Cf. also the *11,600* [horses?] received from the Medes according to Gadd, *Iraq* 16, pl. 46 III 51 (p. 177)
	92[***0***]	oxen	boot.	cf. p. 54 no. 51
	100,225	sheep	boot.	cf. *ibid.* no. 52
	6,054	camels	boot.	cf. p. 55 (cf. n. 48 there)

2 Three possibilities are distinguished, identified by the terms in column four: imp.tr. = "imposed tribute" (*šakānu*); rec.tr. = "received tribute" (*maḫāru*); boot. = "booty carried off" (*našû, waṣû* Š, *šalālu*). Only horses, oxen, sheep and camels are taken into consideration.

3 Rost, *Tigl. III*, Pl. XXXVII:15 (and p. 72) erroneously gives 2 *me*.

Sennacherib	***7,200***	horses		"total" booty of the 1st campaign cf. p. 58
	5,233	camels[4]		
	200,100	oxen		
	800,600	sheep		
Esarhaddon	***30,418***	sheep	tr.?	cf. p. 175 n. 68 (line 17)

METALS (BOOTY AND TRIBUTE)

Tiglath-pileser I	***30*** t	copper bars	boot.	AKA, 59:1 (RIMA 2, 19)
Tukulti-Ninurta II	***20*** m	gold	rec.tr.	two quantif.: Schramm, BiOr 27, pl. III:70; pl. IV:16 (RIMA 2, 175 and 176:98)
	3 t	silver	rec.tr.	Schramm, BiOr 27, pl. III:70 (RIMA 2, 175)
	130? t	bronze	rec.tr.	Schramm, BiOr 27, pl. IV:16 (RIMA 2, 176:98)[5]
	32 t	tin	rec.tr.	*ibid.*
	2 t	iron	rec.tr.	*ibid.*, line 18 (RIMA 2, 176:100)
Ashurnasirpal II	***2*** t	gold	rec.tr.	cf. p. 128 n. 61
	20 t	silver	rec.tr.	two quantif.: AKA, 366:65, 368:73 = Le Gac, *Asn.*, 106 and 108; the second passage also in ICC 44:31 (RIMA 2, 218 and 217:65 = 227:49)
	100 t	tin	rec.tr.	cf. p. 128 n. 61
	300 t	bronze	rec.tr.	cf. p. 47 no. 12 and n. 10 there
	300 t	iron	rec.tr.	cf. p. 128 n. 61
Shalmaneser III	***3*** t	gold	rec.tr.	3 R, 7 II 22 (KB 1, 160)
	100 t	silver	rec.tr.	*ibid.*
	300 t	bronze	rec.tr.	*ibid.*
	300 t	iron	rec.tr.	*ibid.*
Adad-nerari III	(?) ***100*** t	gold	rec.tr.	cf. p. 51 no. 34 - alternatively, ***20*** t (cf. *ibid.*)
	2,300 t	silver	rec.tr.	*ibid.*

4 Cf. also the [X] *lim* camels taken from the Arabs (VS 1, p. 75:23 = Luckenbill, *Senn.*, 92:23).
5 Both the "30" and the "talent" signs are unclear — Schramm (p. 153) reads *140* talents.

	3,000 t	bronze	rec.tr.	*ibid.*
	5,000 t	iron	rec.tr.	*ibid.*
Tiglath-pileser III	***150*** t	gold	rec.tr.	cf. p. 51 no. 37
	2,000 t	silver	boot.?	two different groups of booty or tribute, cf. Wiseman, *Iraq* 18, pl. XXIIf obv. 24′ and rev. 26′ (Tadmor, *Tigl. III*, Summ. 9 - fragm. inscr.)
	500 t	bronze	tr.?	ICC, 51b:10 = Tadmor, *Tigl. III*, Pl. 8 (Ann. 12)
Sargon II	***34*** t ***18*** m	gold	boot.	cf. p. 54
	164 t ***26*** m [*10*] s	gold	"total,"[6]	cf. p. 57 no. 63
	2,100 t ***24*** m	silver	boot.	cf. p. 52 no. 41
Sennacherib	***30*** t	gold	imp.tr.	cf. p. 80 n. 151
	800 t	silver	imp.tr.	cf. *ibid.*

DISTANCES

Tiglath-pileser I	***1 bēru*** + ***20*** MAL	Weidner, AfO 21, Tf. 8:2′
Tiglath-pileser III	***70 bēru***	Tadmor, *Tigl. III*, Summ. 1:24; Summ. 3:25 (cf. Pl. 47f:25; Rost, *Tigl. III*, Pl. XXXIII:[6] and Pl. XXXI:[8])[7]
Sargon II	***30 bēru***	Lie, *Sargon*, 66:443 and *passim* (cf. p. 133); TCL 3, 14:75 (another quantification)
Esarhaddon	***140 bēru***	cf. p. 63 no. 83
Ashurbanipal	***100 bēru***	5 R, 8:91 and 3 R, 24:79 (Streck, *Asb.*, 70); 3 R 35 no. 6 III 34 (Streck, *Asb.*, 204:13; cf. Weippert, WO 7, 79:31)

[6] Given to the gods of Babylonia over three years. Note also the fragmentary passage Gadd, *Iraq* 16, pl. XLIV (p. 175) I 24f, mentioning *177* talents [of gold?] and *730* talents *22* shekels (?) [of silver?] used for the building of a temple.

[7] Layard's copy in ICC, 18:24 is erroneous (having ŠÚ instead of DIŠ), as are Rost's transliterations and translations, which give in both cases "LX" (*60*, cf. 46:24 and 52:39; ARAB 1, 785 and 813 have respectively "80 (?)" and "60"). Cf. Tadmor, *Tigl. III*, 124 n. *ad* line 24.

Highest Quantifications of Selected Objects in the Urartian Royal Inscriptions

	Išpuini	Menua	Argišti I	Sarduri III
Cities		*9* (28:IV)	*60* (80:6,VI)	*200* (103:15,VI)
People	*[2]0,483* (7:V)	*5[X,XXX]* (cf. 18:V)	*52,675* (80:1,VII)	*46,600* (103:16,XII)
Soldiers (Urartian)	*[2]2,704* (*ibid.*)			*352,011* (103:A,II)*
Cavalry	*9,274* (7:III)		*3,551* (77a)	*3,600* (*ibid.*)*
Chariots	*106* (*ibid.*)			*92* (*ibid.*)
Horses	*1,120* (7:V)	*1,753* (18:V)	*4,426* (82 r. II)	*10,408* (103:A,III)
Oxen	*13,54[0]* (6a:VIII)	*7,616* (*ibid.*)	*35,015* (80:1,VII)	*40,353* (103:15,XVI)
Sheep	*[2]05,000* (7:V)	*15,320* (cf. 18:V)	*126.[000]* (80:2,IX)	*214,700* (*ibid.*)
Camels	*365* (*ibid.*)		*184* (80:6,VII)	*115* (103:8,VIII)
Gold			*41* m. (82 r.IV)	*40* m. (103:9,V)
Silver			*37* m. (*ibid.*)	*800* m. (*ibid.*)
Bronze			*20,000* m. (*ibid.*)	*7,077* m. (103:A,III)*

References to König, HCI, no. inscr. and §. Cf. Zimansky, *Ecology and Empire*, 58 for a more comprehensive list of quantifications.

NOTES

• Išpuini: inscription no. 7 according to M. Salvini's edition in Pecorella - Salvini, *Tra lo Zagros e l'Urmia*, 57ff (for the number of soldiers cf. also Zimansky, *loc. cit.*, 56 n. b to Table 6).

• Menua: numerals of people and sheep are given according to Zimansky (*loc. cit.*, 58). Argišti I: the number of sheep according to König is 12 *a-ti-bi* 6 LIM [...] = *126,000* (+ *X*?). Cf. also 10 *a-ti-bi* 1 LIM 8 ME 29 = *101,829* sheep in 80:1, VII. For the bronze, König reads [2] *a-ti-bi*, but the digit is not confirmed by Melikišvili (UKN, 234:21). Cf., however, another 10 LIM minas of bronze in König, HCI, 82 r. V.

• Sarduri III: the quantifications marked with * are given by a text of difficult interpretation (see above p. 108 n. 260). Alternately, *50* chariots are mentioned in 104:VI and 102 r. VI and *3,500* horses in 103:15, XVI.

"EXACT" NUMBERS

The quantifications expressed in "exact" numbers, in chronological order (as far as possible), are reported here. Since the "exactness" of a number is in relation both to its magnitude and to the presence of lower order digits (as described on p. 6), I have established the following categories to take into account (a) numbers higher than *1,000*, having unit(s); (b) numbers higher than *10,000*, having ten(s) and/or unit(s); (c) numbers higher than *100,000*, having hundred(s) and/or lower order digits.

A bar indicates a variant (non-"exact" variants are placed in parentheses). The quantified term is reported without the plurality sign (MEŠ).

Tukulti-Ninurta II		
1.	***2,702***	HORSES, Schramm, BiOr 27, pl. V:48 (RIMA 2, 178: 130)
Ashurnasirpal II		
2.	***47,074***	PEOPLE (ERÍN + MÍ), Wiseman, *Iraq* 14, 44:141 (RIMA 2, 293)
3.	***69,574***	(PEOPLE), *ibid.* (line 149)
Shalmaneser III		
4.	***2,002 > 2,001***	CHARIOTS, cf. p. 50 no. 29
5.	***5,542 > 5,242 > 5,241***	CAVALRY (*pētḫallu*), cf. *ibid.* no. 30
6.	***1,121***	CHARIOTS, cf. p. 50 n. 24
7.	(*16,000* >) ***16,020***	TROOPS (ERÍN *tidūki*), cf. p. 50 no. 27
8.	***110,610***	PRISONERS, cf. p. 93 n. 193 (line 34)
9.	***35,565***	OXEN, cf. *ibid.* (line 37)
10.	***19,690***	ASSES, cf. *ibid.*
11.	***184,755***	SHEEP, cf. *ibid.* (line 38)
Tiglath-pileser III		
12.	***1,223***	PEOPLE, 3 R, 9:33 = Tadmor, *Tigl. III*, Pl. XVI:12 (Ann. 19)
13.	***6,208***	MEN, 3 R, 9:45 = Tadmor, *Tigl. III*, Pl. IX:5 (Ann. 13)
14.	***13,52***[***0***?]	[PEOPLE?], ICC, 72b:9 = Tadmor, *Tigl. III*, Pl. XVI (Ann. 24)[8]
15.	***72,950***	PEOPLE, cf. p. 182
Sargon II		
16.	***27,280 / 27,290***	PEOPLE, cf. p. 52 no. 38
17.	***9,033***	PEOPLE, Botta, 71:4′ = Winckler, *Sargon*, no. 3 (Lie, *Sargon*, 8:57)

[8] The context is very fragmentary — however, it is likely that the numeral refers to deportees.

18.	***25,212***	SHIELDS of copper (*a-ri-at* URUDU), Thureau-Dangin, TCL 3, line 392
19.	***305,412***	DAGGERS of copper (GÍR URUDU), *ibid.*, line 394
20.	(*6,110 > 6,170 >*) ***20,170***	PEOPLE, cf. p. 54 no. 49
21.	***1,235 > 1,285 > 100,225***	SHEEP, cf. *ibid.* (n. 52)
22.	***8,609***? ***> 4,609***	HORSES, cf. p. 55 no. 56
23.	***18,430***	PEOPLE, Botta, 65b:6′ and 4, 113:6′, cf. also 156:3′ = Winckler, *Sargon*, nos. 31 and 42, cf. no. 28 (Lie, *Sargon*, 44:279 and n. 8)
24.	***12,062***	PEOPLE, Botta, 66a:16′ = Winckler, *Sargon*, no. 32 (Lie, *Sargon*, 52)
25.	***90,580***	PEOPLE, cf. p. 55
26.	***6,054***	CAMELS, cf. *ibid.*
27.	***1,604 / 1,804***	talents of SILVER, cf. p. 57 no. 64
28.	***16,280***	CUBITS, cf. p. 140 n. 112 and 141 n. 120
Sennacherib		
29.	***11,073***	ASSES, cf. p. 58
30.	(*5,230 >*) ***5,233***	CAMELS, cf. *ibid.*, no. 67
31.	***80,050 >*** (*80,100 /*) ***200,100***	OXEN, cf. *ibid.*, no. 68
32.	***800,100 > 600,600 / 800,600***	SHEEP, cf. *ibid.*, no. 69
33.	***200,150***	PEOPLE, cf. p. 114 n. 296
34.	***12,515***	(CUBITS), CT 26, 30:62 (Luckenbill, *Senn.*, 111; Heidel, *Sumer* 9, 166:68)
35.	***21,815***	CUBITS, *ibid.* (two lines after); cf. p. 62
Esarhaddon		
36.	***1,586***	GARMENTS (TÚG MAŠ.MAŠ), cf. p. 175 n. 68 (line 14)
37.	***30,418***	SHEEP, *ibid.* (line 17)
38.	***19,323***	ASSES, *ibid.* (line 18)
Ashurbanipal		
39.	***1,535 / 1,635*** (*/ 1630 / 1830*?)	YEARS, cf. p. 66 no. 90

Note that:
- no. 39 is a *Distanzangabe* and nos. 28 and 34-35 are measures of perimeters (of cities, cf. p. 152), thus their "precision" is, so to say, justified;
- nos. 4-5, 7, 16, 20-21, 22, 27, 30-32 and 39 have variants (usually other "exact" numbers);
- nos. 1, 4-5, 8-11 are "totals" relative to the whole reign (up to the date of the inscription) and have no parallels, being reported by one inscription only, and consequently cannot have variants;
- nos. 2-3 come from the "banquet stela" (cf. p. 141), an unparalleled document, and therefore also cannot have variants; the second number is a sum, including also the first. Nos. 18-19 come from the "letter to the God" of Sargon (cf. p. 128).

LISTS OF "SUMMED" NAMES

(A) In the first table are reported the instances of lists of names followed by their number (e.g. "the cities X, Y, Z, ..., these *8* cities I conquered"). The various columns show, respectively: no. of the occurrence; number of names actually listed, if different from their "total" (? appears when this cannot be ascertained); object (Akkadian term or logogram); possible presence of the term *napḫar* ("total"); number-sum reported; reference (to transcription and translation).

	(names)	term	"total"	sum	reference
Shalmaneser I					
1.		KUR	-	***8***	RIMA 1, 183:32-36 (ARI 1, 527)
Tiglath-pileser I					
2.		KUR	-	***16***	RIMA 2, 21:58-65 (ARI 2, 30)
3.		LUGAL	(PAP)	***23***	RIMA 2, 21:71-83 (ARI 2, 30), cf. p. 92 n. 189
4.	?	URU	PAP	***14***	RIMA 2, 58:8′-14′ (ARI 2, 137)
Ashurnasirpal II					
5.	7	É.GAL	-	***8***	*Iraq* 14, 33 = RIMA 2, 289:25f
Shalmaneser III					
6.		URU	-	***6***	KB 1, 160:16f (ARAB 1, 601)
Shamshi-Adad V					
7.		*māḫāzu*	PAP	***27***	KB 1, 176:45-50 (ARAB 1, 715)
8.		URU	-	***3***	Weidner, AfO 9, 92:21f
Tiglath-pileser III					
9.		URU	P[AP]	***29***	Postgate, *Sumer* 29, 53:26-33 = Tadmor, *Tigl. III*, 114
10.	?	*nagû*		***19***	Tadmor, *Tigl. III*, Ann. 19:4-9; Ann. 26:1-5 (cf. ARAB 1, 770)[9]
Sargon II					
11.		*nagû*	-	***6***	Lie, *Sargon*, 16:98f (ARAB 2, 11)
12.		URU *birtu*		***10***	Lie, *Sargon*, 36:216ff (ARAB 2, 26f)[10]
13.		*nasīku*	-	***8***	Lie, *Sargon*, 45 n. 9 (ARAB 2, 31)
14.	?	URU	-	***5***	Lie, *Sargon*, 46:286f
15.		URU	-	***2***	Lie, *Sargon*, 46:287f
16.	?	URU	-	***3***	Lie, *Sargon*, 46:289
17.	?	URU	-	***6***	Lie, *Sargon*, 46:290f
18.	?	URU	-	***6***	Lie, *Sargon*, 46:292f
19.	?	URU	-	***7***	Lie, *Sargon*, 46:294f
20.	?	*nagû*	-	***6***	Lie, *Sargon*, 48:1 (ARAB 2, 32)

---

[9] Fragmentary passage: it appears that some cities (URU) and mountains/lands (KUR) are summed up as *19* districts (*nagû*). Another fragmentary passage mentions some cities and some lands, which are summed as "[x +] *8* provinces," cf. Tadmor, *Tigl. III*, Summ. 9:20′ff.

[10] The list follows the number.

21.		*nasīku*	-	***5***	Lie, *Sargon*, 48:4f (ARAB 2, 32)
22.		*nasīku*	-	***4***	Lie, *Sargon*, 48:326 (ARAB 2, 32)
23.		URU	-	***14***	Lie, *Sargon*, 50:11f (ARAB 2, 32)
24.		*nasīku*	-	***6***	Lie, *Sargon*, 52:1f (ARAB 2, 34)
25.		KUR	-	*sibitti* (*7*)	TCL 3, 8:28f (ARAB 2, 143)[11]
26.	13	URU	-	***12***	TCL 3, 16:87ff (ARAB 2, 151)
27.	?	URU	-	***21***	TCL 3, 38:235-239 (ARAB 2, 163)
28.		URU	-	***7***	TCL 3, 42:270ff (ARAB 2, 165)
29.	29	URU	-	***30***	TCL 3, 44:281-286 (ARAB 2, 166)
30.		URU	-	***5***	TCL 3, 46:304f (ARAB 2, 167)
31.		URU	-	***15***	Gadd, *Iraq* 16, 186:50-58
32.		ABUL	-	***8***	Weissbach, ZDMG 72, 182:42ff (ARAB 2, 85)

Sennacherib

33.	26	URU	*napḫar*	***33***	Luckenbill, *Senn.*, 52:36ff (ARAB 2, 261)[12]
34.		URU	*napḫar*	***8***	Luckenbill, *Senn.*, 53:40f (ARAB 2, 261)
35.		URU	*napḫar*	***39***	Luckenbill, *Senn.*, 53:42ff (ARAB 2, 261)
36.		URU	*napḫar*	***8***	Luckenbill, *Senn.*, 53:48f (ARAB 2, 261)
37.		ABUL	PAP	***7***	Luckenbill, *Senn.*, 112:74ff (ARAB 2, 397)[13]
38.		ABUL	PAP	***3***	Luckenbill, *Senn.*, 112:87ff (ARAB 2, 397)
39.		ABUL	PAP	***5***	Luckenbill, *Senn.*, 113:94ff (ARAB 2, 397)
40.	36?	URU	-	***34***	Luckenbill, *Senn.*, 39:61ff (ARAB 2, 248); dupl. Ling-Israel, *Fs Artzi*, 233:53ff
41.		URU	-	***18***	Luckenbill, *Senn.*, 79:8ff (ARAB 2, 332)[14]

Esarhaddon

42.		LUGAL	-	***8***	Borger, *Asarh.*, §27 ep. 17A (ARAB 2, 520)
43.		LUGAL	-	***12***	} Borger, *Asarh.*, §27 ep. 21A:55-63 and 63-71
44.		LUGAL	-	***10***	} (cf. also §67:31ff - ARAB 2, 690)[15]

Ashurbanipal

45.		URU	-	***8***	Piepkorn, *Asb.*, 50:34-37 (ARAB 2, 851)
46.		LUGAL	ŠU.NIGIN	***22***	Streck, *Asb.*, 138:24-46 (ARAB 2, 876); Bauer, IWA, 14 II 39-46 + cf. p. 18 (K 1848)

---

11 One mountain is mentioned earlier, line 18.

12 Nos. 33-36: the cities are added at the end of the passage (*napḫar* 88 URU, line 50), that is *33* + *8* + *39* + *8*. The first group of cities includes only 26 names: possibly one line, including 7 names, had been omitted by mistake (cf. discussion in Eph'al, *Ancient Arabs*, 40 n. 106 and *id.*, "«Arabs» in Babylonia in the 8th Century B.C.," JAOS 94 (1974), 111 n. 17).

13 Nos. 37-39: on the gates of the "palace without rival" other versions exist: cf. p. 62 no. 77.

14 Lists the names of *18* villages through which *18* channels (ÍD) had been excavated.

15 Nos. 43-44: here too, at line 72, PAP 22 LUGAL^{MEŠ}-*ni* are summed (in Esarhaddon). Ashurbanipal's list (no. 46, cf. BAL^{2}, 93) is similar to that of his predecessor: cf. Cogan, *Fs Tadmor*, 122f and here above, p. 150. Cf. also here below, *sub* B, nos. 7 and 8.

(B) VERSIONS WITHOUT LIST — In this table are shown those occurrences of (A) represented by the number alone in different texts, the list of names being missing. The columns here show: no. of the occurrence; (no. listed in scheme A); number-sum; number in the second version (sometimes it has varied); reference.

VERSION WITH LIST OF NAMES AND NUMBER - VERSION WITH NUMBER ONLY

Tiglath-pileser I					
1.	(3.)	***23***	-	***30***	cf. p. 46 no. 6 (ARI 2, 69, 80)
Sargon II					
2.	(27.)	***21***	-	*21*	Lie, *Sargon*, 24:140 (ARAB 2, 20)
3.	(28.)	***7***	-	*7*	Lie, *Sargon*, 24:142 (ARAB 2, 20)
4.	(29.)	(29) ***30***	-	***30***	Lie, *Sargon*, 24:144f (ARAB 2, 20)
5.	(30.)	***5***	-	*5*	Lie, *Sargon*, 26:146 (ARAB 2, 20)
Esarhaddon					
6.	(42.)	***8***	-	*8*	Borger, *Asarh.*, §27 ep. 17B (ARAB 2, 537); MacGinnis, SAAB 6, 4 and 16
7.	(43-44.)	***12+10***	-	*22*	Borger, *Asarh.*, §27 ep. 21B (ARAB 2, 697)
Ashurbanipal					
8.	(46.)	***22***	-	*22*	Streck, *Asb.*, 8:69 (ARAB 2, 771)

INDEX

The following list is conceived mainly as an index, and reports all the above mentioned quantifications of some selected categories — categories that have been established rather arbitrarily — taking into consideration especially practicality and historical relevance.[1] They are subdivided into two groups:

• MILITARYMEN, including prisoners and Assyrian military contingents.

• BOOTY, including tribute.

Note that prisoners are listed under "MILITARYMEN", except when they are mentioned together with (quantified) booty, in this case being listed under "BOOTY" in order to respect the context.[2]

To obtain a general framework of the subject, some quantifications never mentioned in the work are also included, the list resulting is thus complete, except for a small number of doubtful cases or very fragmentary passages. The mark *** indicates that a very long tribute / booty list has been (intentionally) abridged, reporting only some commodities (animals and metals - except metal objects).[3]

Entries are listed chronologically, as far as possible.

The columns show: 1) year/campaign (if known) - 2) number - 3) object - 4) action (verb) - 5) context (geographic, etc.) - 6) references (and assorted notes) to latest edition; id = same context, p. = refers to this work (reference only to a single page, not to multiple pages).

Remarks and abbreviations for each column:

1) AY = accession year; ? = chronological context unclear; l = chron. cont. unclear, but late in the reign; - = outside any chron. cont. (e.g. appendix to the inscr.); year/campaign according to the main or latest annalistic inscriptions[4]

2) * = varying numbers (only the oldest version or the lowest numeral is reported); *x* = a missing digit (but certainly present); [*x*] = a lacuna *possibly* containing digits; t = talent(s); m = minas; s = shekels; h = homer(s)

---

1 A list of quantifications of cities would need much space, and would not be very useful. All quantifications of hunted animals are mentioned within the pertinent paragraph.

2 Kings, princes or commanders of any kind (*rab kiṣri*, etc.) are *not* listed, unless included in booty or if the quantification is discussed in the work. They are mostly quantified in such contexts as "*5* chieftains of the Puqudu ... heavy tribute brought to Dur-Sharruken" (ARAB 2, 32), or "*7* of my officials I sent to avenge him" (ARAB 2, 65), which are difficult to assign to a category ("MILITARYMEN" or "BOOTY").

3 In addition, any reference to non-quantified objects has been omitted, as well as some uncertain quantifications included in booty lists appearing in fragmentary passages.

4 E.g. the Khorsabad annals for Sargon and prism "A" for Ashurbanipal. For Ashurnasirpal II, cf. p. 29.

3) Objects. For certain Akkadian words I have employed specific conventional English equivalents:

oxen (incl. cattle) GU4(.NÍTA)
cavalry *pēthallu*
men LÚ (*amēlu/û*)
people UN (*nišē*)
captives *šallatu*
enemies LÚ.KÚR
killed *dīktu*
dead souls *šalmat niši*
soldiers (LÚ) ERÍN (= *ṣābē*)
troops ERÍN.ḪÁ
fighters *mundaḫṣi*
fighting men ERÍN *mundaḫṣi*
warriors ERÍN *taḫāzi*
combatants GAZ (= *tidūki*)
combat troops ERÍN *tidūki*
armed men *muqtablu*
men-at-arms ERÍN *muqtabli*
hordes *ummanu/āte*
captive warriors *šallat* LÚ *quradi*
infantry zūk (*šēpē*)
picked men *zīm pāni*
elite troops *gunni*
bow(men) (ERÍN) GIŠ.PAN
shields GIŠ *arīte*
shield bearers *nāš kabābi*
lancers *nāš išmarê*
princes *zēr (šarri)*

The other translations are obvious.
c = URUDU, copper (object)
b = ZABAR, bronze (object)

4) Verbs. The actions to which the objects were subjected are also expressed with conventional translations:

crossed [with] *ebēru*
marched to *alāku*
called up *dekû*
equipped *rakāsu*
sent *šapāru*
fought with *šanānu* Gt
clashed *maḫāṣu* Gt
killed *dâku*
slain *napāṣu* D
fallen (in a river) *maqātu*
slaughtered *maqātu* Š
+ *ina* GIŠ.TUKUL
cut down *palāqu*
burnt *ina* IZI.MEŠ GÍBIL
impaled *zaqāpu*
beheaded *batāqu* D (+ GÚ)
blinded *napālu*
flayed *kâṣu* (+ KUŠ)
escaped *naparšudu*
captured *ṣabātu* (D), DIB
alive + TI(.LA)
deprived of *ekēmu*
seized *tamāḫu* D
took *leqû*
counted (as booty) (*ana šallatiš*) *manû*
lifted *našû* (referring both to the Assyrian king - "I took away" - or to the enemy - "he brought to me")
brought out *waṣû* Š
brought back *târu* D
brought *wabālu* (Š: had brought)
carried off *šalālu*
uprooted *nasāḫu*
collected *kaṣāru*
settled *wašābu* Š
received *maḫāru*
fixed *kânu* D
imposed *emēdu*
established *šakānu*, GAR

5) some selected contextual references are indicated
l = land (KUR); c = city (URU); i = individual (I); p = people (LÚ); r = river (ÍD)

7) id = *idem*, see line above, same context. References to this work are selective. They are given only to a single page, not to multiple pages. The terms relating to tribute are indicated in parentheses, if appearing (*m.* = *madattu*; *n.* = *nāmurtu*; *b.* = *biltu*).

MILITARYMEN

Arik-den-ili

	33 chariots	?	i Esini	p. 22 (fragm.)
	90 chariots	crossed	(r?)	id (fragm.)
	600 men	killed	c Ḫi-...	id
	254,000 men	killed		id, p. 182

Shalmaneser I

	14,400 elite troops	blinded + carried off	l Ḫanigalbat	p. 150

Tukulti-Ninurta I

	28,800 soldiers	uprooted	(l) Ḫatti	p. 150 (wide area)

Tiglath-pileser I

AY	(*) *20,000* men-at-arms	fought with	l Muški	p. 6, 46
	6,000 hordes	captured	l Muški	p. 9 (out of the above)
	30 chariots	took	marched to l Išdiš	RIMA 2, 16:65 (Assyrian forces)
(1)	*4,000* soldiers	-	l Kasku, l Urumu	RIMA 2, 17:100 (heard of the arrival of the Assyrian king)
	120 chariots	took		id, cf. p. 181
?	*4,000* (soldiers)	took	l Urumu, l Abešlu	RIMA 2, 33:21; 42:20; cf. 53:24 (same as above?)
(2)	*6,000* hordes	clashed	l Papḫu	RIMA 2, 19:9
(3)	*120* chariots	seized	l Nairi	p. 181
(4)	*20,000* hordes	clashed	l Qumānu	p. 6, 181
	300 families	received	c Kipšuna	p. 10 n. 53 (uprooted by the king of Qumānu)
(4?)	*2,000* captives	?	l Qumānu	(together with the above mentioned *20,000*), cf p. 182
?	(*) *12,000* troops	conquered?	l Muški	p. 46 (= the *20,000* of the AY?)

"Broken obelisk" (Ashur-bel-kala)

	1,000 soldiers	?	l?	p. 181 (fragm.)
	4,000 captives	uprooted	?	id
	3,000? captives?	brought out	c Erešu l Ḫabḫu	RIMA 2, 102:17 (fragm.)

Tukulti-Ninurta II

	2,702 horses	equipped	-	p. 153, 167, 170 n. 48, 181 (comprehensive Assyrian forces)

Ashurnasirpal II

AY	*200* combat troops	slaughtered	l Tummu	RIMA 2, 197:52; 241:75
	260 fighting men	slaughtered	l Ḫabḫu	RIMA 2, 198:64 (cf. 242:85); cf p. 28 n. 43
(2)	* *600* fighting men	slaughtered	c Kinabu	p. 46
	3,000 captives	burnt		id
	50 combat troops	slaughtered	c Mariru	RIMA 2, 201:111
	200 captives	burnt		id

	332 soldiers	killed	l Nirbu	RIMA 2, 201:111; cf p. 29
	3,000 combat troops	slaughtered	c Tēla	RIMA 2, 201:115; cf p. 177 n. 85
(3)	*1,460* fighting men	killed	c Babitu	p. 28 n. 43, 177 n. 85
	1,200 troops	uprooted	i Nūr-Adad (l Dagara)	cf *ibid.*
	320 combat troops *300* troops	slaughtered uprooted	c Bāra	RIMA 2, 204:32; 245:103 id
	326 armed men	slain	c Bunāsi (i Muṣaṣina)	RIMA 2, 204:36; cf p. 29
	172 combat troops	killed	c Larbusa (i Kirteara)	p. 9, cf p. 29
	50 soldiers	killed	c Bāra	RIMA 2, 205:45; 245:18
(4)	*800* fighting men	slaughtered	c Ammali (i Araštua)	RIMA 2, 206:55; 246:41
	50 fighters *20* soldiers	killed captured alive	i Ameka (c Parsindu)	RIMA 2, 207:71; 247:84 id
	500 armed men	slain	c Mesu	RIMA 2, 208:83
(5)	*2,800* combat troops	slaughtered	c Matiatu	RIMA 2, 209:89; 249:9; 258:45
	300 combat troops	slaughtered	c Bunnu?, Maṣula	RIMA 2, 259:53
	800 fighting men * *500* soldiers	slaughtered impaled	c Pitura l Dirru	RIMA 2, 210:107; 250:75; 260:74 id; p. 47
	700 combat troops	slaughtered	c Kūkunu	RIMA 2, 210:110; 281:85
	* *40* soldiers	captured alive	l Dirru	p. 47
	1,000 men-at-arms *200* soldiers *2,000* prisoners	slain captured alive carried off	c Arbakku l Ḫabḫu	RIMA 2, 211:114; 251:96; 260:81 id id
	1,500 troops	uprooted	i Amme-ba'li (Bīt-Zamāni)	RIMA 2, 261:95 ("Arameans")
	900 combat troops? *2,000* captives	slaughtered carried off	c Šūra	RIMA 2, 262:103 id
(6)	*70* soldiers *50* cavalry *3,000* combat troops	fallen captured	c Sūru (i Kudurru l Suḫu) (i Nabû-apla-iddina)	RIMA 2, 213:18 RIMA 2, 213:19; cf p. 9 n. 49 id; cf p. 177 n. 85
(7?)	*470* combat troops * *20* (alive)	slaughtered captured, impaled	l Laqû, Suḫu	RIMA 2, 214:32 id, p. 47
	6,000 troops *6,500* armed men	called up slaughtered	l Suḫu, Laqû, Ḫindānu	p. 177 n. 85 (attack the Assyrians) id, p. 181
	1,000 combat troops	killed	c Kipinu (i Azi-ili)	RIMA 2, 215:39
	500 troops	uprooted	i Ilâ l Laqû	RIMA 2, 215:43
(8?)	*800* armed men * *2,400* armed men	slain uprooted	c Kaprabu (l BḪt-Adini)	RIMA 2, 216:53 id, p. 48
(18)	*600* fighting men *400* soldiers *3,000* captives	slaughtered captured alive brought out	c Damdammusa (i Ilānu)	p. 9 id id
	1,400 fighting men * *580* soldiers *3,000* captives	slaughtered captured alive brought out	c Udu (i Labṭuru)	RIMA 2, 220:111 id, p. 48 id
?	[*x*+]*200* combat troops	blinded?	?	RIMA 2, 266:5′ (fragm.)
	174 soldiers *12* - [*x*+]*153* - *20?* ?	alive captured? flayed beheaded impaled?	c Mal?ḫānu?	p. 9 n. 49 id id id

	50,000	hordes	called up	l Meḫru	p. 181 (Assyrian)

Shalmaneser III

1	*300*	fighters	slaughtered	c Burmar'ina	Mahmud - Black, *Sumer* 44, 140: 50; KB 1, 156:35; cf p. 30 fn. 53
	1,300	combat troops	slaughtered	l Paqarruḫbūni	Mahmud - Black, *Sumer* 44, 141: 6; KB 1, 156:39
	2,800	combatants	killed	l Ḫattina	p. 48 n. 17
	* *4,600*	captives	carried off		id, p. 48
	22,000	troops	uprooted	l Ḫatti	p. 93
?	*40,400*	soldiers	uprooted	l Ḫatti	p. 94 (total?)
?	*87,500*	troops	uprooted	l Ḫatti	p. 94, 169 (total?)
3	* *3,400*	fighters	slaughtered	i Arramu l Urartu	p. 49, 95 n. 201, 155
	3,000	captives	carried off	c Šilaia (l Ḫubuškia)	p. 155
(3?)	*18,000*	enemies	killed (mātu Št)	?	p. 49 n. 18
4	* *17,500*	troops	uprooted	i Aḫuni (Bīt-Adini)	p. 49, 95
6	*1,200*	chariots	-	i Adad-idri [l] Damascus	p. 103 (offer battle at Qarqar)
	1,200	cavalry			id
	20,000	soldiers			id
	700	chariots		i Irḫulina l Hamāt	id
	700	cavalry			id
	10,000	soldiers			id
	2,000	chariots		i Aḫabbu (l Israel)	id
	10,000	soldiers			id
	500	soldiers		l Gue	id
	1,000	soldiers		l Muṣre (= Egypt)	id
	10	chariots		l Irqanate	id
	10,000	soldiers			id
	200	soldiers		i Matinu-Ba'il c Arwad	id
	200	soldiers		l Usanate	id
	30	chariots		i Adunu-ba'li l Ši'ana	id
	10,000	soldiers			id
	1,000	camels		i Gindibu' (l Arabia)	id
	x,000	soldiers		i Ba'sa l Ammon	id
	* *14,000*	combat troops	slaughtered	i Irḫulena, c Qarqar	p. 49, 107
11	*10,000*	combat troops	slaughtered	i Adad-idri i Irḫulena	p. 136
14	*120,000*	troops	crossed	r Euphrates	p. 107
18	* *16,000*	combat troops	slaughtered	i Hazael (Damascus)	p. 50
	1,121	chariots	deprived of		id
	470	cavalry			id
-	* *2,002*	chariots	(set up)		p. 50, 107 (Assyrian forces)
	* *5,542*	cavalry			id

Shamshi-Adad V

3	*6,000*	combatants	killed	c Uraš (l Gizilbunda)	KB 1, 180:13
	1,200	fighters	captured	(i Pirišāte)	id
	2,300	combatants	killed	(i) Ḫanaṣiruka l Media	KB 1, 180:32
	120	cavalry	brought back		id (cf p. 50 fn. 29)
	1,070	fighters	slaughtered	i Munsuartu l Araziaše	KB 1, 182:39
4	*330*	fighters	killed	c Datēbir + Izduya	KB 1, 184:16*0*
	500	fighters	killed	c Qiribtu	KB 1, 184:20
	13,000	fighters	slaughtered	c Dūr-Papsukkal	p 181
	3,000	(alive)	captured		id

	5,000 dead souls	cut down	i Marduk-balatsu-iqbi	p. 177 n. 85
	2,000 (alive)	captured		id
	100 chariots	deprived of		id, cf p. 181
	200 cavalry			id
	650 combatants	killed	?	Weidner, AfO 9, 92:11
	30 cavalry	deprived of		id
	1? chariot			id
(5)	*30,000* captives	?	c Dēr	p. 168, 182

Tiglath-pileser III

(1)	*10,000* -	settled		Tadmor, *Tigl. III*, Ann. 9:10
	x,000			id
	5,000		l Mazamua	id
(3)	*72,950* [people?]	?	i Sarduri (l Urartu)	p. 181 (fragm.)
(8)	*83,000* [people?]	settled	in c Tuš[ḫan]	p. 101, 181 (fragm.)
	1,223 people	settled	in l Ulluba	id
	12,000 people	brought	[to l Ḫatti]	Tadmor, *Tigl. III*, Ann. 19:17 (from c Sarragitu)
	600 captives	settled	in c Kunalia, etc.	p. 100 (from c Amlate)
	5,400 captives			id (from c Dēr)
	x captives	settled	in c Ṣimirra, etc.	id (from l Bīt-Sangibūti)
	1,200 men			id (from Illila)
	6,208 men			id (from Nakkaba, Buda)
	* *588* men	settled	in prov. c Tu'imme	id, cf p. 51 (from Buda, c Duna)
	252 men			id (from Bila)
	554 men			id (from Banīta)
	380 men			id (from Nergal-andil-māti)
	460 men			id (from Sangilla)
	x [men]			id (from Illila)
	458 captives			id (from l Bīt-Sangibūti)
	555 captives	settled	in c Til-Karme	id (from l Bīt-Sangibūti)
(13)	*800* people	carried off	c Ḫadara	Tadmor, *Tigl. III*, Ann. 23:14ff
	750 captives	?	c Kurūṣṣa	id
	550 captives	carried off	c Metuna	id
	625 captives	?	c ?	Tadmor, *Tigl. III*, Ann. 18 and 24: 4ff (fragm., cf p. 103 fn. 239)
	226 ?		?	id
	650 captives		c Ku-...	id
	4XX ?		?	id
	656 captives		c Sa-...	id
	13,520 ?		?	id (total, incl. the above?)
(9?)	*6,500* people	carried off	l Media	p. 91, 128 (wide area)
(13?)	*9,400* -	killed	[i Samsi l Arabia	Tadmor, *Tigl. III*, Summ. 8:24′
1	*10,000* soldiers	?	[i Samsi l Arabia]	Tadmor, *Tigl. III*, Summ. 9 rev. 22 (fragm.)
1	*55,000* people	carried off	i Nabû-ušabši, c Sarrabānu	p. 91, 156
	30,000 people	carried off	c Tarbaṣu + Yaballu	cf *ibid.*
	40,500 people	carried off	l Bīt-Ša'alli, c Dūr-Baliḫāya	cf *ibid.*
	155,000 people	carried off?	c Tarbaṣu, Yaballu, etc.	cf *ibid.*

Sargon II

1	* *27,280* people	carried off	l Samaria	p. 52, 80
	* *200* chariots	collected	l Samaria	p. 52 (annexed to the Assyrian army)
	x+7 people	uprooted	l Babylonia	Lie, Sargon, 6:22 (fragm.)

2	* *200* chariots	collected	l Hamath	p. 56, 106 (annexed to the Assyrian army)
	600 cavalry			id
	6,300 men	settled	l Hamath	p. 10 n. 54
	9,033 people	carried off	c Rapiḫu	p. 188 no. 17
4	*7,350* people	counted as booty	c Šinuḫtu	Lie, Sargon, 10:70
(4)	*30* chariots	counted as booty	c Šinuḫtu	Winckler, Sargon, 102:28
5	*50* chariots	collected	i Pisiri c Karchemiš	p. 52, 153 (annexed to the Assyrian army)
	* *200* cavalry			id
	3,000 foot soldiers			id
7	*4,200* people	carried off	l Tuaiadi	p. 7
	4,000 picked men	received	c Anzaria, etc.	Lie, Sargon, 20:112
	4,820 people			id
	2,350 people	carried off	c Kimirra l Bīt-Ḫamban	Lie, Sargon, 20:116
8	* *260* princes	captured	c Parda l Urartu	p. 53 no. 46
	1,000 cavalry, bowmen, etc.			TCL 3, 50:320; Lie, Sargon, 26:150 (personal guard)
9	*2,200* picked men	received	l Karallu	Lie, Sargon, 28:168
	100 chariots	took	i Ambaris l Bīt-Buritiš	p. 106
10	*5,000* captive warriors	brought	c Til-Garimmu	Lie, Sargon, 34:213; Winckler, Sargon, 112:81
12	*600* cavalry	given	p Gambulu, c Dūr-Atḫara	Lie, Sargon, 44:276 (p Marduk-apal-iddina gave to p Gambulu)
	4,000 soldiers			id
	18,430 people	carried off	c Dūr-Atḫara	p. 188 no. 23
	7,520 men	carried off	c Samʾuna and Bāb-Dūri	Lie, Sargon, 52:16 (p Elamites)
	12,062 people		c Agurumu	id (p 188 no. 24)
13	*1,000* soldiers	deprived of	i Mitâ l Muški	Lie, Sargon, 66:447 (by the gov. of l Que)
	2,400 soldiers	carried off	i Mitâ l Muški	Lie, Sargon, 66:450 (by the gov. of l Que)
	1,000 picked men	brought	to c Samʾuna	Lie, Sargon, 68:451 (by the gov. of l Que)
	150 chariots	collected	l Bīt-Yakin	p 56 (given to an Assyrian officer)
	1,500 cavalry			id
	20,000 bowmen			id
	* *10,000* shield bearers - lancers			id
(14)	*4,500* bowmen	escaped	p Elamites	Lie, Sargon, 74:3

Sennacherib

1	*80,000* bowmen	sent	l Elam	p. 171 n. 54, 175 n. 65 (help to Merodach-baladan)
(8)	*150,000* warriors	slaughtered	c Ḫalulē	p. 116, 175
-	* *10,000* bows	collected		pp. 89, 167 n. 38, p. 169 n. 45, p. 170 n. 48 (total - whole reign)
	* *10,000* shields			id

Ashurbanipal

	16,000 bows	?	?	p. 176 n. 76 (fragm.)

BOOTY

Arik-den-ili

	100 sheep	brought	?	RIMA 1, 126:3′ (fragm.)
	100 oxen			id

Tiglath-pileser I

AY	*180* c kettles	(I) lifted	i Kili-Tešub (l Katmuḫu)	p. 123, 125
	5 b bathtubs			id
	60 c kettles	(he) lifted	i Šadi-Tešub c Urattinaš	p. 123, 125 (*b.* + *m.*)
	120 men			id
(2)	* *60* c kettles	brought out	c Murattaš	p. 46, 123
	30 t copper bars			id, cf p. 181
	25 gods	brought out	l Sugu	p.
(3)	*1,200* horses	fixed	l Nairi	p. 181 (*m.*)
	2,000 oxen			id
	1 h lead ore	fixed	c Milidia l Ḫanigalbat	RIMA 2, 23:39 (*m.*)

Tukulti-Ninurta II (cf. p. 97)

3 t silver	received	i Ilī-ibni l Suḫu	p. 132, 177 n. 86 *** (*n.*)
20 m gold			id
18 tin bars			id
10 m gold	received	i Amme-alaba l Ḫindānu	RIMA 2, 175:76ff *** (*n.*)
10 m silver			id
2 t tin			id
30 dromedaries			id
50 oxen			id
30 asses			id
14 ducks			id
200 sheep			id
200 sheep	received	i Mudadda l Laqû	RIMA 2, 175:85 (*n.*)
30 oxen			id
200 sheep	received	l Ḫamatāya (l Laqû)	RIMA 2, 176:86 (*n.*)
50 oxen			id
200 sheep	received	i Ḫarānu l Laqû	RIMA 2, 176:88 (*n.*)
30 oxen			id
3 m gold	received	i Mudadda c Sirqu	RIMA 2, 176:90f; cf p. 97 *** (*n.*)
7 m silver			id
x t tin			id
x00 sheep			id
x40 oxen			id
20 asses			id
20 birds			id
3 m gold	received	i Ḫarānu l Laqû	RIMA 2, 176:93f; cf p. 97 (*n.*)
10 m silver			id
6 t tin			id
500 sheep			id
1x0 oxen			id
20 asses			id

	20 m	gold	-	[l Ḫamat]āya (l Laqû)	RIMA 2, 176:98ff; cf p. 8 n. 40, p. 97 *** (*n.*)
	20 m	silver			id
	32 t	tin			id
	130? t?	bronze			id
	2 t	iron			id
	1,200	sheep			id
	100	oxen			id
	200	sheep	received	[c Usalā]	RIMA 2, 177:103 (*n.*)
	30	oxen			id
	10 m	silver	received	c Dūr-katlimmu	RIMA 2, 177:106f; cf p. 97 *** (*n.*)
	[*x*+]*14* m	?			id
	x t	ir[on]			id
	11 t	tin	[received?]	[c Qatnu?]	RIMA 2, 177:109f; cf p. 97 (*n.*)
	[*x*+]*100*	ducks + geese			id
	3 m	bronze?	-	c Šadikannu?	RIMA 2, 177:113f; cf p. 97 (fragm.)

Ashurnasirpal II

(5)	*40*	chariots	received	l Bīt-Zamāni	p 128, 127 fn. 57 *** (*m.*)
	460	horses			id
	2 t	silver			id
	2 t	gold			id
	100 t	tin			id
	* *200* t	bronze			id, p. 47
	300 t	iron			id
	2,000	oxen			id
	5,000	sheep			id
(9?)	*4* m	silver	received	i Ḫabinu c Tīl-abni	RIMA 2, 217:63f (*m.*)
	400	sheep			id
	10 m	silver	established		id (*m.*)
	20 t	silver	received	i Sangara l Ḫatti	RIMA 2, 217:65ff; cf p. 181 (*m.*)
	100 t	bronze			id
	250 t	iron			id
	200	girls			id
	20 t	silver	received	i Lubarna l Patinu	RIMA 2, 218:73ff, 227:49ff; cf p. 181 (*m.*)
	1 t	gold			id
	100 t	tin			id
	100 t	iron			id
	1,000	oxen			id
	10,000	sheep			id, p. 181
	1,000	garments			id
	10	singers			id

Shalmaneser III

AY	*2*	camels	received	i Asû l Guzana	p 123 (*m.*)
2	*3* t	gold	received	l Patinu	KB 1, 160:22ff; cf p. 99
	100 t	silver			id
	300 t	bronze			id
	300 t	iron			id
	1,000	b vessels			id
	1,000	garments			id
	20 t	purple			id
	500	oxen			id
	5,000	sheep			id
	1 t	silver	established		id (*m.*)
	2 t	purple			id
	200	cedar logs			id

	10 t silver	received	i Ḫayānu (l Sam'al)	KB 1, 162:24ff; cf p. 99
	90 t bronze			id
	30 t iron			id
	300 garments			id
	300 oxen			id
	3,000 sheep			id
	200 cedar logs			id
	10 m silver	established		id (*m.*)
	100 cedar logs			id
	10 m gold	received	i Aramu (Bīt-Agusi)	KB 1, 162:27; cf p. 99
	6 t silver			id
	500 oxen			id
	5,000 sheep			id
	3 t gold	received	i Sangara c Karchemiš	KB 1, 162:27ff; cf p. 99
	70 t silver			id
	30 t bronze			id
	100 t iron			id
	20 t purple			id
	500 weapons			id
	100 daughters			id
	500 oxen			id
	5,000 sheep			id
	1 m gold	established		id
	1 t silver			id
	2 t purple			id
	20 m silver	received	i Qatazilu l Kummuḫu	KB 1, 162:29f; cf p. 99
	300 cedar logs			id
3	*7* camels	received	i Asāu l Gilzānu	p. 123
20	*110,610* captives	-	-	p. 93, 112, 86 n. 168, 96 n. 205 (total of whole reign)
	82,600 killed			id
	9,920 horses, mules			id
	35,565 oxen			id
	19,690 asses			id
	184,755 sheep			id

Shamshi-Adad V

4	*3,000*? ?	carried off	c Di'bina	p 34 (chieftains?)

Adad-nīrāri III[5]

	* *2,000* t silver	received	i Mari' l Damascus	p 51 (Rimaḫ and Saba'a stelae) (*m.*)
	[*x*+]*100* t gold			id
	1,000 t copper			id
	* *2,000* t iron			id
	3,000 garments			id
	2,300 t silver	received	in c Damascus	cf *ibid.* (Nimrud slab) (*m.*)
	20 t gold			id
	3,000 t bronze			id
	5,000 t iron			id

Tiglath-pileser III

(1)	*10* t gold	received	?	Tadmor, Tigl. III, Ann. 10:6 (fragm.) (*m.*)
	1,000 t [silver]			id

---

5 The subdivision between the version(s) given by the stelae and that of the Nimrud slab is *not* an indication that two different episodes are involved (cf. above p. 51 n. 31).

2	*300* t	lapis lazuli	?	i Mannu-kīma-ṣabē c Bīt-abdadāni	Tadmor, *Tigl. III*, Ann. 12:10 (fragm.) (*m.*)
	500 t	bronze			id
(7?)	*3?* t	gold	?	i Raḫianu [c Damascus]	p. 35 n. 72 (fragm.)
	300? t	silver			id
	200? t	?			id
	20 t	ladunu resin			id
	3xx(?)	?			id
	300 t	silver	?	i Tutammû c Kinalia	Tadmor, *Tigl. III*, Ann. 25:9′ (fragm.)
	100 t	[bronze?]			id
(9)	*13x*	horses		c Bīt-Ištar (l Media)	p. 83 n. 158 ***
(11)	*8,650*	people	carried off	c Ḫista, Ḫarbisinna, etc.	Tadmor, *Tigl. III*, Ann. 5:1f
	300	mules			id
	660	asses			id
	1,350	oxen			id
	19,000	sheep			id
(11)	*900*	people	carried off	c Daiakanša, etc.	p 7, 9, 174
	150	oxen			id
	1,000	sheep			id
(15)	*x9,000*	people	?	p Arameans?	Tadmor, *Tigl. III*, Ann. 8:2 (fragm.)
	x,500	oxen			id
(3?)	*30* t	gold	brought back	l Bīt-Agusi	Tadmor, *Tigl. III*, Summ. 9 obv. 24′ (fragm.)
	2,000 t	silver			id, cf p. 181
(8?)	*20* t	[gold]	?	[l Unqi]	Tadmor, *Tigl. III*, Summ. 9 obv. 26′
(9?)	*5,000*	horses	took	l Media	p. 83 n. 158 *** (total?)
(12?)	*800* t	silver	?	[i Ḫanū]nu c Gaza	p. 80 n. 148 (fragm.)
(13?)	*10?* t	gold	received	i Peqaḫ (Israel)	cf *ibid.* (fragm.)
	1,000? t	silver			id
?	[*x*+]*100* t	silver	uprooted	[c Aškelon?]	cf *ibid.* (fragm.)
1	*1,x00*	people	deprived of	i Samsi l Arabia	p. 85 (fragm.)
	30,000	camels			id
	20,000	oxen			id
	5,000	spices			id
1	*10* t	gold	[received?]	i Wassurme l Tabal	Tadmor, *Tigl. III*, Summ. 7 rev. 15
	1,000 t	silver			id
	2,000	horses			id
?	*20* t	[gold]	?	i Hir]am (c Tyre) c Maḫalab	Tadmor, *Tigl. III*, Summ. 9 rev. 7
1	*150* t	gold	[received?]	i Mētenna c Tyre	p. 51
1	*50* t	gold	?	[i Mētenna c Ty]re?	cf *ibid.* (fragm.) (*ta*[*mārtu*])
	2,000 t	silver			id

Sargon II

5	*2,100* t	silver	carried off	m Pisiris l Karchemiš	p. 52 (*kišittu*, afterwards placed in a palace)
	11 t	gold			id
(5)	*12*	horses	brought (našû)	i Šilkanni l Egypt	Tadmor, JCS 12, 78:10 (*t.*)
8	* *6,110*	people	counted	c Muṣaṣir	p. 54, 112, 129
	* *12*	mules			id
	* *380*	asses			id
	* *525*	oxen			id
	* *1,235*	sheep			id, p. 73, 86, 173
	34 t *18* m	gold	carried off	c Muṣaṣir	p. 54, 129 (from the palace treasure)
	* *160* t *2½* m	silver			id ***

	[*x*+]*4* t *3* m	gold	carried off	c Muṣaṣir	p. 129 (from the temple treasures) ***
	162 t *20* m	silver			id
	3,600 t	bronze			id
(9)	*33,600*	people	?	c Ḫubaḫna (l Ellipi)	Gadd, *Iraq* 16, 177:51 (fragm.)
	11,600	[horses?]	?	c Ḫubaḫna (l Ellipi)	p. 184 (fragm.)
9	* *4,609*	horses	received	l Medes	p. 55 (wide area)
12	*1* t *30* m	silver	fixed	c Dūr-Atḫara	Lie, Sargon, 46:284 (*nadan šatti*, yearly tribute)
	2,000 h	barley			id
13	*90,580*	people	received	(c Dūr-Yakin, i Marduk-apal-iddina)	p. 55, 177 (carried off by the army)
	* *2,080*	horses			id
	* *610*	mules			id
	6,054	camels			id
	[*x*+]*40*	sheep			id
	1,000	horses	?	i Marduk-apal-iddina	Winckler, Sargon, no. 55:14′ [tribute accepted]
	800	asses			id

Sennacherib

1	*1*	ox	fixed	c Ḫirimmu	p. 132 n. 79
	10	sheep			id
	10 h	wine			id
	20 h	dates			id
	208,000	people	carried off	(p Arameans)	p. 58, 90, 113, 173 (total of campaign)
	7,200	horses, mules			id
	* *11,073*	asses			id
	* *5,230*	camels			id
	* *80,050*	oxen			id
	* *800,100*	sheep			id
3	*200,150*	people	counted as booty	i Hezekiah l Judah	p. 114, 173
	30 t	gold	had brought	i Hezekiah (l Judah)	p. 80, 174 (*m.*)
	800 t	silver			id
?	*x,000*	camels	deprived of	p Arabs	p. 185

Esarhaddon (cf. also p. 175 nn. 69 and 71)

	65	camels	fixed	i Tabûa (p Arabs)	p. 175 (added to earlier *m.*)
	10	asses			id
	10 m	gold	imposed	i Yautaʿ (p Arabs)	p. 63, 175 (added to earlier *m.*)
	* *100*	choice gems			id
	50	camels			id
	* *100*	bags of spices			id
	6 t *19* m	gold	?		p. 175 n. 68, 189 (fragm.) *** (*m.*?)
	1,586	garments			id
	30,418	sheep			id
	19,323	asses			id

Ashurbanipal (cf. also p. 176 n. 76)

2	*2*	obelisks	took	c Thebes	p. 124 (weight: *2,500* t)
5	*30*	horses	added	i Ualli (l Mannea)	p. 14, 175 (*m.*)

OTHER QUANTIFICATIONS MENTIONED IN THE WORK[6]

Adad-nerari I

10 / 14 bricks, thickness of a wall, p. 45

Shalmaneser I

41 / 51 cities destroyed and burned, p. 13, 45
8 lands and their fighting forces, conquered, p. 12

Tukulti-Ninurta I

40 kings of Nairi, p. 139

Tiglath-pileser I

60 kings, vied with (whole reign), p. 92 fn. 190, 139
23 / 30 kings of Nairi, *60* kings incl. those who came to their help p. 46, 92
42 lands conquered (first *5* camp.), p. 88
28 crossings of Euphrates, p. 89
6 / 17 cities at the foot of Mt. Bešri, p 46

Adad-nerari II

"for the *Xth* time", p. 19, 25

Tukulti-Ninurta II

470 wells (annals, line 43), p. 27 n. 38

Ashurnasirpal II

40 / 50 cities of Dirru, 5th year, p. 47
250 cities of Nairi, 5th year, p. 181
20 / 30 cities, neighborhood of Bunasi, p. 48
20 / 30 cities, neighborhood of Hudun, p. 48
47,074 (+ others, total: *69,574*), people invited to the inauguration of the royal palace, p. 141

Shalmaneser III

4 / 14 cities, neighborhood of Sugunia, p. 48
97 + *100* cities of Sangara, 11th year, p. 178
86 / 89 / 99 cities conquered, 11th year, p 49
11 strongholds (Mt. Kašiyari), 5th year, p. 126 n. 51
12 kings of the seacoast / of Ḫatti, p. 134
20 / 24 kings of Tabal, 22nd year, p. 50
17th *palû* (year of reign), p. 148, cf also p. 163
18th crossing of the Euphrates, 19th year, p. 50, 136

6 This section is very selective. For the animals hunted see chapter 4 §9.

Shamshi-Adad V

255 / 256 cities of the neighborhood of Nimittišarri, p. 50
11 + 200 cities of Uspina, p. 14
Cities (in general), p. 34, 168

Sargon II

3 / 12 + 23 / 24 / 84 cities of Mitatti of Zikirtu, 8th year, p. 53
55 / 56 cities of Ursâ, 8th year, p. 15, 53 n. 40
21 + 140 / 146 cities of Bīt-Sangibūti, 8th year, p. 54, 130
430 cities conquered, "total" of the 8th camp., p. 132
24 / 25 cities of Ḫubaḫna (9th year), p. 55
12 / 22 fortresses of Ullusunu the Mannean, p. 53
250 / 260 nobles of the family of Ursâ, 8th year, p. 53
5 / 6 chiefs (*nasikāte*) of the land Yadburu (12th year), p. 55
7 kings of Yā', p. 133
1 charioteer, *2* cavalrymen, *3* outriders, losses of the 8th camp., p. 139
4,610 talents, weight of the *8* lions set up in Dur-Sharruken, p. 56, 152
154 / 164 tal. gold + *1,604 / 1,804* tal. silver, precious metals presented to the temples of Babylonia, p. 57
30 bēru, distance from Dilmun, p. 133
5 / 6 bēru distance between mount Uauš and mount Zimur, p. 53
16,280 cubits, perimeter of Dur-Sharruken, p. 140

Sennacherib

88 and *820* (+ var.) cities of Babylonia, 1st campaign, p. 57, 114
34 cities, near Marubišti and Akkuddu, p. 16
46 cities of Judah, 3rd campaign, p. 15
33 / 35 cities of Ukku, 5th campaign, p. 58
"Palace without rival" (É.GAL *šānīna lā išû*) in Nineveh, p. 59 (on the Perimeter and Gates of Nineveh, cf also p. 152)
11,400 talents, weight of the *8* lions set up in Nineveh, p. 152

Esarhaddon

22 kings of the seacoast, p. 150, 191f
15 days, distance between Išḫupri and Memphis, p. 78
120 / 140 bēru, distance from Bāzu, p. 63
580 / 586 years, time-span between Adad-nerari I and Esarhaddon, p. 63
70 (or *11*) years, desolation of Babylon, p. 140

Ashurbanipal

22 kings of the seacoast, p. 150, 191f
60 nobles escaping from Elam, p. 64 n. 96
17 magnates and *85* (+ var.) princes of Elam (8th camp.), p. 64
60 bēru / a month + *25* days, length of time during which Elam has been devastated, p. 65
1,535 / 1,630 / 1,635 years in which the goddess Nanna has been in Elam, p. 66
Market prices in Assyria, p. 64
1½ shekels of silver, price of camels in Assyria after the Arab campaign, p. 179

State Archives of Assyria Studies

VOLUME I
Neuassyrische Glyptik des 8.-7. Jh. v. Chr.
unter besonderer Berücksichtigung der Siegelungen
auf Tafeln und Tonverschlüsse
by Suzanne Herbordt
1992

VOLUME II
The Eponyms of the Assyrian Empire 910–612 BC
by Alan Millard
1994

VOLUME III
The Use of Numbers and Quantifications
in the Assyrian Royal Inscriptions
by Marco De Odorico
1995